T0275880

IMPROVE

IMPROVE

The Next Generation of Continuous Improvement for Knowledge Work

GEORGE ELLIS
Vice President Innovation, Envista Business System Office
Envista Holdings Corporation, Brea, CA, United States

Butterworth-Heinemann is an imprint of Elsevier
The Boulevard, Langford Lane, Kidlington, Oxford OX5 1GB, United Kingdom
50 Hampshire Street, 5th Floor, Cambridge, MA 02139, United States

Library of Congress Cataloging-in-Publication Data
A catalog record for this book is available from the Library of Congress

British Library Cataloguing-in-Publication Data
A catalogue record for this book is available from the British Library

ISBN: 978-0-12-809519-5

For information on all Butterworth-Heinemann publications
visit our website at https://www.elsevier.com/books-and-journals

Publisher: Joe Hayton
Acquisitions Editor: Brian Guerin
Editorial Project Manager: Isabella Silva
Production Project Manager: Poulouse Joseph
Cover Designer: Victoria Pearson

Typeset by SPi Global, India

Contents

Foreword

As I returned to Danaher to join the senior team leading the divestiture of a $2.8 billion business spinning onto the public markets, we needed to replicate the coveted and legendary Danaher Business System Office (DBSO) and the Danaher Business System (DBS); it was our cultural way of life at Danaher. The proven system of how we did our work and a rally for a culture never to be satisfied remained at its core. Many companies have tried emulating the "tools" or process of the Danaher Business System. As we have learned over the years, it's all about the people who drive a culture and change in an organization. That is how I met George Ellis. He was part of the founding team of the Envista Business System Office, from our heritage of Danaher.

During our time of building of the Envista Business System Office, George was tapped to lead half of our growth pillar as the Vice President of Innovation. He held the position on the team with other veterans leading commercial and lean manufacturing from Danaher, some having more than 28 years of experience with DBS and a long history with the company. In his role, he was responsible for implementing and developing project management, launching excellence and innovation tools that would get our R&D teams in control of the process, delivering value, and capturing a greater return on investment. Efficient process and project management drive innovation, as they allow teams to spend time and energy on creating solutions to problems that our customers are having instead of thinking about minor details of when the next task or deliverable is due. Time is a common currency we all trade in; no matter what your country or business is, when you waste time it costs you money. When a project or a product is late, the team can't move on to developing and servicing the next problem for your customers.

Lean thinking and a lean culture works; the data support it. However, it will take years to move an organization forward. Winning lean cultures by definition are never satisfied; there is no destination. Don't get frustrated, start now. This book will give you the confidence to move your organization or team forward using good lean thinking and process rigor. If you aren't new to the concept of lean, this book will reaffirm your commitment with updated ideas and thinking around managing projects and teams. George is an experienced R&D leader. He is a great teammate who sets a high bar, and a big thinker, backed by a career of performance. I trust you will enjoy this book as much as I have.

John Marotta
Senior Vice President
Envista Holdings Corporation, Brea, CA, United States

Endorsements

Continuous improvement is at the foundation of any high-performance team. In its purest form, a continuous improvement mindset enables individuals to constantly learn throughout their lives – guided by intellectual curiosity and personal humility. Such an open minded disposition allows these individuals to be the best versions of themselves. As a result, one can harness both analytical and emotional intelligence, creating a team member more aware of her/his environment, more respectful of other's points of view, and open to both constructive and positive feedback. George embodies continuous improvement and has dedicated his 35 year career to improving both individuals and teams by unleashing their true potential through the power of learning. He has spent five years capturing that approach in this book.

Amir Aghdaei
CEO, Envista Holdings Company

Praise for earlier work -

Everyone recognizes the importance of sustainable innovation for any growing business. Less well understood is the vital role outstanding project management plays in innovative businesses. Our approach to project management at Danaher evolved greatly over the last two decades and played no small part in our success. George Ellis nicely captures the "state of the art" and demonstrates how process alone is not enough – true "Total Leadership" in project management differentiates the winners from the runners-up. If you want to win the innovation game, read his book.

Lawrence Culp, Jr.,
Former CEO, Danaher Corporation

Preface

June 2015, Athens, Greece. I sit at an outdoor café while my family shops for mementos of our vacation. Peaceful protesters march along the narrow street, waving signs and shouting slogans. People alone and in small groups enjoy the sun. Except me. I am tapping on the only laptop in sight, straining to understand what precisely "lean thinking" means. Having practiced it for 15 years at one of the foremost companies in lean thinking, I knew I could explain it if I just had some time. I stole away for an hour, sketched out a few thoughts, and was sure I had the core of it. I didn't, but I wouldn't know that for more than a year.

That would be the first of many hours stolen from vacations, weekends, and evenings in the pursuit to describe what lean thinking is and how to apply it to knowledge work, the work that professionals do. A week at St. Thomas. Two weeks in Myrtle Beach. A ski vacation in Traverse City. And virtually every weekend for 5 years. Over the last 30 years, I've written a few books and each took about a year; this one took five. Twice I wrote nearly the full manuscript and discarded it.

This brings to mind two questions: why is lean thinking so difficult to describe, and why did I care enough to dedicate perhaps 5000 hours of my free time to do it? Let's begin with the latter. The passion for lean thinking derives from the good it does. The effect that resonates with me most is how it changes people: those who were used to losing learn to win. Professionals struggling just to get though tedious work cut waste in order to make time to invest in themselves, doing more of the work they enjoy and doing it better. My deepest satisfaction comes when I see in my mind's eye the faces of those who used to feel they didn't measure up but now know that they do.

Lean thinking brings many other benefits. The organization sees dramatic increases in work quality and productivity. Customers, clients, patients, and users have better experiences. Leadership capability grows naturally in teams of professionals. These are the benefits that pay the bills. We must improve customer experience and organizational performance if that organization is to invest. And lean demands investment, especially mindshare. The benefits of lean thinking will grow into a handsome reward for those that invest. But the passion for it comes from the way it coaxes every person to grow to the limit of their capabilities.

Now to the first question: why is it so difficult to describe lean thinking? For one thing, lean thinking is part philosophy and part science. And lean has a long and rambling history, starting with Sakichi Toyoda and his inventions on the loom, which we will discuss in Chapter 14. Four decades later, his son Kiichiro started the Toyota Motor Corporation, where Taiichi Ohno would lead the creation of the Toyota Production System, which is the foundation for all lean thinking. Ohno quotes from *The Analects of Confucius* in his books, and certainly the principles like even flow, continuous personal growth, and respect for others are aligned to that philosophy. But lean is pragmatic, bringing in what works and eschewing what doesn't. So, as lean crosses boundaries it constantly changes. It's as difficult to describe as it is to contain.

My journey led me to a structured view that would help peel away some of the mystery of *lean*. The books I read were inspiring, giving examples, success stories, and advice. But they were unsatisfying because they lacked structure. It seemed many were written as a long stream of ideas peppered with techniques and stories. But without a structure, how was I to choose which technique or story applied best to the problem I was grappling with today? Most didn't tell me and many went so far as to advise the reader against even trying, recommending I find an expert or sensei who would lead me through a years-long apprenticeship. I didn't find that inviting.

My hypothesis was that there was an underlying structure on which the many facets of lean thinking could hang. That would illuminate not just what to do, but also why and how to know it's working. And this is how I spent 5 years writing a book three times. I started by postulating a structure, then wrote 50,000 or 70,000 words building on that structure, and only then could see that the foundation was inadequate. Start again.

My hypothesis today is that all lean thinking reduces to three axioms: simplify, engage, and experiment. *Simplify* is to reduce the waste we see and so increase the value we produce. *Engage* is to focus our effort as a team in the right places. *Experiment* is to learn systematically by first doubting what we think we know, then measuring results so we can see the waste that is hiding in plain sight.

From these axioms, a structure began to form. A five-faceted definition of value for knowledge work (Chapter 3), Ohno's definition of waste captured in the Lean Equation (Chapter 4), and from that to the 8 Wastes of Knowledge Work that form the acronym DIMINISH (Chapter 5):

Discord
Information Friction
More-is-Better Thinking
Inertia to Change
No-Win Contest
Inferior Problem Solving
Solution Blindness
Hidden Errors

Along the way, I learned many methods from others and invented a few. But every technique, tool, visualization, and workflow in the 100,000 words and 250 illustrations that follow reduces to some combination of simplifying a workflow, engaging those we work with, and experimenting to discover what we don't know.

Simplify, *engage*, and *experiment* has helped me immensely in my work as it did writing this book. It is a lens that I look though when creating, deploying, sustaining, and improving workflows. I hope you will find this approach and this book useful. I also hope you will help me improve. Please write me with your thoughts, comments, and corrections. I respond to every reader who writes. I can be reached at: BetterKnowledgeWorks@Outlook.com.

George Ellis

Acknowledgments

Writing a book is a large undertaking and many people have supported me. My loving and patient wife Regina is first among them. She has tolerated me working on every vacation, every weekend, and quite a few evenings, always encouraging me to spend the time I needed. She endured a month where I took over the living room with more than 1000 sticky notes to organize the book (I have photos!). She listened to ideas, read text, and reviewed figures. She has been as supportive as any wife could be. Her one condition: this must be my last book!

I want to thank my family, especially Gretchen, Hannah, Rachel, Alexia, Brandon, Tucker, and Greg, who have all listened to and encouraged me. They have reviewed material and tried some lean thinking ideas at home, including the Kanban board that we are currently using to renovate a house.

Many thanks to Dave Banyard and Andrew McCauley, who read parts of the manuscript and provided valuable feedback. Every reader will benefit from their insights. I must also thank Tony Ricci who, upon seeing an early manuscript, kindly identified several gaps that would prove difficult to fill. That was a hard day, but it improved the book. And thanks to Robb Kirkpatrick for his insights on canvases, which also changed the book. And many thanks to Juliet Gardner who diligently reviewed this text and provided guidance at many levels.

I want to thank my superiors who have contributed so much to my understanding and growth, starting with my supervisor for 8 years, John Boyland, who spent so many evenings helping me grow, perfecting ideas, and showing me how to follow a leader by his example. Those years sitting a few evenings a month until 7 or 8 p.m. in John's office prepared me for much more than I could have imagined. Thanks also to Todd Brewster and Cathy Biltz, who taught me about improvement, especially during our 3 years of working on our new product development system. And thanks to Scott Davis and Tom England, who helped me as I was developing material. I also want to thank Kevin Layne, one of the strongest critical thinkers I have worked with. I thought of him often when writing Chapter 24. Kevin helped me find my way more times than I can count and guided me find many opportunities to grow.

I'm especially grateful to Ron Voigt who, as President of X-Rite, hired me as VP of R&D. Ron demonstrated so much of what I wanted to be:

consistent, respectful, curious, and thoughtful. He allowed me to lead and let me make my share of mistakes. He is still a model for me. There were many people at X-Rite from whom I learned about lean thinking, especially Olivier Calas, Mike Weber, Brian Harleton, Richard Roth, Mike Moldenhauer, Michael Gloor, Milos Sormaz, Leo Mora, Adrian Fernandez, Mike Galen, Jill Henning, John Meighen, Marcus Jones, Stephanie Kempa, Jeff McKee, and Barry Ben Ezra of AVT. There are so many others I could mention. Thank you all.

The Danaher Business System Office (DBSO) is a small corporate team that deploys lean thinking processes to operating companies. I want to thank the people in that office with whom I was privileged to work closely. I miss Craig Overhage, who passed a few months ago. I want to thank especially Brian Ruggiero, John Fini, Olaf Boettger, and Chad Buckles, who, along with Craig, invested so much energy to help me transition to the Envista Business System Office (EBSO).

I want to thank the people whom I have been privileged to join at Envista, which was spun off from Danaher as an IPO in 2019. My colleagues at Envista have been so supportive of this work, starting with our CEO Amir Aghdaei, who has placed lean thinking at the forefront of our company. Many thanks to my supervisor John Marotta, who guides me as we deploy our own brand of lean thinking to our new organization. I am grateful for all I have learned from my colleagues: Kelly Jones, who regularly and rightly reminds me to keep things simple, and Torben Nielsen, who teaches me something about the commercial world every week. Thanks to Brian Huh, who has enriched my understanding about "carpetland" processes, and many thanks to Yexenia Torres for her focus on keeping structure and order. Finally, I want to thank those I lead in the EBSO Innovation Team: Grainne Dugue, Suneel Battula, and Justin Ong. I treasure our partnership. You tolerate a lot!

CHAPTER 1

30% of what you think is wrong

> *Even a wise man probably is right seven times out of ten but must be wrong three times out of ten.*
>
> ***Taiichi Ohno, creator of the Toyota Production System [1]***

1.1 A good story

Continuous improvement for knowledge work is my favorite topic. I've seen so many times its benefits: customers are delighted, organizational performance improves in virtually every measure, and the experts at the center of it all take deep satisfaction from mastering their craft and growing their leadership skills. I've watched teams transform from *okay* to *amazing*, following a pattern:

At the start, a team of good people trying to do the right thing is stumbling: they are often late, make too many mistakes, and frequently focus on the wrong things. This happens even though they are smart, experienced, and hardworking.

Their transformation begins as we search for hidden waste in their workflows. The brilliant insight of lean thinking is that waste is always there in abundant quantities, most of it hiding in plain sight. The core of what professionals and experts do—what we will call *knowledge work*—is complex, even mysterious, but most of the waste it creates is tedious and predictable.

Next, we install a few techniques carefully targeted at real problems. These techniques are a combination of: **simplify** to cut the waste we see, **engage** the full team, and **experiment** continuously to find waste that's hiding in plain sight. A new visualization method. A regular team standup meeting. An unambiguous call to action. A "Single Point of Truth" to give everyone the same information. Quantifying success and then laying out a path to it. Defining the problem thoroughly before

Improve
https://doi.org/10.1016/B978-0-12-809519-5.00001-6

starting to solve it. Always **simplifying**. Always **engaging**. Always **experimenting**.

A few months later, that team is outperforming its initial level by a margin so wide it's noticed by colleagues, the boss, a few customers, and probably some family members. People in the team smile more and walk taller. At first, many stood back waiting to see results before they bought in. Now almost all are convinced that they can improve, and they want to.

The story never gets old. The people who used to fail now succeed. They always had the ability to succeed. Of course they did. Just learning a profession like medicine, engineering, or finance takes years and only the most capable people qualify. The expertise, diligence, and acumen represented in the average team of knowledge staff take decades to develop. So, the best that can be hoped for within the first months of lean knowledge transformation is to release the capability that is already there. It does. I often hear, "I always knew we could do this." Usually, no one can point to exactly what changed or when, but there's always a lot of "we," such as "We are helping each other now" or "We understand each other's problems because we talk to each other." And people will notice a new common purpose: "We all want the same thing." And they want more opportunities to get better at what they do, to delight more customers, and to deliver better results to their organization.

Lean knowledge releases the capability that was there already.

Continuous improvement was born in the 1940s in the Toyota Production System. Its roots are on the manufacturing floor where it's often called "lean manufacturing." When we apply it to what Peter Drucker called "knowledge work"—engineering, medicine, business, finance, and essentially any professionals working in teams—we'll call it "lean knowledge."

It's surprising that the same thinking works on the factory floor and in the professional office. The mystery of lean knowledge is the simplicity of the remedies it recommends juxtaposed with the complexity of the domains where these people work. Knowledge staff such as doctors, engineers, coders, scientists, and lawyers are smart...really smart. Lean techniques aren't designed to bring them new domain expertise. Lean thinking is based on the premise that it is rarely a lack of expertise that holds back these teams; it's almost always a preponderance of simple failures: failing to define goals clearly, failing to identify errors and resolve them, failing to solve problems at their root, or failing to communicate with colleagues or customers. Lean knowledge targets these sorts of failures.

1.1.1 Dangerous assumptions

Assumptions, as we will use the term here, are unvalidated conclusions that form a foundation for action. Our lives demand assumptions. I'm going to assume my car will start tomorrow morning. This assumption is not a deeply held conviction. I know there likely will be a morning when my car doesn't start. But tonight, I'm not taking any actions to validate (such as asking a mechanic to look at it) or creating any countermeasures (such as arranging for a friend to pick me up). I'll act as if my assumption is true and deal with the problem if it presents. I'll also assume that the road system will function well enough to get me to work. That my card key will open the building door. These and a hundred other assumptions are necessary just to start the workday.

We must make many assumptions just to function at home and at work. The problem is that some of those assumptions in the complex domains of knowledge work are going to be wrong—not dead wrong, but wrong enough to create a great deal of waste. Some of the most dangerous assumptions are among the most common:

- This is how we've always done it and it works fine.
- Someone I trust told me, so I don't need to see it myself.
- It seems logical, so it probably works.
- It's not working because other people are not trying hard enough.

So how often are our assumptions wrong? In the routine parts of life, it may be rare. I don't remember the last time my car didn't start. But assumptions about knowledge work are different. A small misunderstanding of a customer's need or a problem's root cause can render a large effort entirely wasteful. Quantifying how often our assumptions are wrong is difficult, so let's defer to Taiichi Ohno, the creator of the Toyota Production System and one of the first lean thinkers. The opening of this chapter quoted him as saying the wisest of us is right only 7 out of 10 times. Put simply, at least 30% of our assumptions are wrong.

1.1.2 The dilemma of lean knowledge

This is the dilemma of lean knowledge. You must make assumptions to function, but many of those assumptions will be wrong and you have no idea which ones. You understand most when you realize there's much you don't understand. The height of poor understanding is to think you understand everything. Or, as Ohno is often quoted, "having no problems is the biggest problem of all." This dilemma never resolves. It never becomes intuitive to distrust your intuition. And this is why lean knowledge rests so heavily on experimentation. New ideas always require assumptions, some of which you won't even be aware you're making. No matter how smart or hardworking you are, a large portion of your assumptions are going to be wrong. The only way to find them is to experiment: measure the results and compare them to what you expected. This may seem like common sense, but it's an uncommon way to think. This is why a great deal of this book is spent addressing that dilemma with examples and experiences. There is also a healthy dose of references from those who went this way first: the pioneers who applied lean thinking to the manufacturing floor like Deming, Ohno, and Shingo. But more on that in Chapter 2.

1.2 What problem are we trying to solve?

One of the most important aspects of lean knowledge is effective problem solving. So, let's start our journey by defining the larger problem we are trying to solve using two techniques we'll cover in detail later: the *canvas view* from Chapter 8 and formal problem solving from Chapter 12. A problem statement that forms the premise of this book is shown in Fig. 1.1; it is a high-level view of this book. In later chapters, we will zoom into areas of interest again and again. But for now, Fig. 1.1 tells the story.

1.2.1 The need

The problem statement begins with the need, the purpose of taking up the problem at all. In this case, the need is inarguable. Knowledge organizations from product designers to business developers to IT groups to scientists address some of the most important issues their organizations encounter. An organization's health now and in the years to come depends on the solutions that knowledge staff create. New products. New acquisitions. New markets. New channels. Work product for critical clients. Problem solving. All require a great deal of work from some of the organization's smartest, most educated, hardest-working team members.

Organizations aren't getting the performance they need from their teams [2].

Need	Organizations depend on knowledge staff, teams of highly skilled experts and professionals, to deliver high-value, timely solutions to external and internal customers.
Problem	Even when good people are trying to do the right thing, knowledge work frequently disappoints. The work product is often late—sometimes very late. Moreover, finished work product often disappoints customers because of omissions and errors.
Root cause	Hidden waste consumes 80%–90% of the effort in a typical knowledge organization.What is left creates too little value to meet expectations.
Solution	Cut hidden waste. Because there is so much waste, even small increments of waste reduction lead to large increases of value. In addition, new waste reduction can be realized year after year, leading to dramatic increases in value creation over time.
Counter-measures	Eliminating waste with techniques and tools that simplify, engage, and experiment **Simplify** complex and varying workflows, long task queues, ambiguous calls to action, and knowledge transfer. **Engage** knowledge staff through a common vision, meaningful connections among colleagues, creating challenges that can be won, and tenaciously "protecting the brand." **Experiment** continuously. Start with business humility; measure gaps, remedy those gaps, and validate. Repeat.

Fig. 1.1 The premise of this book written as a formal problem statement.

1.2.2 The problem

The problem is that knowledge work frequently causes disappointment to the organization. Why? Normally, the people doing the work are competent and dedicated, but that doesn't guarantee good results. Far from it. From running projects to hiring top talent to simply managing ordinary work, everyone with a few years of experience in a knowledge organization has stories about mediocre results peppered with the occasional spectacular failure.

More than 50% of all engineering projects fail to meet their goals [3].

1.2.3 The root cause

There are multiple root causes for why knowledge work fails to meet expectations. The hypothesis of this book is that a great many of those causes fall into the broad category of hidden waste. Hidden waste is, of course, exceedingly difficult to measure. After all, if you can measure it, it's not hidden. So, the quantification is an estimate, but experts in the field of lean transformation commonly peg the proportion of waste in our workday at about 85%. In other words, about 1 hour in 6 is spent on something that a customer really wanted. The rest is spent working on problems that customers care little about, creating solutions that don't solve problems, passing on mistakes that will have to be corrected later, writing long and angry emails to colleagues, and so on and so on.

> *Most business processes are 90% Waste and 10% Value-added work.*
>
> ***Jeffrey Liker [4]***

About 15 years ago, when I first heard that 85% of our effort generated waste, I thought, "Not us...we're better than that." Today, I'm convinced of it. Consider one definition of waste: the effort it took to do something of value minus the effort it would have taken had it been done *perfectly*. Occasionally, we do something almost perfectly—we catch every substantial mistake, we listen to everything the customer or client asked for, and our leadership coordinates every person to get something genuinely complex out the door on schedule. It happens that way sometimes, but more often we push and strain and get things done, wondering why it's so hard. That frustration comes from waste like mistakes, false starts, arguments, finding out you've been working on the wrong problem, weak solutions, and indecision. From that perspective, comparing the *actual effort* to the *perfect effort*, 85% can seem optimistic!

1.2.4 The solution

The obvious solution to this problem is to reduce hidden waste. The good news about waste being so large is that small increments of waste reduction deliver large opportunities to increase value. Consider this example: assume that waste in our knowledge organization is 85%; what if we could reduce that by just 5% in 1 year? That may seem a small gain: 80% or 85% waste both seem dismal. But let's look at what happens when we use that 5% to increase value. If we go from spending 15% of our day creating value to 20%, that's an increase of one part in three. That's enormous! Now, let's say we install a culture where every year we chip away a few percent of waste. In 3 or 5 years, a team can double the value it creates per year!

> *If we could convert all the waste to value-creating time, we would increase our development throughput by a factor of four...*
>
> ***Ward and Sobek [5]***

waste so she remains successful. But over time, most people wear down. Perhaps, in a few months, Gretchen will push herself a little less and start asking, "Why is it so hard to get something done here?"

1.4 Six ways this book is unique

There are a lot of books on management techniques and some that apply lean thinking to knowledge work. So, what makes this book different? Here are six ways it is unique:

1. This book is aimed at expert teams with deep respect for their knowledge and experience. There is no attempt to bring "check the box" thinking to fields that require decades to master. Rather, we will seek to untangle tedious problems precisely so the experts can apply more of their expertise to the areas where it is most needed. And the material is wholly focused on knowledge work, avoiding overused analogies to the factory, which can be confusing or even off-putting to experts.
2. It presents skills and techniques common to all knowledge work. Included are examples from IT, product development, sales, management, and other areas. This book isn't focused on a narrow slice of knowledge work such as software development or project management.
3. It provides dozens of simple visual tools that anyone can implement in their existing framework, all of which work in virtual and face-to-face interactions. This includes a number of tools unique to this book, such as:
   - the Engagement Wheel (starting in Chapter 7);
   - the Test Track and Success Map (Chapter 9);
   - the Error-Proofed Fever Chart (Chapter 18); and
   - the canvas view in many forms.
4. It introduces the **Knowledge Work Improvement Canvas** (**KWIC**, pronounced "quick") a single-page view to manage improvement cycles systematically. The KWIC is applied to 10 common knowledge workflows in the later chapters. It will let you create and sustain a broad range of improvements for your organization.
5. It presents a systematic use of Stop-Fix alarms[a] to provide reliably clear calls to action for teams dealing with complex issues.

[a] Stop-Fix alarms in lean knowledge are similar to Andon/Autonomation in lean manufacturing.

6. It brings structure, which can be lacking in lean thinking for knowledge work where tools are often presented one after the other. Without structure, readers struggle to apply the thinking to the problems that present in their organization. This book presents a well-structured approach based on the following:
   - **Simplify** to reduce effort needed to accomplish something.
   - **Engage** to focus effort in precisely the right place.
   - **Experiment** to learn at astonishing speed.

1.5 Who is this book for?

This book is written for anyone who wants to improve the work product quality, response time, and engagement of their professional and expert teams, especially:

- those who practice knowledge work and want to build a culture of continuous improvement: doctors, attorneys, software developers, engineers, managers, salespeople, architects, scientists, product managers, marketers, business developers, chemists, investors, accountants, analysts, and domain experts of every stripe;
- project and initiative managers who want to deliver better results in their projects and initiatives; and
- managers and thought leaders who want to guide the teams they lead to improve.

1.6 Structure of the book

This book is organized into three main sections: Fundamentals (Chapters 2–6), Understanding the 8 Wastes of Knowledge Work (Chapters 7–14), and Improving 10 Key Workflows (Chapters 15–24).

1.6.1 Section I. Fundamentals (Chapters 2–6)

Chapter 2: A brilliant insight

Today it's clear that lean thinking works in manufacturing environments. More than 60% of industrial companies have adopted lean manufacturing, starting with automotive and moving to almost every industry on earth. But does this thinking apply to knowledge work? This chapter explores how lean thinking has been applied to health care, product development, and software development to establish that the general principles work.

Chapter 3: Creating value from knowledge work

Define the kinds of value that knowledge work creates following the axiom that *value* is what the customer is willing to pay for.

Chapter 4: The lean equation

Some portion of our effort goes to create value; everything else we do is defined as *waste*. This leads to the lean equation: Effort = Value + Waste. This chapter explores how the lean equation presents itself in knowledge work and the many insights it provides.

Chapter 5: DIMINISH: Recognizing the 8 Wastes of Knowledge Work

The 7 Wastes of Manufacturing have a long history, starting in the 1950s with Taiichi Ohno, but they are focused on the factory floor, capturing waste like inventory and worker motion. Using the acronym DIMINISH, this chapter introduces the 8 Wastes of Knowledge Work, which are more relevant to the workflows of professionals and experts.

Chapter 6: Simplify, engage, and experiment

Cut waste structured along three dimensions: simplify, engage, and experiment.

1.6.2 Section II. Understanding the 8 Wastes of Knowledge Work (Chapters 7–14)

Each chapter in Section II discusses methods to reduce one of the 8 Wastes of Knowledge Work, detailing examples of the waste and offering a few tools to reduce it.

Chapter 7: Reduce Waste #1: Discord

Discord is the waste created when people don't align to the organization as a whole or to the people and groups within. It's marked by people pulling in competing directions, failing to share information, and dropped handoffs between groups.

Chapter 8: Reduce Waste #2: Information Friction

Information Friction is the waste that arises when information is present somewhere in an organization, but unavailable to or inaccurate for all those who need it. The lack of information availability creates bad decisions, failure to generate or respond to a call to action, and mistakes in work product.

Chapter 9: Reduce Waste #3: More-is-Better Thinking
More-is-Better Thinking is the waste of grinding away at a task without a plan to succeed. It creates busyness, sometimes to the point of exhaustion, but little value.

Chapter 10: Reduce Waste #4: Inertia to Change
Inertia to Change is the waste that occurs when people act to impede improvement. It is usually unintentional, such as when people experience a natural aversion to change. On rare occasions, it may be intentional, for example, when people actively work against change to protect their position or status.

Chapter 11: Reduce Waste #5: No-Win Contests
No-Win Contest is the waste created by asking people to do something they cannot do well within the allotted time. Left without the option of success, they will do something. Give people two jobs to do in the time they can do only one well and they will almost always fail: most will either do one well and leave the other undone or do both, but poorly.

Chapter 12: Reduce Waste #6: Inferior Problem Solving
Inferior Problem Solving is created when smart, hardworking, well-meaning people jump to conclusions before they understand either the problem or a proposed countermeasure.

Chapter 13: Reduce Waste #7: Solution Blindness
Solution Blindness is the waste created when we proceed with solutions, even well-crafted solutions, either ignoring contradictory information collected since the solution was developed, or failing to take reasonable steps to collect such information.

Chapter 14: Reduce Waste #8: Hidden Errors
Hidden Errors is the waste encountered by creating knowledge work product with undetected mistakes and omissions. We will begin with our core assumption that most waste comes from good people trying to do the right thing. So, rather than blaming people for making errors, we will focus on asking how our workflows allowed these errors to pass undetected.

1.6.3 Section III: Improving 10 Key Workflows (Chapters 15–24)

Chapter 15: Standardize workflow

Apply the principles of standard work to knowledge workflows.

Chapter 16: Workflow improvement cycle

Manage a cycle of improvement using the Knowledge Work Improvement Canvas (KWIC).

Chapter 17: Workflow—Checklists and expert rule sets

Build checklists and expert rule sets, both of which grow a group's capability in a domain of knowledge by recognizing primary guidelines the team uses, recording them and then refining and augmenting them over time. These checklists and expert rule sets also provide a natural mechanism to teach people new to the organization.

Chapter 18: Workflow—Problem Solve-Select

The Solve-Select workflow is a modification of Chapter 12's formal problem solving aimed at problems that are too complex to be resolved with a single team. This workflow has one group driving through the expert issues ("solve team") working hand-in-glove with a leadership group ("select team").

Chapter 19: Workflow—Visual management for initiatives and projects

Visualize schedule for modest-sized projects. Project management is one of the most commonly identified areas of dissatisfaction with knowledge work stakeholders. It's frequently stated that most projects are late or fail to deliver to expectations or both. These techniques are effective at helping a cross-functional team coordinate effort to deliver high-quality work on time.

Chapter 20: Workflow—Visual management with buffer

Augment Chapter 19 with the concept of project buffer derived from Critical Chain Project Management (CCPM). Buffer is an added complexity, so it's not for every project. But its benefits are substantial, so these techniques are normally applied where they are most needed: with large projects.

Chapter 21: Workflow—Kanban and Kamishibai: Just-In-Time Rationalization

Create a resource-rationalized workload for large groups of small tasks to avoid oversubscription and multitasking. There are three main tools: (1) "Ruthless Rationalization": taking on only tasks that are within the capacity of the team;

(2) a Kanban board for irregular tasks (that is, tasks that are highly varying in size and timing); and (3) a Kamishibai board for tasks that repeat at a set cadence like monthly project reviews or quarterly performance reviews.

Chapter 22: Workflow—Putting out "fires"

Modify the formal problem solving of Chapter 12 to put out a "fire." Whatever discipline of knowledge work you practice, you're probably sometimes faced with a sudden demand for a large amount of resources when something goes wrong. A competent and diligent response to these events increases engagement of the team and builds confidence in the organization.

Chapter 23: Workflow—Visualizing revenue gaps

Visualize revenue shortfalls and the countermeasures to make them up.

Chapter 24: Workflow—Leadership review of knowledge work

Improve leadership review of knowledge work, a powerful tool to catch defects and mistakes that are too expensive to find by experimentation. Here, people with high business acumen and experience guide a team of domain experts.

References

[1] T. Ohno, Taiichi Ohno's Workplace Management, Special 100th Birthday Edition, McGraw Hill Education, Kindle Edition, 2012, 1.
[2] S.M. Johnson Vickberg, K. Christfort, Pioneers, Drivers, Integrators, and Guardians, HBR, March–April, 2017.
[3] Huthwaite Innovation Institute, www.barthuthwaite.com (landing page), viewed October 10, 2019.
[4] J.K. Liker, The Toyota Way. 14 Management Principles Form the World's Greatest Manufacturer, McGraw-Hill, 2004, 8.
[5] A.C. Ware, D.K. Sobek II, Lean Product and Process Development, second ed., Lean Enterprise Institute, 2014, 33.
[6] R. Maurer, One Small Step Can Change Your Life: The Kaizen Way, Workman Publishing Company, 2004, 31–32.
[7] E. Goldratt, J. Cox, The Goal: A Process of Ongoing Improvement, third ed., North River Press, 2004, 332.

CHAPTER 2

A brilliant insight

If you are going to do kaizen[a] continuously, you've got to assume that things are a mess.

Taiichi Ohno [1]

[a]Japanese for "continuous improvement."

2.1 Introduction

This chapter presents evidence that lean thinking applies to knowledge work—that is, it works across domains, across a range of functions, and for organizations small and large. The discussion starts with a brief history of the birth of lean in Toyota in the middle of the 20th century. Using lean health care, the discussion then bridges from manufacturing to knowledge work; health care is unique in lean thinking because it has large portions of both operationally demanding procedures (those ideal for lean manufacturing techniques) and knowledge workflows that are among the most demanding of individual expertise. The discussion then moves to other proven areas of lean knowledge:

- lean product development derived mostly from the auto design industry;
- lean startup, a method of managing entrepreneurship;
- Critical Chain Project Management (CCPM), one of the first structured approaches to lean knowledge work outside the auto industry; and
- Agile Software Management, lean thinking applied to software product and IT development.

Improve
https://doi.org/10.1016/B978-0-12-809519-5.00002-8

2.2 The birth of lean thinking

Lean thinking took its first foothold in the Western automotive industry in the 1990s due in large part to the book *The Machine That Changed the World* [2]. Lean has dug in more deeply since; today, lean factories are the standard in manufacturing organizations of virtually every kind. For example, top lean manufacturing organizations as of 2014 included marquee companies: Toyota, Ford, John Deere, Textron, Intel, Caterpillar, and Nike [3]. And, as lean has strengthened its presence in manufacturing, it has begun to transform knowledge work, for example, in lean health care [4], lean product development [5], lean entrepreneurship [6], lean organization management [7], lean sales [8], lean legal [9], lean accounting [10], and lean nonprofit management [11]. But the story of lean begins in Japan nearly 100 years ago.

Taiichi Ohno

Taiichi Ohno developed the production systems "that helped make the Toyota Motor Company one of the most powerful automobile producers in the world" [12]. He is sometimes called the father of lean thinking. His methods were used in Toyota, where they "helped transform Toyota from a small car maker near bankruptcy in the late 1940s into the third-largest auto maker" [12]. His wisdom in lean thinking is broad, which is why he's quoted here so often. His contributions earned praise, including this small sample:

"He ranks among the production geniuses of the 20th century," said Michael A. Cusumano, Asst. Prof. at the Massachusetts Institute of Technology, an author on Japanese manufacturing.

Norman Bodek, President of the Productivity Press (who translated Mr. Ohno's books into English), said, "His contribution to modern manufacturing ranks with the work of Henry Ford" [12].

Lean thinking began at the Toyota Motor Company in the 1930s and 1940s. In the years after World War II, Japan's economy was nearly destroyed. Toyota was producing cars in small numbers. For example, in the early 1950s, Toyota was averaging a few hundred automobiles per month, where as Ford and Chevy were each producing that many on a daily basis [13–15]. Toyota was forced to develop techniques that cut costs and raised quality without the massive economies of scale enjoyed by its Western rivals.

Unable to spend its way to higher efficiency through tooling and other capital-intensive methods, Toyota began a journey that it is still on: continuously improving using the reduction of *hidden* waste to fund increased focus on customer value. Every company knows that there is a certain amount of hidden waste within its walls. The brilliant insight of Toyota was that there was so much *hidden* waste that the company could be transformed by reduction of this. Of course, it was hidden—ordinary thinking will remove revealed waste. But it was genius to recognize that most of what every person does every day creates waste.

The brilliant insight of Toyota was there was so much hidden waste that reducing waste would transform the company.

The techniques they created were eventually captured in the Toyota Production System (TPS). The birth of TPS is largely attributed to Taiichi Ohno, the Head of Production at Toyota after World War II [16]. In his later years, Ohno wrote about his experiences. He recognized that if waste were reduced, all performance indicators would improve: cost would be reduced, quality defects would decrease, and delivery time would shorten; customers would be more satisfied and so would the employees. And the organization would be more profitable.

W. Edwards Deming

Dr. W. Edwards Deming was an early practitioner of lean thinking, but in Japan rather than in his US homeland. Deming is best known for his pioneering work in Japan in quality and operations. Starting in 1950, he taught methods for improvement with a high focus on how people worked together. One of his best-known methods is called the Deming Cycle or Plan-Do-Study-Act (PDSA), an improvement on the Shewhart Cycle or Plan-Do-Check-Act (PDCA) [17].

He was a contributor to the post-war turnaround of industry in Japan. "Deming's theories were discovered by a few American managers in the early 1980s as they struggled to catch up to such Japanese companies as Toyota and Matsushita, which had adopted Deming's methods" [18]. "Dr. Deming's role as the architect of Japan's post-World War II industrial transformation is regarded by many Western business schools and economists as one of the most significant achievements of the 20th

Continued

W. Edwards Deming—cont'd

century" (LA Times, October 25, 1999). He is often called the "father of the third wave of the industrial revolution" [19].

"The Japanese rewarded Deming for his efforts in 1951 by establishing a corporate quality award in his honor. The Deming Prize is often referred to as the Nobel Prize of Japanese business" [20].

"The patron saint of Japanese quality control, ironically, is an American named W. Edwards Deming, who was virtually unknown in his own country until his ideas of quality control began to make such a big impact on Japanese companies" (Akio Morita, co-founder of Sony) [21].

TPS is a system of continuous improvement based on standard work augmented by Kaizen ("incremental improvement") to bring sustainable performance increases. *Standards* and *improvements* work together: the standard creates a foundation upon which the improvement can build. Then, yesterday's improvement becomes tomorrow's standard and the cycle repeats in small increments. From a distance, these small increments appear to meld into a continuous flow of improvements; hence, the result is called *continuous improvement*.

TPS is said by Toyota to rest on two concepts [22, 23]:

- *Just-In-Time* Inventory Management minimizes inventory to just what is needed delivered to just where it's needed just when it's needed.
- *Autonomation* or *Jidoka*, in this text, called Stop-Fix.

Stop-Fix guides the development of machines and processes to identify defects quickly and then stop the process until the conditions that caused those defects have been corrected. It is "Stop-Fix" that led to the famous behavior at Toyota whereby assembly workers would pull a cord to stop the line when they found a defect.

At Toyota assembly, workers stop the line

Granting that level of authority to assembly workers was unheard of at the time—not just because the rest of the world's car makers didn't think the people putting together their products had enough judgment but because stopping the line was expensive [24].

Stop-Fix alarms are as applicable in knowledge work as they are on the factory floor. Stop-Fix is perhaps just common sense: in our work, we define the errors we cannot tolerate and then don't tolerate them. While it may seem obvious, few people appear to grasp how to apply this simple principle. Stop-Fix alarms will be discussed throughout this text.

So, does lean manufacturing work? There is virtually no doubt that it does. The first example is Toyota, a one-time insignificant automobile manufacturer which is today probably the most successful car company in the world. But that story took more than half a century to develop. What about in just 1 year? That's how long it took to bring substantial improvement to the GM factory in Freemont, CA, which would come to be known as NUMMI.

Shigeo Shingo

Another pioneer of lean thinking, Shigeo Shingo, also helped transform Toyota. For 50 years [25], Shingo taught lean thinking and helped countless companies in Japan and around the world to improve. "His disciplined, methodical approach and careful notes have provided the basis for 18 books" [26] on lean manufacturing and quality. He is most famous for the process "Single Minute Exchange of Dies" (SMED), developed in 1969, which is a combination of mindset and methodology to drive down the time to change over a factory line from hours to minutes.[b] For example, Toyota used it to convert their factory lines to produce different models of cars. SMED is known to reduce inventory by 90%, shorten lead times from months to hours, and reduce defects by 97%. But "shortening setup times...is only a small part of Mr. Shingo's achievements" [26]. He also wrote "Study of Toyota Production System from an Industrial Engineering Standpoint," the first English book to present just-in-time (JIT) production and Kanban [26]. He also developed "poka-yoka," a now commonly used systematic method for mistake proofing in the factory [27].

[b]"Single minute" is literally a number of minutes that can be represented with a single digit; in other words, under 10 minutes.

2.2.1 The story of NUMMI

One of the most compelling success stories for continuous improvement is the 1984 Toyota/GM joint venture at the New United Motor Manufacturing, Inc. (NUMMI) plant in Freemont, CA. The NUMMI plant started as the worst in the GM system according to a wide variety of measures [28, 29]. "Productivity was among the lowest of any GM plant, quality was abysmal, and drug and alcohol abuse were rampant both on and off the job" [30]. Morale was stunningly low, with workers admitting to sabotaging the cars they made: "They'd intentionally screw up the vehicles, put Coke bottles or loose bolts inside the door panels so they'd rattle and annoy the customer. They'd scratch cars" [31]. But just 1 year of continuous improvement imported from Toyota transformed NUMMI. After the first year, NUMMI became perhaps the best auto plant in America. Quality improved as did employee morale. According to Jeffrey Liker, one of America's most admired advocates of continuous improvement, "The best measure they use is how many defects there are per 100 vehicles. And it was one of the best in America. And it was the same for the Toyota cars that were made in California as the Corollas that were coming from Japan, right from the beginning" [32].

Lean manufacturing works. In their 1990 bestseller, "The Machine That Changed the World," Womack et al. compared Japanese auto manufacturers to US and European manufacturers [33]:

- productivity higher by 82%;
- quality better 49%;
- space for assembly lower by 27%; and
- inventory days lower by 92%.

Table 2.1 gives a brief overview of lean manufacturing.

Can you overuse lean manufacturing analogies?

Lean has gotten something of a bad name since it was coined in the 1990s. For one thing, it's become a euphemism to reduce head count [34]: "We cut jobs 10% to lean our organization." That view is antithetical to lean thinking, which relies on creating trust for the long-term relationships needed to transform an organization.

For knowledge work, there's a more subtle problem with analogies from lean manufacturing that are, at best, not helpful in teaching knowledge staff and are, at worst, off-putting:

- Using Japanese words where equivalent English words are available.
- Using analogies of machines or processes that are largely unfamiliar to knowledge staff.
- Devaluing tacit knowledge. Tacit knowledge is understandably avoided on the factory floor, but is core to how experts and knowledge staff contribute value.
- Using analogies of factory work, which often can be learned in a few weeks or months, to knowledge staff who have invested decades building their unique expertise.

Doubtless there are many analogies between the factory floor and knowledge work, but many are confusing or create unnecessary barriers to knowledge staff. Thus, in this book, those analogies will be kept to the minimum. But we will use the term "lean" because it is judged to be the best term to connect the core thinking here to the heritage of focusing first on customer value.

Table 2.1 Lean manufacturing overview.

Lean manufacturing	
What is it?	A management philosophy founded on creating smooth flow, minimizing waste, and respect for the worker.
When was it developed?	The Toyota Production System (TPS) led by Taiichi Ohno and Shigeo Shingo starting in post-World War II Japan.
How does it use lean thinking?	• Focus on the slice of things we do that create value for our customers. The rest is waste and should be ceaselessly slashed. • The way we work is captured in standards which are improved in small increments. • The foundational pillars are Just-In-Time, producing and sourcing only what is needed when it is needed and Autonomation or "Stop-Fix" where problems we cannot tolerate are identified and corrected before defects are passed forward.
Is it used?	Lean manufacturing has revolutionized manufacturing starting in the automotive industry in the West in the 1980s and spreading to almost every industry.
Does it work?	Yes. One example: Toyota went from making a few cars a day to being the world's largest automotive company, which is largely credited to TPS.

2.3 What is knowledge work?

The term *knowledge work* is generally attributed to Peter Drucker in his 1959 book "Landmarks of Tomorrow." It typically refers to work done by professionals and experts who are said to "think for a living." Knowledge work is part of virtually every organization of any size. Fig. 2.1 shows examples of knowledge work domains. In many organizations, the need for knowledge work is core to the mission: a medical practice depends on doctors and nurses

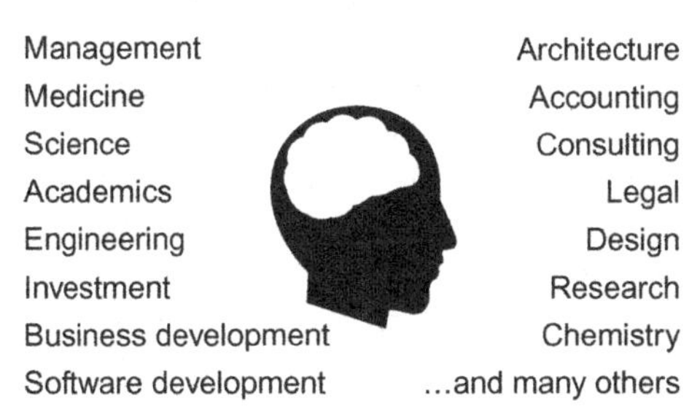

Fig. 2.1 Examples of knowledge work.

Domain knowledge

- The single most defining characteristic of knowledge work
- Usually requires formal education at or above a 4-year diploma
- High levels of expertise are acquired only with years of practice

Fig. 2.2 Domain knowledge is central to knowledge work.

and an electronics company relies on engineers, scientists, and technical salespeople. In others, such as a distribution business, even if knowledge work is not at the core, it's a necessary support—management and accounting are two professions that are seen in most organizations.

Most of us think about domain knowledge (or *expertise*) as the central requirement for knowledge work: med school for doctors, law school for attorneys, and university training for most other professions. Domain knowledge is certainly a requirement for knowledge work. These skills are specific to the domain in which they are practiced. They are difficult to acquire—expensive, time-consuming, and requiring high cognitive skills. And the initial acquisition is only the start. Domain knowledge takes years of experience to master (Fig. 2.2). Malcom Gladwell famously marked the time to acquire expertise at 10,000 hours of working in the core of the field, something that takes at least a decade for most people [35].

It has become a kind of truism in the study of creativity that you can't be creating anything with less than 10 years of technical-knowledge immersion in a particular field. Whether it's mathematics or music, it takes that long.

Mihaly Csikszentmihalyi [36]

Start by assuming that people are not the problem

Approach every problem assuming knowledge staff are able and willing to do the right thing. Why? These people have worked hard to get where they are. For example, engineers got through university where the dropout rate is about 50% [37], compared to Marine Corp boot camp training with a dropout rate under 15% [38]. The above assumption is almost always right and demands leaders evaluate themselves honestly rather than blaming others. So, hold the assumption that the system is the problem until the facts demand another conclusion.

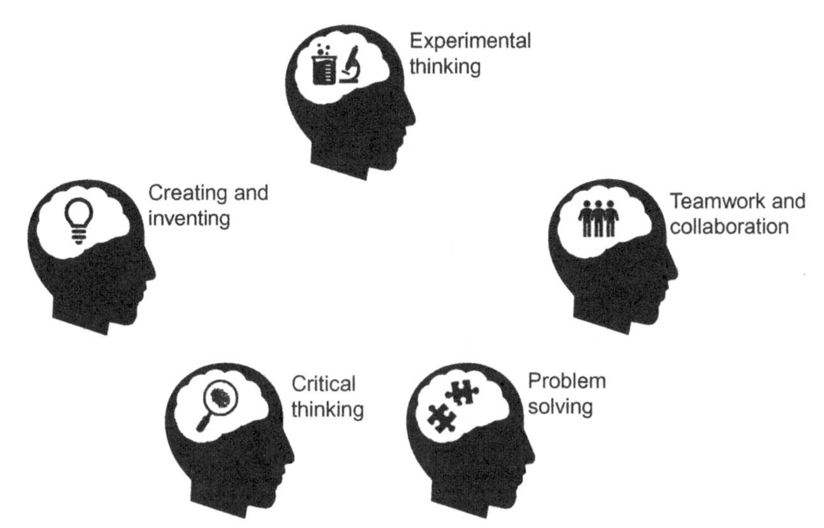

Fig. 2.3 This book will focus on five skills, which are shared across the many disciplines of knowledge work.

Even though domains vary, knowledge work of all stripes shares common skills, especially when the work is done in a team. Five of the most common shared skills, shown in Fig. 2.3, are as follows:

- Experimental thinking: being able to create open-minded hypotheses and measure outcomes to identify good alternatives.
- Teamwork and collaboration: the social skills to work with colleagues to accomplish a shared goal.
- Problem solving: diligently understand problems, find root causes, and create effective solutions.
- Critical thinking: evaluate a chain of logic to validate sound reasoning. Prized in management reviews where leaders, people of high general acumen, evaluate the work of a domain expert against a broad range of characteristics.
- Creating and inventing: developing novel viewpoints or solutions.

2.4 Lean knowledge vs. lean manufacturing

Applying the Toyota Production System outside the shop floor can be done, but this takes some creativity.

Fujio Cho, President of Toyota Motor Corporation [39]

Don't try to bring lean manufacturing upstream to product development. The application of Lean in product development and manufacturing are different. Some aspects may look similar, but they are not! Be leery of an expert with experience in lean manufacturing that claims to know product development.

Jim Womack's advice to Harley-Davidson [40]

Can the principles of lean manufacturing be applied to knowledge work? Yes, but there are several pitfalls to avoid. As Stats and Upton stated in their 2011 Harvard Business Review article, one issue is that lean knowledge is much less repetitive:

> *...attempts to apply lean approaches to knowledge work have proved frustratingly difficult. Most in the business world believe that knowledge work does not lend itself to lean principles, because, unlike car assembly, it is not repetitive and can't be unambiguously defined [41].*

If processes are repeated only rarely, it stands to reason that the benefit of optimizing will be diminished. Of course, no task is repeated exactly in any domain, whether you are assembling a refrigerator or finding a new area of the world to expand distribution into. But on the factory floor, the first step is usually to minimize variation. If we have two parts going together that now don't fit, it's likely because one of the parts has an undesirable dimensional variation. In knowledge work, variation is expected. Every region of the world we open to distribution will be different; laws, languages, and customs will come in new combinations and we will always rely on experts to cope with that variation. On the factory floor, we seek to eliminate tacit knowledge everywhere possible; in knowledge work, tacit knowledge is the foundation of our work.

The methods of lean manufacturing have been optimized for a domain with very different characteristics...[M]anufacturing deals with predictable and repetitive tasks, homogeneous delay costs, and homogenous task durations [42].

Reduced repetition is just one of many differences between factory and knowledge work; knowledge work has orders of magnitude higher reliance on tacit knowledge, takes longer to execute (often years to complete complex initiatives), and has obscure workflows. Yet there is an important similarity: knowledge work has subprocesses that repeat almost exactly, including:

- approval processes;
- progress reporting;
- customer service contracts;
- formal problem solving;
- tracking metrics and key performance indicators (KPIs); and
- patent and trademark applications.

The list goes on. It is the subprocesses that support knowledge work where waste can be identified and cut. They repeat even when the work they support does not. A design engineer creating a new strut may repeat the same inefficient purchase processes or useless design reviews many times, even though the strut design will vary each time. And these repetitious tasks occupy a great deal—often most—of the precious time of knowledge staff. Stated in the language of lean health care, "Lean is not necessarily about clinical care, but about reducing inefficiencies in processes needed for that care, by transforming waste into value" [43].

Lean applies in every business and every process. It is not a tactic or a cost reduction program, but a way of thinking and acting for an entire organization [44].

The Lean Enterprise Institute

2.5 Lean health care

Those who implemented lean health care were among the first to face the question of whether lean thinking could be applied outside of manufacturing. "[I]n the minds of many, the health industry is different. This is certainly true as to its history, technology and culture. However, the decisive factors in what works and what does not are the managerial processes, which are alike for all industries" [45]. This led Teich and Faddoul to the following conclusion: "This is the reasoning that allows the principles of lean production and management to be applied in healthcare, despite these being originally developed for application in other industries" [13]. Or, as Sloan et al. put it, "While healthcare provision and manufacturing operations may differ in many ways, the prerequisites for successful change in each have been

Table 2.2 Results of lean health care conversion at the Virginia Mason Medical Center, 2003–10 [47].

Improvement after lean conversion	Over 7 years
Labor/productivity increase	45%–75%
Cost reduction	25%–55%
Throughput increase	60%–90%
Space reduction	35%–50%
Lead-time reduction	50%–90%

found to be remarkably similar" [46]. Results back this up. Table 2.2 shows results from an early study done at the Virginia Mason Medical Center (VMMC) demonstrating that lean health care delivers real results.

A few other points about lean health care:

- Dr. Gary Kaplan, Head of Seattle's Virginia Mason Medical Center in 2001, applied continuous improvement to health care. The Virginia Mason Medical Center started losing money in 1998. The hospital staff applied basic techniques to trace the path a patient travels during a procedure: "it was this immensely powerful visual experience for the entire team" that they realized how time was wasted. What eventually emerged from that continuous improvement transformation was the "Virginia Mason Production System," which has been copied by dozens of hospitals. "A recent study of the nation's hospitals put Virginia Mason now in the top one percent in safety and efficiency for two years running" [48].
- Teich and Faddoul cite other gains at VMMC: "hospital increased profit margins, decrease in deaths, and decrease in the number of medication errors...85% reduction in how long patients wait for a lab result, increased productivity by 93%, and lowering inventory costs by $1 million" [13].
- Toussaint and Barry cite a 2-year study with events in eight hospitals. Results as of spring 2012 include on-time starts increased from 50% to 70%, operating room (OR) cases per month increased from 329 to 351, OR turnaround time from 60 minutes to less than 40 minutes, cases rescheduled due to late starts decreased from 21% to 4.4%, and same-day surgery cancellations reduced from 7% to 3% [49].
- Mark Dean cites numerous other examples of successes in health care including Jewish Hospital increasing throughput in a lab by 28%, while improving patient satisfaction by 50%, and the University of Pittsburgh Medical Center Shadyside going from two patient falls per day to zero patient falls in 3 months [50].

Table 2.3 Summary of lean health care.

Lean health care	
What is it?	Applying lean thinking to health care starting with operational issues like patient flow and physical inventory
When was it developed?	Became popular around 2000, most famously with the Virginia Mason Medical Center (VMMC)
How does it use lean thinking?	• Uses standard factory lean thinking in medical settings, often starting with value-stream mapping, a staple of lean manufacturing • Seeks to deliver maximum value to patient • Defines defects in a way that is relevant to medicine such as patient falls, wrong-site surgery, and medication errors
Is it used?	VMMC is one of the leading voices with the Virginia Medical Production System (VMPS). The VMMC's website shows the Center has run 3000 improvement events: https://www.virginiamason.org/vmps

- Dr. Lisa Yeri is the Director of Hepatobiliary Pathology and Medical Director of Continuous Improvement within the Division of Clinical Transformation at the Cleveland Clinic. In the article *A Collaborative Approach to Lean Laboratory Workstation Design Reduces Wasted Technologist Travel*, Dr. Yerian and her coauthors describe how they applied lean principles to improve workflow during the design phase of the Cleveland Clinic's pathology lab [51, 52].
- Author and surgeon Atul Gawande describes the lean principle of checklists for complex workflows (see Chapter 17, *Workflow—Checklists and expert rule sets*). He describes research showing that small avoidable oversights occur 75,000 times each year in US operating rooms. When surgical staff used simple checklists before surgery, major complications dropped by 36%, and deaths fell by 47% [53].

Table 2.3 shows a summary of lean health care.

2.6 Lean product development

Lean product development is a collection of lean techniques, some taken directly from the Toyota product development centers such as "Big Rooms" (Obeya) [54]. The many methods are only loosely coupled and are deployed differently at different organizations. Therefore, collecting categorized information is difficult. Table 2.4 shows an overview of lean product development and several success cases follow.

Table 2.4 Summary of lean product development.

Lean product development	
What is it?	A collection of techniques adapted from lean manufacturing to product development. (Alternatively, Toyota's "Set-Based Concurrent Engineering" method.)
When was it developed?	Introduced in the West in the 1990s, especially by Don Reinertsen and publications about the Toyota design center
Examples of how lean product development uses lean thinking	• Reduce waste through standardization • Define customer value to separate waste • Heavy dependence on cross-functional teams • Creating smooth flow by keeping knowledge work queues small and moving work in small "batches"—for example, dividing a large project into a series of smaller projects • Rigorous standards to reduce variation and maximize flexibility
Does it work?	Reduces development time by 4 ×, reduces impact of quality problems by 10 ×, and increases innovation by 10 ×[a]

[a]Allen C. Ware, Durward K. Sobek II, *Lean Product and Process Development*, second ed., Lean Enterprise Institute, 2014, p. 13.

- The Delphi Rochester Technical Center claimed a throughput increase of 200% over a period of 3 years by partly implementing lean development—with plenty of additional gains still in sight [55].
- According to his 2015 article, Jeffrey Liker says that Ford implemented lean product development in auto body design and, over the 5 years starting in 2004, "surpassed benchmarked levels of performance for quality, lead time, and cost" [56]. For example, Ford reduced lead time by 40%, improved quality by 35%, increased dimensional accuracy by 30%, and "dramatically improved craftsmanship and body fit and finish that is now among the very best in the world." During that time, Ford went from the worst automaker to the best according to J.D. Power measures of body exterior quality. "Finally, morale as measured by Ford's annual internal survey also improved by about 30% during this same time period."

2.7 The lean startup

When Eric Ries wrote his New York Times bestseller "The Lean Startup," it created a revolution in product development. Ries advises organizations to create a series of experiments to validate the value proposition of a new product in small increments. He explains that when a company spends 12 months creating a phone book-sized product specification and then another 12 months implementing it, the company goes 24 months before the customer sees how the product will work in their environment. Instead, he recommends the use of the "Minimal Viable Product" or MVP [6], the first variant of a product that delivers real value to the customer [57] (Table 2.5).

The lean startup highlights numerous enterprises that gained startling success using lean thinking on its website: http://theleanstartup.com/casestudies

Table 2.5 Summary of the lean startup method.

Lean startup method	
What is it?	Application of lean thinking to rapidly understand customer needs concurrently with product development
When was it developed?	Credited to Eric Ries starting in 2008, coming into wide use after his New York Times best-selling title, "The Lean Startup" in 2011
Examples of how the lean startup method uses lean thinking	• High-cadence experimentation to validate the value proposition of a product in the smallest possible increments • Value measurement from early in development • Focus on a small number of critical issues at each stage
Is it used?	It's difficult to measure, but Eric Ries came into wide notoriety in 2011 as a best-selling author, publicizing marquee clients like General Electric and Procter & Gamble. The Lean Startup Company (of which Ries is a co-founder) holds annual conferences on the topic
Does it work?	Yes, based on innumerable success cases, a few of which are below

- Zappos: Nick Swinmurn, founder of Zappos, used an experimental approach to build the world's largest online shoe store. He saw a need for a store with a wide selection. But, rather than relying on traditional marketing and then making large investments in an infrastructure of warehouses and distributors, he used experiments to validate that customers would value an online shoe store. He started by photographing the inventories of local shoe stores, agreeing to pay full price for any shoes he sold online. His experiment had him interacting with customers: transacting payments, providing support, and dealing with returns. By building a tiny business rather than just asking customers what they wanted (or, more accurately, what they thought they wanted), he gathered reliable data about how customers behave—what they really wanted. For example, he could test how discounts affected the buying behavior of online shoppers. The approach started slowly, but was ultimately successful. In 2009, Zappos was acquired by Amazon.com for $1.2 billion [58].
- Dropbox: Drew Houston, CEO of Dropbox, founded the Lean Startup blog (https://leanstartup.co/blog). After that, his company introduced products with smaller, faster iterations to discover what customers wanted. In 15 months, Dropbox went from 100,000 to 4 million registered users.
- Wealthfront provides access for investment managers to top hedge funds and money managers. They use continuous deployment ("smooth flow" of new products and services) in a highly regulated environment. Using lean techniques, this company manages more than $200 million and processes more than $2 million per day.
- General Electric: Mark Little (GE) and Eric Ries describe how the lean startup principles, often pigeon-holed as a software technique, dramatically shifted thinking in GE's power generation division [59].
- Procter & Gamble's CTO said in 2018, "We were spending a lot of money, sometimes before we really should have." So the company began several lean startup experiments. These were so successful that Procter & Gamble decided to take these to the enterprise [60].
- Steve Blank gives more examples of early successes [61].

Lean start-up practices aren't just for young tech ventures. Large companies, such as GE and Intuit, have begun to implement them.

Steve Blank [61]

Lean Advisors list two dozen success stories from widely varying domains: health care, education, government, and manufacturing. See http://www.leanadvisors.com/lean-success-stories. The lean startup will be featured in Section 13.5.

2.8 Critical Chain Project Management

CCPM was created by Dr. Eliyahu Goldratt in the 1990s in large part because of poor schedule performance of the most common method in industry, Critical Path Method (CPM). CCPM changes both the planning and execution of projects to provide more focus on tasks that must be completed efficiently for the project to be done on time. Today, CCPM is applied across a wide range of project types including many examples in product development. An overview of the method is shown in Table 2.6, with several success cases listed below.

CCPM has delivered impressive results to many organizations. Here are a few testimonials [62]:

- Harris Corp. finished a $250 million wafer fabrication plant in 13 months against an original projected schedule of 18 months.
- FMC Energy Systems project on-time delivery went from below 50% to above 90% when using CCPM.

Table 2.6 Summary of Critical Chain Project Management.

Critical Chain Project Management (CCPM)	
What is it?	Improves traditional project management (Critical Path Management) to rapidly identify and focus the team on schedule gaps. Manages projects using schedule buffer (see Chapter 20)
When was it developed?	Started in 1997 when Eliyahu Goldratt published "The Critical Chain"
Examples of how CCPM uses lean thinking	• Unambiguous measure of whether the project is meeting schedule requirements coupled with clear calls to action to address gaps • Heavy dependence on cross-functional teams • High-cadence team meetings
Is it used?	Hundreds of companies have used CCPM with many success cases published
Does it work?	Many well-known companies have published success stories, especially showing sharp improvements in projects completed on schedule (typically 90%)

- The Boeing T45 training simulator development saw a 20% cost reduction, substantial quality improvement, and 1.5 months reduction in schedule [63, 64].

There are many more. Goldratt's website states, "Tens of thousands of people worldwide depend upon Goldratt's business knowledge to successfully manage their organisations," and then lists numerous cases from BAE Systems to Habitat for Humanity [65]. The book *Advanced Multi-Project Management* lists almost 400 examples [66, 67] of companies deploying the method.

2.9 Agile software development

Agile is targeted at software development projects, though it has been applied successfully to some hardware projects. Three of the most common components of Agile [68] are Scrum, Scrumban, and eXtreme Programming.

- Agile Scrum divides large projects into short *sprints*, which are iterations that sum to the full project. Each *sprint* is a sort of mini-project, lasting just a few weeks and ending with a new product that could be released. The team executes one *sprint*, and then defines the next; team members loosely follow a plan to the project end, refining the project direction at the start of every *sprint*. Every project management method accepts the possibility of the endpoint changing as the project proceeds, though it's normally an undesired event. Scrum expects—even welcomes—the redirection and builds its processes to ensure that this happens on a regular basis.
- Scrumban is a lightweight version of Scrum based on the Kanban project management method of Chapter 21. It is a good fit for most IT work as well as simpler product development projects.
- *eXtreme Programming* (XP) is a set of standard processes for software development. Where as Scrum defines interactions of the team (when they meet, what roles people take on), XP defines how the team does their work (how they code, how they test, how they validate).

The first widely accepted definition of Agile is a short document resulting from a meeting of leading software developers in 2001: the Agile Manifesto [69]. As shown in Fig. 2.4, the document presents four principles of relative value, each favoring agility over rigidity, although the more rigid items still retain significant value.

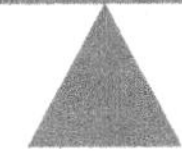

Fig. 2.4 The defining principles of Agile.

Agile brings real benefits

Continuous delivery improved quality, increased productivity, and improved the employee experience.

Ashish Pandey, Technical Lead, Cisco Cloud and Software IT Team [70]

Agile works, as these examples illustrate:

- Agile innovation has revolutionized the software industry. Now it is poised to transform nearly every other function in every industry [71].
  - John Deere uses Agile to develop new machines.
  - Saab uses Agile to create fighter jets.
  - National Public Radio employs Agile for new programming.
  - Mission Bell Winery employs Agile methods for wine production and warehousing; it even uses Agile to run its leadership team.
- In his article "Agile: The World's Most Popular Innovation Engine," Steve Denning stated that in one survey, more than 60% of respondents said Agile Scrum works [72]. For example, Cisco claimed a 40% fall in critical and major defects, a 16% reduction in their "defect rejected ratio (DRR)," a 14% improvement in "defect removal efficiency," and improved employee satisfaction by cutting after-hours work and reducing the numbers of meetings and calls by a quarter.
- According to the Standish Group Chaos Report of 2015, Agile methods succeed 39% and fail 9% of the time, compared to traditional project management methods, which succeed 11% and fail 29% of the time [73]. This study measured more than 10,000 projects between 2011 and 2015.

This is a small slice of success stories for Agile. It has also earned accolades from the Government Accounting Office (GAO) [74], Spotify [75], Microsoft [76], LEGO [77], British Telecom [78], and many more organizations (Table 2.7).

Table 2.7 Summary of Agile Scrum software management.

Agile Scrum software management	
What is it?	A nimble way to develop software based on daily team meetings, frequent feedback from users, and adding code in small increments with constant validation
When was it developed?	Around 2000
Examples of how Agile Scrum uses lean thinking	• Delivers code in small increments with constant validation • Presents software prototypes to users throughout a development project • Heavy dependence on cross-functional teams • Visualizes workflow (esp. "Kanban" board) • Standardized interactions
Is it used?	Agile is widely used in software/IT development and continuing to grow in popularity
Does it work?	According to one survey of 8000 projects, Agile teams are 25% more productive teams that are 50% more likely to deliver on schedule with 75% fewer software defects[a]

[a]http://www.deltamatrix.com/why-are-agile-teams-25-more-productive.

2.10 Personal experience

I have spent many years working for one of the most lean thinking companies in the world, the Danaher Corporation. Danaher transformed itself, starting in the mid-1980s, using lean thinking. Early on, the organization created the Danaher Business System (DBS) modeled after the Toyota Production System. Danaher has been, by any measure, a most successful business, growing fast and outperforming the stock market by a wide margin for decades. According to the corporate website: "Success at Danaher doesn't happen by accident. We have a proven system for achieving it. We call it the Danaher Business System (DBS), and it drives every aspect of our culture and performance" [79].

My journey began in the 1980s as a design engineer at Kollmorgen, which would be acquired by Danaher in 2000. My first substantial application of lean thinking was as part of a team creating and sustaining a lean product development process. I was then fortunate in having opportunities to lead R&D teams, culminating in the role of VP of Global Engineering for X-Rite, an operating company of Danaher, from 2015 to 2018.

As VP of R&D, I spent what time I could in lean knowledge, but leading a large group of people involves many other demands. In early 2018, I decided to focus the rest of my career on lean knowledge conversion. It took more than a year, but I was able to realize that goal as VP of Innovation for Envista Holdings Corporation. Envista is a company created in 2019 by Danaher beginning as three operating companies from Danaher's dental segment. Today, every minute of my day is spent deploying, improving, and sustaining our own brand of lean knowledge, the innovation processes of the Envista Business System, which is based on the Danaher Business System [80].

Over my career, I've been able to watch lean knowledge conversion work with many benefits, a few of which are:

- New products generate >100% of planned revenue.
- Late projects cut nearly 4×.
- Voluntary severance down by two-thirds.
- Engagement up: number of people reporting as neutral or disengaged cut in half.

I can state with confidence: lean knowledge works!

2.11 Conclusion

This chapter has sought to provide a brief history of lean thinking and present evidence that lean knowledge works, and that it dramatically improves organizations in a broad range of domains. The remainder of the book will present a structured approach to lean thinking as it applies to knowledge work, along with dozens of tools and techniques to help you accelerate your journey in lean knowledge.

References

[1] T. Ohno, Taiichi Ohnos Workplace Management, Special 100th Birthday Edition, McGraw-Hill Education, 2012, 166. Kindle edition.
[2] J.P. Womack, D. Jones, D. Roos, The Machine That Changed the World: The Story of Lean Production—Toyota's Secret Weapon in the Global Car Wars That is Now Revolutionizing World Industry, Free Press, 2007 (Reprinted).
[3] Global Manufacturing, Top 10: Lean Manufacturing Companies in the World, Global Manufacturing, June 11, 2014.https://www.manufacturingglobal.com/top-10/top-10-lean-manufacturing-companies-world.
[4] M.L. Dean. Lean Healthcare Deployment and Sustainability, McGraw-Hill, 2013.
[5] G. Ellis, Project Management for Product Development: Leadership Skills and Management Techniques to Deliver Great Products, Butterworth-Heinemann, 2016.

[6] E. Ries, The Lean Startup: How Today's Entrepreneurs Use Continuous Innovation to Create Radically Successful Businesses, Crown Business, 2011.
[7] M. Rother, Toyota Kata: Managing People for Improvement, Adaptiveness and Superior Results, McGraw-Hill, August 4, 2009.
[8] R.J. Pryor, Lean Selling: Slash Your Sales Cycle and Drive Profitable, Predictable Revenue Growth by Giving Buyers What They Really Want, Author House, 2016.
[9] K. Grady, Ken Grady on Applying Lean Thinking to the Practice of Law, https://tlpodcast.com/category/lean-thinking/, February 8, 2018.
[10] J.E. Cunningham, O.J. Fiume, Real Numbers: Management Accounting in a Lean Organization, Management Times Press, 2003.
[11] https://npengage.com/nonprofit-management/lean-implementation/.
[12] O. Holusha, T. Ohno, Whose Car System Aided Toyota's Climb, Dies at 78, NY Times, May 31, 1990.https://www.nytimes.com/1990/05/31/obituaries/taiichi-ohno-whose-car-system-aided-toyota-s-climb-dies-at-78.html.
[13] S.T. Teich, F.F. Faddoul, Lean management—the journey from Toyota to healthcare, Rambam Maimonides Med. J. 4 (2) (2013) e0007.https://www.ncbi.nlm.nih.gov/pmc/articles/PMC3678835/.
[14] https://www.toyota-global.com/company/history_of_toyota/75years/data/automotive_business/sales/sales_volume/japan/1950.html.
[15] https://auto.howstuffworks.com/ford6.htm.
[16] K. Shimokawa and T. Fujimoto. The Birth of Lean. Lean Enterprise Institute, Inc., 2009, Kindle edition, location 54.
[17] http://www.apiweb.org/circling-back.pdf.
[18] C. Lazzareschi, W. E. Deming, Quality Control Guru, Dies at 93, LA Times, December 21, 1993. Alsohttps://www.latimes.com/archives/la-xpm-1993-12-21-mn-4178-story.html.
[19] https://deming.org/deming/deming-the-man.
[20] C. Lazzareschi, W. E. Deming, Quality Control Guru, Dies at 93, LA Times, December 21, 1993.
[21] A. Morita, E.M. Reingold, M. Shimomura, Made in Japan: Akio Morita and Sony Autobiography, Dutton, 1986.
[22] https://global.toyota/en/company/vision-and-philosophy/production-system/.
[23] K. Shimokawa and T. Fujimoto. The Birth of Lean. Lean Enterprise Institute, Inc., 2009, Kindle edition, location 207.
[24] J. Kurtzman, Toyota's Problems Start at the Top, https://hbr.org/2010/03/toyotas-problems-start-at-the-1, March 11, 2010.
[25] S. Shingo, The Sayings of Shigeo Shingo: Key Strategies for Plant Improvement, English Translation, Productivity Press, 1987, p. 161.
[26] S. Shingo, The Sayings of Shigeo Shingo: Key Strategies for Plant Improvement, English Translation, Productivity Press, 1987, p vi.
[27] S. Shingo, The Sayings of Shigeo Shingo: Key Strategies for Plant Improvement, English Translation, Productivity Press, 1987, p xvii.
[28] J. Shook, How to Change a Culture, Lessons from NUMMI, vol. 51, MIT Sloan Management Review, Winter 2010 No 2.
[29] https://youtu.be/v8bNBtqfmrM.
[30] P. Adler, Time-and-Motion Regained, HBR January–February 1993.
[31] http://www.thisamericanlife.org/radio-archives/episode/561/nummi-2015.
[32] https://www.thisamericanlife.org/561/nummi-2015.
[33] J.P. Womack, D.T. Jones, D. Roos, The Machine That Changed the World, Simon and Schuster, 2007, p. 92 derived from Figure 4.7 by averaging Japanese worldwide and comparing to average of North America and Europe.

[34] K.L. Sim, J.W. Rogers, Implementing lean production systems: barriers to change. Manag. Res. News 32 (1) (2008) 37–49, https://doi.org/10.1108/01409170910922014.
[35] M. Gladwell, Outliers: The Story of Success, Little, Brown, & Company, 2008.
[36] M. Csikszentmihalyi, Flow: The Secret to Happiness, TED, February 2004.https://www.ted.com/talks/mihaly_csikszentmihalyi_on_flow?language=en.
[37] http://dondodge.typepad.com/the_next_big_thing/2008/11/50-of-us-engineering-students-dropout—why.html.
[38] http://usmilitary.about.com/od/joiningthemilitary/l/blbasicattrit.htm.
[39] J. Liker, The Toyota Way: 14 Management Principles From the World's Greatest Manufacturer, McGraw-Hill, 2004, 269.
[40] D. Oosterwal, The Lean Machine, Productivity Press, 2010, pp. 131–132.
[41] B. Staats and D.M. Upton, Lean Knowledge Work, Harvard Business Review, October 2011, https://hbr.org/2011/10/lean-knowledge-work
[42] D.G. Reinertson, The Principles of Product Development Flow: Second Generation Lean Product Development, Celeritas Publishing, 2009, p. 2. Kindle edition.
[43] C. Berczuk, The Lean Hospital, The Hospitalist 2008(6) (June 2008). https://www.the-hospitalist.org/hospitalist/article/123698/lean-hospital.
[44] https://www.lean.org/WhatsLean/.
[45] A. Manos, M. Sattler, G. Alukal, Make healthcare lean, Qual. Prog. 39 (2006) 24.
[46] T. Sloan, A. Fitzgerald, K.J. Hayes, Z. Radnor, S. Robinson, A. Sohal, Lean in healthcare—history and recent developments, J. Health Organ. Manag. 28 (2) (2014) https://doi.org/10.1108/JHOM-04-2014-0064.
[47] C. Kenney, Transforming Health Care: Virginia Mason Medical Center's Pursuit of the Perfect Patient Experience, CRC Press, 2010 with an introduction by Don Berwick.
[48] D. Weinberg, US Hospitals Turn to Toyota for Management Inspiration, Voice of America, May 2, 2011.http://www.voanews.com/content/us-hospital-turns-to-toyota-for-management-inspiration–121165799/163553.html.
[49] J.S. Toussaint, L.L. Berry, The Promise of Lean in Health Care, Mayo Clinic, 2013. https://www.mayoclinicproceedings.org/article/S0025-6196(12)00938-X/fulltext.
[50] M.L. Dean, Lean Healthcare Deployment and Sustainability, McGraw-Hill, 2013, p.6.
[51] L.M. Yerian, J.A. Seestadt, E.R. Gomez, K.K. Marchant, A collaborative approach to lean laboratory workstation design reduces wasted technologist travel, Am. J. Clin. Pathol. 138 (2012) 273–280.
[52] D. Drickhamer, Transforming Healthcare: What Matters Most? How the Cleveland Clinic is Cultivating a Problem-Solving Mindset and Building a Culture of Improvement, Lean Enterprise Institute, (May 28, 2015) https://www.lean.org/common/display/?o=2982.
[53] A. Gawande, The Checklist Manifesto, Henry Holt and Company, 2009.
[54] J. Liker, The Toyota Way: 14 Management Principles From the World's Greatest Manufacturer, McGraw-Hill, 2004, pp. 156–157.
[55] A.C. Ware, D.K. Sobek II, Lean Product and Process Development, second ed., Lean Enterprise Institute, 2014, p.31.
[56] J.K. Liker, J. Morgan, Lean product development as a system: a case study of body and stamping development at Ford, Eng. Manag. J. 23 (1) (2015) 16–28.
[57] http://ramlijohn.com/a-landing-page-is-not-a-minimum-viable-product/.
[58] E. Ries, The Lean Startup: How Today's Entrepreneurs Use Continuous Innovation to Create Radically Successful Businesses, Crown Business, 2011, p.58.
[59] https://www.youtube.com/watch?v=DQGVRgVAlpU.
[60] https://leanstartup.co/how-procter-gamble-uses-lean-startup-to-innovate/.
[61] S. Blank, Why the Lean Start-Up Changes Everything, Harvard Business Review, May 2013.

[62] R. Roberts, Critical Chain Project Management: An Introduction, Stottler Henke Associates, 2013.
[63] Souders, P. "(Boeing) T45 Undergraduate Military Flight Office Ground based Training System", Project Flow 2011, https://www.youtube.com/watch?v=ZEucNl8vpfI.
[64] http://www.realization.com/pdf/casestudy/Case-Study-Airline-Boeing.pdf.
[65] http://www.goldratt.co.uk/Successes/pm.html.
[66] G.I. Kendall, K. Austin, Advanced Multiproject Management, J. Ross Publishing, 2012.
[67] http://www.chaine-critique.com/fr/Les-resultats-de-la-Chaine-Critique-33.html.
[68] G. Ellis, Project Management for Product Development: Leadership Skills and Management Techniques to Deliver Great Products, Butterworth-Heinemann, 2016, 223f.
[69] K. Beck, M. Beedle, et al., Manifesto for Agile Software Development, Available at: http://agilemanifesto.org/, 2001.
[70] https://www.scaledagileframework.com/cisco-case-study.
[71] D.K. Rigby, J. Sutherland, H. Takeuchi, Embracing Agile, HBR, May 2016.
[72] S. Denning, Agile: The World's Most Popular Innovation Engine, Forbes.com https://www.forbes.com/sites/stevedenning/2015/07/23/the-worlds-most-popular-innovation-engine/#50186d2d7c76, July 23, 2015.
[73] https://www.standishgroup.com/sample_research_files/CHAOSReport2015-Final.pdf.
[74] US Government Accounting Office (GAO), Software Development: Effective Practices and Federal Challenges in Applying Agile Methods, (July 27, 2012) p.8. Available at:http://www.gao.gov/products/GAO-12-681.
[75] Sunden, J., "Agile at Spotify", 2013. Available at: http://www.slideshare.net/JoakimSunden/agile-at-spotify.
[76] Bjork, A., "Scaling Agile Across the Enterprise," Microsoft Visual Studios Engineering Stories, 2015. Available at: http://stories.visualstudio.com/scaling-agile-across-the-enterprise/.
[77] https://www.youtube.com/watch?v=TolNkqyvieE.
[78] http://www.methodsandtools.com/archive/archive.php?id=43.
[79] https://www.danaher.com/how-we-work/danaher-business-system.
[80] A. Aghdaei, CEO Envista, as quoted at "Danaher announces new dental company to be named Envista Holdings Corporation," Danaher Press Release, June 27, 2019. Available at http://investors.danaher.com/2019-06-27-Danaher-Announces-New-Dental-Company-To-Be-Named-Envista-Holdings-Corporation.

CHAPTER 3

Creating value from knowledge work

> *Lean thinking defines value as providing benefit to the customer; anything else is waste [1].*

3.1 Introduction

In this chapter, we will look at how a knowledge organization creates value for its customers. The connection of effort to value is complicated because of the large variety of value that knowledge work creates and the indirect ways in which end customers pay for them. But the connection is certain and those organizations that increase it enjoy substantial benefits over time: better customer and client experience, more engaging work, and ultimately better financial performance.

3.2 A definition of value

The term *value* has many meanings from a bargain ("that's good value") to an indication of worth ("I value your input") to estimating a monetary equivalent ("we value those assets at $300,000"). For our purposes, value will have a specific definition: what customers are willing to pay for goods or services. Notice that value is what customers are willing to pay, whereas price is what they actually pay; customers pay no more than the value they place on a product or service, but if offered the opportunity, they will pay less.

3.3 Who is the customer to knowledge work?

Since value is based on the perception of the customer, we must know who the customer is. On the factory floor, the customer is usually defined as the

Improve
https://doi.org/10.1016/B978-0-12-809519-5.00003-X

end customer: the person or organization that purchases products. For example, an automotive factory makes cars; the value added to each automobile is what customers are willing to pay less what that car costs to make.

Knowledge work provides indirect value to end customers, so their willingness to pay may be indeterminate. Who is the customer for the finance team to close the books each month? We could say it's the end customer, because a well-run financial group ultimately reduces the costs of producing products and services, but that connection is distant. If a company cannot pass a tax audit, it won't be providing anything for very long. But can a financial team really connect what they do to end-customer value in a measurable way? As an alternative, we could define the general manager as the customer for compliance in a tax audit. We call this an *internal customer*.

> *Development exists to create operational value streams. Operations is our customer. We should listen to and serve operations just like external customers [2].*

A broad definition of a customer would include the person or people who specify, fund, direct, and approve work. Unfortunately, that can be a large group of people. So, we'll normally start with something narrower: the primary customer for knowledge work is probably the person who determines when the work product has met the need. For an attorney in a legal firm, the customer is normally the client, an external customer. But for a product development team, it's probably the general manager and his or her staff. A common industry standard is that the cost of research and development (R&D) should be a certain fraction of sales. For example, Craft.co, a company that "organizes financial, operating and human capital data from thousands of sources," estimates that average R&D costs across a range of industries vary from 3% to 16% [3]. There are similar measures for other functions, like cost of sales [4]. It is much more likely that the general manager would evaluate these measures than an external customer. Where you sit in the organization determines in large part whom your primary customer is; if your work product is never seen by an external customer, an internal customer may provide a more effective arbiter of the value you offer.

3.4 What is a knowledge organization?

An *organization* is a group of people aligned to accomplish a defined set of purposes. The Ford Motor Company is an organization. So is the 15-person sales team at a Ford dealership in Saginaw. So is the state government of Michigan. A "knowledge organization" here is an organization with the primary mission of delivering knowledge work, which is creating and transforming information.

> *We're a knowledge company. Everything we do comes out of the heads of people who work here.*
>
> ***Commonly attributed to Jim Goodnight, CEO, SAS***

Organizations with more than a dozen or so people are usually built in layers. In my role at X-Rite, there were perhaps six layers: I led the R&D organization (Layer 3) for X-Rite (Layer 2), an operating company of the Danaher Corporation (Layer 1). The R&D team contained a global firmware/software design group (Layer 4), and within that a local firmware/software design group operated out of Kentwood, MI (Layer 5), which in turn contained a firmware design group (Layer 6).

3.4.1 Examples of value

Value creation is often divided into two facets (see Table 3.1):

- direct: what the customer is willing to purchase; and
- support: what work must be provided to make that purchase possible.

At the auto repair shop, customers are willing to pay for a brake pad being replaced; most are unwilling to pay a separate line item for booking an appointment. But people are willing to pay for appointments in the form of overheads, which is one reason why customers willingly pay more for the repair than the sum of the parts and the mechanic's wage—so they will pay for appointments and heating and annual taxes, just not as separate line items. This distinction between direct and support is important in much of lean manufacturing, but it's less important in knowledge work. While a slice of knowledge work is for direct value creation—for example, physicians caring for patients and attorneys representing clients—most of it is indirect.

3.5 Value $\neq$ effort

It's easy to think that value is proportional to effort, but actually the two are only loosely tied together. Value as perceived by customers is usually based on the market rates for similar products and services. If the going rate to change front brakes is $150 and your shop charges $300, you'll probably

Table 3.1 Examples of value in a few common businesses.

	Total value	
Business	**Direct value**	**Value support**
Auto repair	Replacing worn brakes	Booking appointment Taking customer information
Manufacturing	Assembling a product a customer will purchase	Inventory of parts Invoicing customers
Doctor's office visit	Making a diagnosis Prescribing medication	Setting up appointment Invoicing insurance

start asking questions. Let's say the owner tries to charge you based on effort with an explanation like, "We didn't have a good mechanic, so we had to change your brakes three times to get it right. So, I had to charge you double the usual." You're probably going to say something like, "Your bad mechanic is not my problem." Customers value products and services at the market rate, not according to the effort or cost required to produce them.

3.6 Five facets of value

The specific elements of knowledge work value will vary across industries. Patient recovery rates, trial win rates, and patents generated are each specific to narrow slices of knowledge work. For the general case, we need to broaden measures. Here, we will categorize value of knowledge work into five major facets as shown in Fig. 3.1:

Fit for purpose: Work that's done well enough to meet the need.

Profitable: Financial reward for the organization resulting from work product.

On-schedule: Providing a product or service when the customer needs it.

Innovative: Products and services that meet real customer needs in creative and differentiated ways.

Protected and compliant: Meeting requirements of regulatory agencies and standards that customers value; protecting intellectual property (so the competitors have more to comply with!).

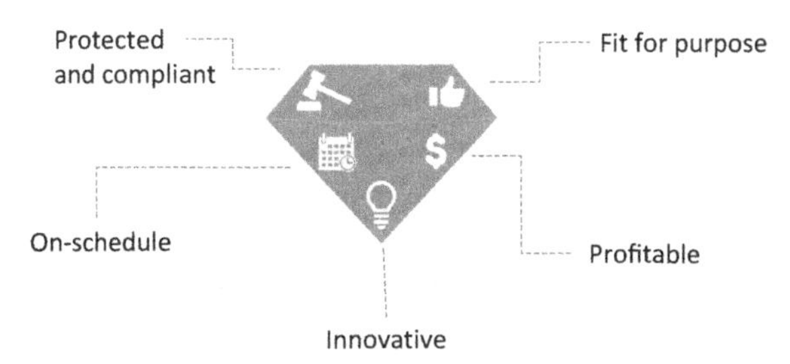

Fig. 3.1 Knowledge work delivers value in five key categories.

These are a derivative of Shingo's list of values, which he called "Goals of Improvement" [5]: easier, better, faster, cheaper, but rebalanced for knowledge work. Let's review each of the five facets of knowledge work value in detail.

3.6.1 Fit for purpose

Fit for purpose is defined in the *MacMillan Dictionary* as "something good enough to do the job it was designed to do" [6]. As shown in Fig. 3.2, it describes knowledge work that is free of errors and omissions that disappoint customers. Errors and omissions reduce the value of work product in multiple ways. The most obvious is the simple cost of repeating work to correct mistakes. After that can come financial penalties for poor work—repeating certifications, design recalls, and damage that customers suffered from the errors. Further, the loss of customer confidence may remove revenue opportunities in the future. Last, the internal "brand" of the knowledge organization can be damaged, eroding trust and willingness of stakeholders to take risks on the team.

- Good enough to do the job it was meant to do
- Avoiding errors and omissions that anger customers and require time and energy to correct
- Avoiding damage to the "brand" from shoddy work
- *Example measures: # returns, # complaints, % win ratio, defective parts-per-million (ppm)*

Fig. 3.2 The first facet of value in knowledge work: Fit for purpose.

3.6.2 Profitable

The most measurable facet of value is usually profitability: higher revenue and lower cost (see Fig. 3.3), which captures the direct economic benefit to the organization from knowledge work. Knowledge work can create at least two financial advantages for the organization. The products and services of the organization can be made more desirable by good knowledge work: better design, easier to use, more reliable, more functional, etc. More desirable products lead to greater sales and allow for higher prices, which creates more margin. The second effect is increasing the efficiency of the organization to provide the goods or services for the markets where they are consumed. This results in lower costs and thus higher margins. Both of these allow the organization to enjoy a higher return on investment (ROI) on knowledge work.

3.6.3 On-schedule

The third facet of knowledge work value is on-schedule performance, shown in Fig. 3.4. This includes the concrete need to finish according to the schedule that meets customer expectations. In addition, the schedule is also a leading indicator of other facets of value. For example, a patent increases profitability to the extent that it prevents others from producing something they otherwise would. So, having the patent application filed on time is a leading indicator; measuring the economic value of preventing a competitor from copying a product may lag that on-time patent application by many years.

On-schedule value is of special interest in continuous improvement because it's usually the first of the five facets that will show measurable results. It's often the focus early in continuous improvement because it gets

- Increasing revenue with superior products and services
- Reducing costs of products and services
- Increasing return on investment (ROI) of knowledge work

Example measures: currency, % margin, % growth, % return on investment

Fig. 3.3 Profitability is the facet that represents the ultimate measure of value in knowledge work, manufacturing, or almost any other organization.

Fig. 3.4 On-schedule is a separate facet of value: If work is late, the organization will devalue it.

traction with stakeholders quickly. When a CEO sees work product go from 30% on-schedule to 85% in 3 or 6 months—something not uncommon, as discussed in Chapter 2—trust usually increases.

3.6.4 Innovative

The fourth facet of value is innovative, which is commonly defined as *invention that solves a valuable problem.* As shown in Fig. 3.5, we will use that definition but add that the organization can provide the innovative product or

Fig. 3.5 The most engaging facet of value for professionals and experts is usually innovation.

service. Here, innovation must provide value to the customer and to the organization providing it.

Innovation: the process of translating an idea or invention into a good or service that creates value or for which customers will pay [7].

Innovation is the primary driver of prosperity, job growth, social responsibility, environmental sustainability, and national security [8].

For knowledge staff, innovation is often the most engaging of the five facets. For most of us, novel solutions to real problems are fascinating. This engagement spreads well beyond the inventors of these novel solutions to the people who build, sell, repair, and support innovative products and services.

3.6.5 Protected and compliant

Protected and compliant (Fig. 3.6) are probably the most difficult facet of value to monetize. Compliance with government regulations and commonly expected standards is usually binary: if you don't comply, you don't sell. It's often referred to as "table stakes" because there's no choice for ethical companies.

- Complying with applicable laws and regulations from governments and regulatory agencies
- Protecting intellectual property through copyrights, patents, trademarks, and trade secrets
- Fair use of intellectual property of others

Example measures: # compliance issues, compliance cost, # patents

Fig. 3.6 The last facet of value is ensuring the organization competes fairly and enjoys protections it deserves from competitors.

In a similar category is intellectual property (IP), both respecting the IP of competitors and generating IP for the organization. The results of failing to comply can be devastating once discovered. Failing to protect IP can injure an organization's competitive position for many years. Like compliance, the value of protection is high, but difficult even to estimate. Yet a great deal of knowledge work must be invested in this facet of value, especially in highly regulated businesses like transportation and medical care.

3.7 Conclusion

Knowledge staff create value in highly varying ways, from heads-down, diligent work to comply with the regulations to exciting moments of creating an innovative solution. And value traverses from the concrete metrics of delivering something to the schedule to sometimes a years-long journey to generate financial benefit. There are many things knowledge work does that cannot be monetized in the time frame necessary to aid a decision; for example, patents often reach their full value a decade after the patent is issued and, even then, are usually difficult to monetize. So, we will proceed understanding that our purpose is to create value, even though the quantification of value will often be subjective.

References

[1] E. Ries, The Lean Startup: How Today's Entrepreneurs Use Continuous Innovation to Create Radically Successful Businesses, Crown Business, 2011, p. 48.

[2] A.C. Ward, D.K. Sobek II, Lean Product and Process Development, second ed., Lean Enterprise Institute, 2014, p. 23.

[3] What Industry Spends the Most on Research and Development? https://craft.co/reports/s-p-100-r-d.

[4] AccountingTools, Cost of Goods Sold, https://www.accountingtools.com/articles/2017/5/4/cost-of-goods-sold.

[5] S. Shingo, The Sayings of Shigeo Shingo: Key Strategies for Plant Improvement, Productivity Press, 1987, p. 35.

[6] Fit for purpose—definition and synonyms, https://www.macmillandictionary.com/us/dictionary/american/fit-for-purpose.

[7] Business Dictionary, www.businessdictionary.com.

[8] Curtis Carlson, former president of Stanford Research Institute (SRI) and author of Innovation (Crown, 2006). https://www.practiceofinnovation.com/2015/04/.

CHAPTER 4

The lean equation

The core idea is to maximize customer value while minimizing waste [1].

4.1 Introduction

Chapter 2 presented Taiichi Ohno's brilliant insight that only a small portion of our effort creates value and the rest creates hidden waste. The lean equation takes us one step further: that we should separate all we do—every effort in every day—into two unambiguous classes, those that create value and those that create waste. This chapter will present the lean equation, the simple truth that effort is the sum of value and waste. It begins with a discussion about waste followed by differences between effort and value, two elements that are often confused. This leads to a discussion about opaque workflows, which are common in knowledge work. It ends by exploring the consequence of waste taking such a large portion of our effort: if we want a sustained increase in value, we should start not by increasing effort, but rather by cutting waste.

4.2 Waste is a villain

[T]he hardest waste to correct is the waste of time because time does not litter the flow like wasted material.

Henry Ford [2]

Waste is a villain. It frustrates every worthy purpose of your organization. It subverts your efforts to delight customers, grow employees, and satisfy investors. Have you ever heard a dedicated, well-meaning, but frustrated person in your organization ask, "Why is it so hard to get anything done around

Improve
https://doi.org/10.1016/B978-0-12-809519-5.00004-1

here?" It's likely that he or she is, at that moment, losing a noble but ill-fated contest with hidden waste that has deep roots in the organization. Your model of waste may be that most of it comes from a few incompetent, unmotivated, or malicious employees: "Get rid of the bad ones and waste will go away." In this view, waste is someone's fault and perhaps your job is to find that someone. But waste is rarely something a few bad people inflict on the others. The overwhelming amount of waste in organizations is created by good people trying to do their jobs well. And most of it hides in plain sight.

> *We can't see something if we don't know to look for it. When people aren't aware of this possibility, they may end up unconsciously overlooking problems.*
>
> ***Shigeo Shingo [3]***

Waste includes every activity of your organization that is not required to provide products and services your customers value. In this view, there are two types of activities an organization performs: those that create value and those that create waste. This is the lean equation, a simple relationship that is foundational for continuous improvement (see Fig. 4.1).

To illustrate the difference between value and waste, Table 3.1 has been modified to create Table 4.1. Because the distinction between value generation and value support is less important in knowledge work, they have been combined into one column, *total value*. A column has been added for waste, which is all the effort that does not create value. For example, the first row shows auto repair; waste includes the effort that goes into:

- repeating work because the first repair was wrong;
- doing the repair right, but having to fix incidental damage like the mechanic scratching the fender;

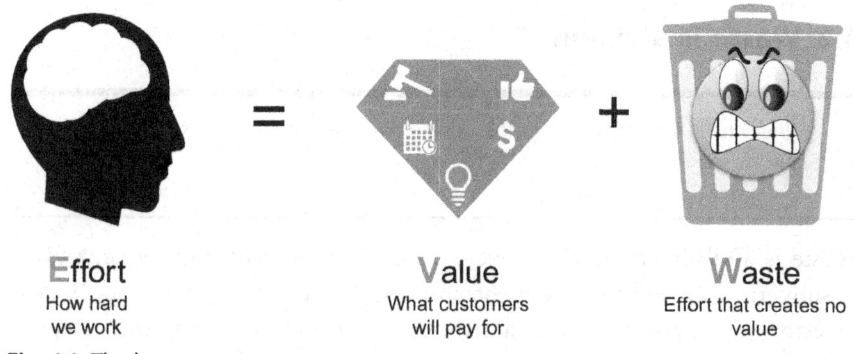

Fig. 4.1 The lean equation.

Table 4.1 Table 3.1 augmented to contrast value and waste.

Business	**Total value**	**Waste**
Auto repair	Replacing worn brakes Booking appointment Taking customer information	Calming an angry customer because the car had to come back to the shop Fixing the fender the mechanic scratched
Manufacturing	Assembling a product a customer will purchase Inventory of parts Invoicing customers	Providing a replacement when the first one didn't work. Customer support for overly complex products
Doctor's office visit	Making a diagnosis and prescribing medication Setting up appointment and invoicing insurance	Second visit due to bad diagnosis Dealing with insurance provider refusal due to missing information

- time and energy spent calming angry customers; and
- time and energy spent attracting new customers to replace those who got so angry they found another garage.

Table 4.1 shows just three businesses, but every business, and every organization within that business, creates waste. When the people in that organization fail to measure that waste, it remains hidden. Sure, the individual instances may be prominent when they happen. But they will quickly recede in memory. This is what is meant by "hidden in plain sight."

Most people will think they naturally comprehend waste, but that's rarely the case. Consider our auto shop owner. Let's say incidental damage during repair was a major cost in his shop. Would he know what those costs were? It's likely he would think he knew, but could he quantify it? Is it a $2k/year or a $20k/year problem? If it's not quantified, each person in the shop will have a different scale in mind. This results in misalignment: maybe the owner overestimates it, but the manager underestimates. The two may have arguments when the owner sees the manager as careless and the manager sees the owner as nitpicking. The waste will cause frustration, but people will habituate to it, and over time, it will be accepted, allowing it to hide in plain sight.

The lean equation is more mindset than math. It visualizes a counterintuitive truth: some of what we do is of great value and the rest has no value. Natural intuition seems to confuse effort and value, leading to a "more is better" attitude: if we try harder, things will get better. That's the topic of the next section.

4.3 Effort vs value

However dedicated the members of an organization are and however hard they try to do what's right, most of what they do is waste. It's commonly stated that 85% of the energy of an organization is spent generating waste. If that portion seems impossibly large, consider a thought experiment: first imagine something of value that your organization delivers to its customers, clients, or patients. Think of something specific: a product you make or a service you provide. Consider these two cases:

- **An ordinary day:**

  First think of the total effort your organization invests to deliver that product or service: all the competent people you must have at the ready, the materials you must manage, the suppliers you engage, the support system you provide, the invoicing, the selling, and time to deal with returns and refunds. And this is the effort needed if the first attempt goes well. Now add the time spent doing something twice because someone did something wrong the first time. Now add the time spent waiting because material or information wasn't ready. Now add the number of times a confused customer accidentally specified the wrong thing or couldn't get something to work or just misunderstood confusing instructions. Add damages from injuries to end users, clients, or patients. Add in the effects of employees and staff getting worn out dealing with these problems day after day. Effort is the sum of these and all the other supporting activities needed to satisfy your customers.

- **A day of pure value:**

  Now, imagine what it would be like if your organization was perfect for one day. No wrong or unclear instructions. Instant access for everyone to totally accurate information. Products and services so easy to understand, you never have a confused customer. No order entry or billing errors. Never selling to a customer who cannot pay. No delivery delays that cause dissatisfaction. No refunds, no repairs, and no replacements needed, and no time spent calming angry customers who needed them. Never sending the wrong part. Not a single mistake. Not one ounce of wasted effort. Everything your organization does delights customers, clients, or patients.

Waste is the difference between the two: everything the organization does less everything it did on the perfect day. Considered this way, I wonder if 85% waste is perhaps optimistic!

Still doubt 85% of your organization's efforts may be invested in creating waste? Remember that any activity you do that is not needed to deliver value is waste. A lawsuit that could have been avoided. Cost of inventory you don't need. Moving parts from one place to another. Employees injured because of unsafe conditions or poor material presentation. Replacing staff who leave due to frustration. Replacing a customer or patient who went elsewhere because your service or product didn't satisfy him or her. It may be counterintuitive that only 15% of your organization's effort creates value. You may be thinking, "Not us...we're better than that." If so, you're not alone. I recall the moment I first heard it and I thought the same.

4.4 Opaque workflows: A place for waste to hide

Knowledge work is complex, and this leads to complex workflows. For example, how does a business developer identify candidate businesses for acquisition? How does an engineering team develop an automotive suspension system? Or a product marketer find a new geography to distribute product to? Or a doctor diagnose a chronic stomachache? All of these are complex workflows and the natural inclination will be for experts to keep the directions in their heads. This creates opaque workflows—workflows where the creation of value is unclear, where everyone doing the work has a different understanding and those outside the team may not have the first idea what happens. As shown in Fig. 4.2, the workflow may produce value, but how or how well it works is unclear.

There are several reasons knowledge workflows are opaque. For one thing, the work is complex and it's difficult to communicate. Knowledge workflow cannot be reduced to unambiguous processes the way most manufacturing instructions can. There may be a general flow, but with plenty of exceptions and exceptions to the exceptions. Adding to the ambiguity, each expert practices his or her craft with his or her own viewpoints and style, drawing on all the personal advantages available. I want my refrigerator assembled according to the work instructions, whether it's made on first shift or second, whether in Illinois or Vietnam. But I want my doctor to apply 100% of his or her skill to me; where he or she knows something the typical

Fig. 4.2 Knowledge workflows are usually opaque.

doctor does not, I don't want him or her to hold back. Manufacturing workers perform according to a standard—knowledge staff must meet a standard and then perform to the highest level their skill can support.

There are often unintentional organizational disincentives to creating workflow transparency. The most obvious is that if others don't understand what you do, they won't ask for change. For the knowledge staff who believe they create little waste, "help" from outside will seem useless. Accepting input from people with less domain knowledge requires what we will call *business humility*: acknowledging our own gaps and learning from all those around us. It's a unique type of humility: *if you know something I don't, help me learn so I can be better.* Business humility is a competitive advantage for those who possess it; for those who don't, transparency feels like a waste of time.

Business humility: acknowledging our own gaps and learning from all those around us.

Another unintentional disincentive to transparency occurs when organizations reward opaque knowledge with higher influence. If the boss is in the dark about what a subordinate does and is fearful that the subordinate may react badly to a decision, the subordinate will have a great deal of influence on decisions that directly affect his or her work.

4.4.1 How waste hides in opaque workflows

Recall our story of Gretchen, the energetic project manager in Section 1.3. That simple example of waste in an approval cycle demonstrated three key concepts about waste:

1. Most waste comes from good people trying to do the right thing.
2. People are generally unaware of the amount of waste their activities generate.

3. The people generating waste are often powerless to improve the workflow.

> *I should estimate that...most troubles and most possibilities for improvement add up to the proportions something like this: 94% belongs to the system (responsibility of management).*
>
> ***W. Edwards Deming [4]***

All knowledge workflows provide rich opportunities for waste: long, angry emails that invite longer, angrier emails; staff limited to tools that are out of date; problems being half understood and solutions being half thought out; and loading more work onto an already overburdened department. The beginning step is to assume that when our workflows are laid bare, they will be full of waste (see Fig. 4.3).

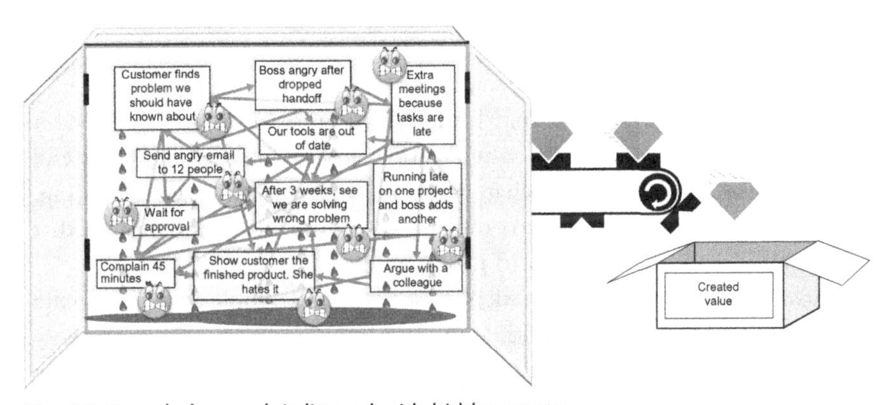

Fig. 4.3 Knowledge work is littered with hidden waste.

4.5 Cut waste by trying harder?

> *No matter how effective it may be to set clear objectives and then strive to achieve them, bursts of effort alone won't do the trick...methods must be improved.*
>
> ***Shigeo Shingo [5]***

If we assume most waste comes from people doing their jobs poorly, our first inclination will be to push harder: to demand more time, to micromanage, or even to ridicule people for poor results. That approach can work in the

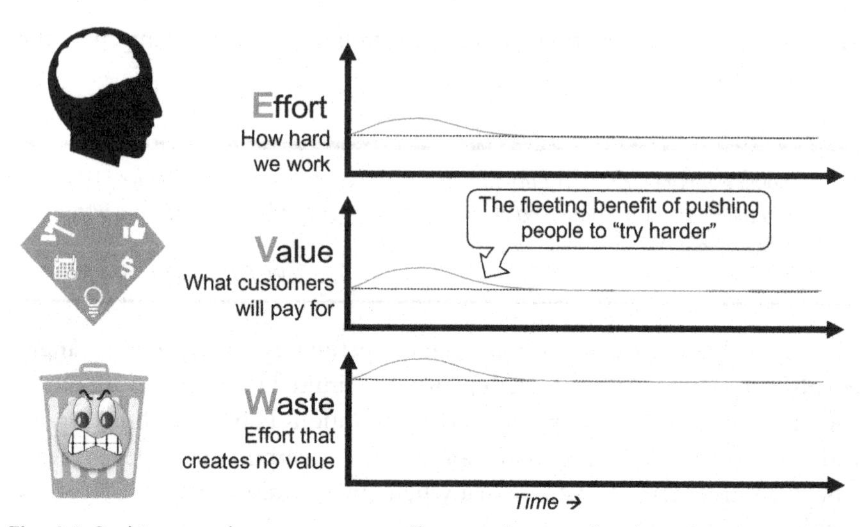

Fig. 4.4 Pushing people to exert more effort may bump value, but it's unsustainable.

short run, as shown in Fig. 4.4. Demanding people "try harder" may create a bump of effort and that effort may bring a temporary bump in value. But it doesn't reduce waste. In fact, when people just try harder in a wasteful system, waste increases. And whatever modest gain is achieved is fleeting. As management's focus moves to other areas over the weeks and months that follow, effort will likely return to its initial level.

Most waste in knowledge work is caused by good people trying to do the right thing. Why do I say "good people?" My experience is that the overwhelming portion of people in knowledge work are good at what they do: they work hard and they want to succeed within the rules. Just getting through college is difficult; then becoming an expert in a field is a long journey. Over more than 35 years, I've been impressed again and again with knowledge staff, with their desire to do the right thing for the customer and the organization. But even with the best intentions, the results can be disappointing because of hidden waste.

4.6 Three ways to expend effort in knowledge work

We can broadly categorize three ways to expend effort in knowledge work. The first is *creativity*, applying expertise to solve, invent, design, and evaluate; this is probably the most valuable time in anyone's day. It also involves some of the most exciting and engaging things knowledge staff get to do: use their expertise to help customers and the organization while mastering their craft. Unfortunately, these moments occupy a relatively small portion of the day–at least for most of us (see Fig. 4.5).

The second way to expend effort is what we'll call *diligence*. Finishing what we started in a moment of invention and creativity: running the needed tests, completing documentation, watching the customer behavior, ensuring all regulatory demands are met, and working through operational issues are just a few examples of the diligent work required to support creativity. For most of us, those moments of creativity are usually accompanied by many hours of diligence to ensure that the result of that creativity brings sustained value. Diligence creates value because it's necessary to deliver products and services to clients and customers. And, as shown in Fig. 4.6, while diligence commonly takes more mindshare than expertise, together they still occupy just a fraction of total mindshare.

The largest portion of mindshare in knowledge work goes to creating waste (see Fig. 4.7). The cost of fixing one or two errors that have escaped to a customer can double the cost to make something. Arguments or petty emails not only waste time to write and read, they cause disengagement when associates feel discord with colleagues and leaders. Working in information "silos" causes misalignment that leaves gaps that injure value. Working on the wrong problem because of a clouded view of customer needs can consume enormous amounts of effort while generating zero value.

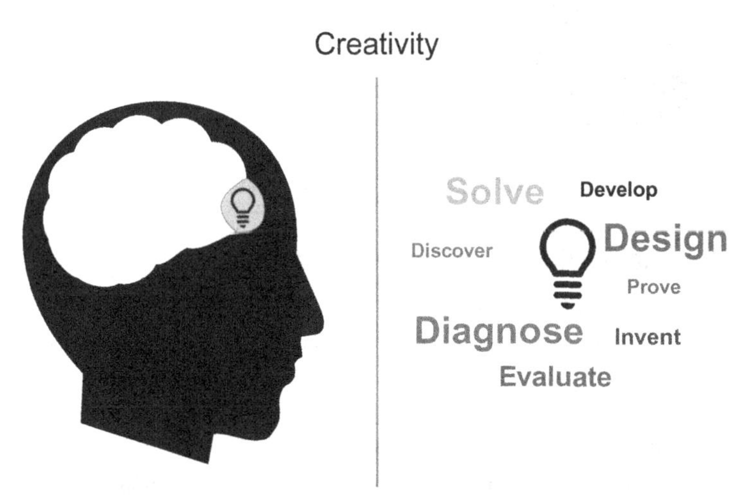

Fig. 4.5 The most satisfying part of knowledge work is when the expert gets to be creative.

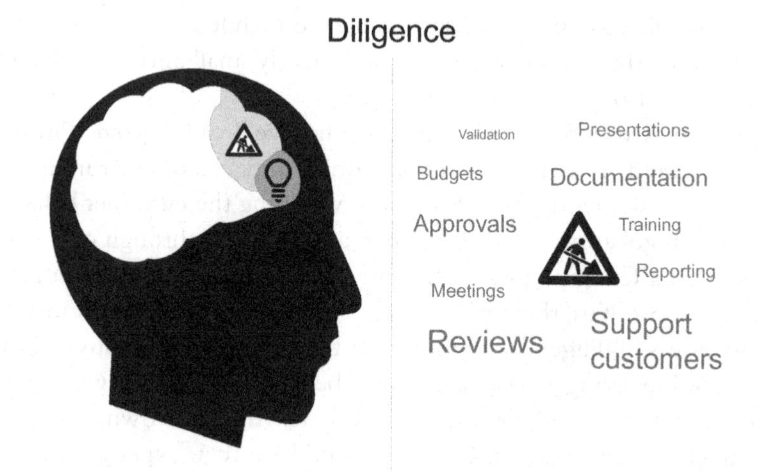

Fig. 4.6 Knowledge work demands diligence—the heavy lifting of value creation.

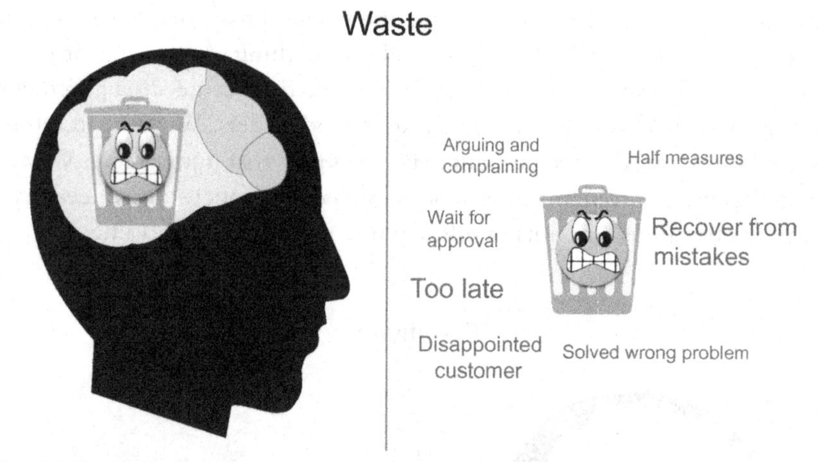

Fig. 4.7 Everything we do that doesn't create value creates waste!

4.6.1 A sustainable way to increase value

There is a better way to elevate value: reduce waste. When we reduce waste, we treat the underlying cause of low value. And this approach is sustainable. Each improvement cycle is an investment into an annuity that pays out in long-term value creation. The payback from each waste-reduction cycle funds future cycles of waste reduction.

Because each improvement reduces waste permanently, or at least for a long period of time, the benefits from the investments accumulate. And they build to a surprisingly large amount in just a few years. Fig. 4.8 shows the results of two assumptions, both of which are well-founded through experience:

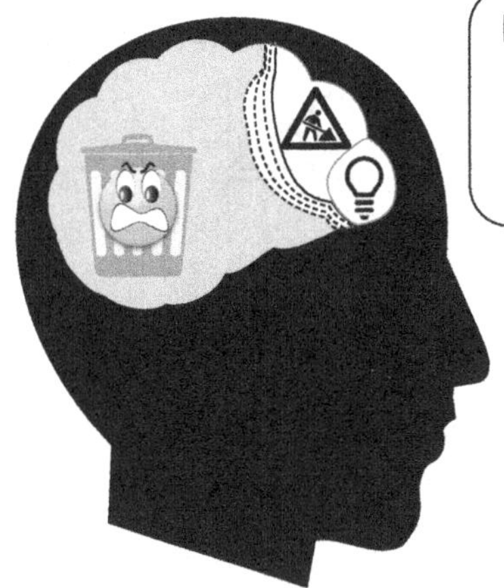

Fig. 4.8 The disproportionately large benefit of small reductions in waste.

(1) Waste in the organization is about 85%.
(2) We can reduce waste by a modest amount, perhaps 5% each year.
These two assumptions and a little math demonstrate a stunning opportunity to increase value. If we can move just 5% of total effort from waste to value, that allows value-creating effort to climb from 15% to 20%; that's an increase in value of one part in three! And whether waste is 85% or 75% or 90%, the result is similar. It depends only on Taiichi Ohno's brilliant insight from Chapter 2: that *most* of what we do creates waste.

Now, let's monetize that benefit for a small knowledge team, say, a staff of seven, again assuming about 85% waste at the start. The annual budget for this team is (let's estimate) about $1M, including compensation, benefits, team expenses, and overhead. (Costs in your organization may vary, so repeat this for your team's budget.) Let's make another conservative assumption: the team is just treading water: if they were able to measure value precisely, it would turn out that value creation is about equal to their budget: $1M. So, about $1M of value is created with 15% of the team's effort, and the 85% of their effort that remains goes to waste.

Now, let's cut waste from 85% to 80%, allowing value to increase to 20% (see Fig. 4.9). If 15% of effort creates $1M in value and we assume the same proportion, then 20% will create about $1.3M ($1M × (20%/15%) = $1.3M). So, for this small team of seven people, cutting waste by just 5% can generate an additional $300k in value in just one year! And the team can make this investment year after year.

So, does this admittedly simplistic math line up with reality? Unequivocally, yes! What is your intuition about the increase in value when

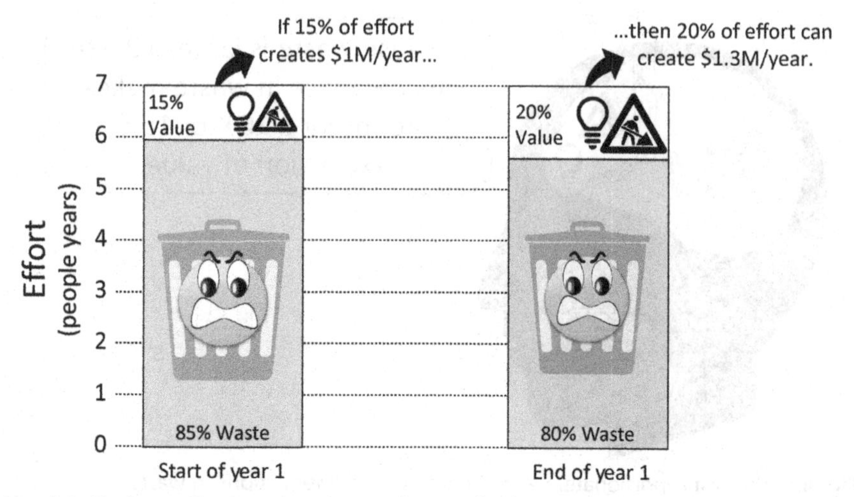

Fig. 4.9 Modest reductions in waste make way for large increases in value.

increasing on-schedule performance from ~30% to 85%, cutting quality defects in half, and reducing voluntary severance by two-thirds? These are all attainable targets in just a few years. Would that double value? Triple it? It's a number we probably cannot measure with precision, but certainly it's large.

Fig. 4.10 demonstrates a sustainable way to increase value, in contrast to just pushing staff harder as shown in Fig. 4.4. Waste can be driven down in

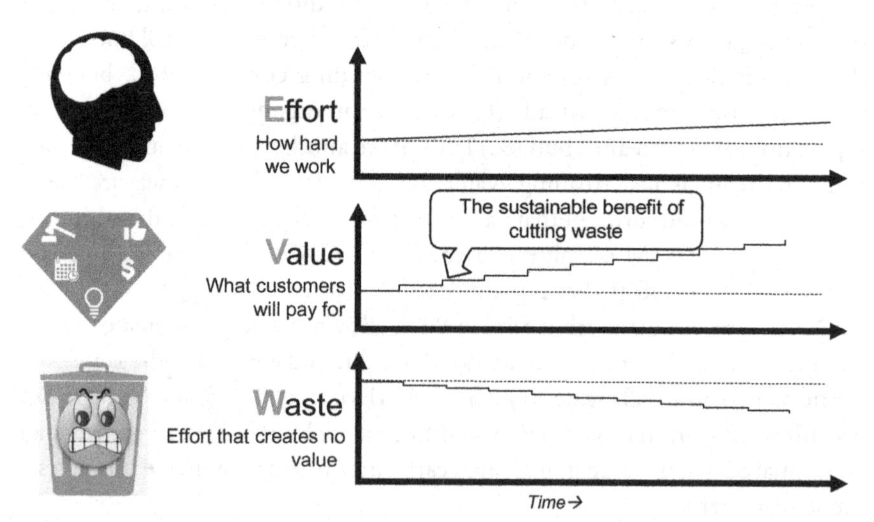

Fig. 4.10 Reducing hidden waste is a sustainable path to creating dramatic increases in value over time.

modest increments year after year, and this increases capacity for value, perhaps doubling value generation over a few years. This model, removing waste to increase value, follows Toyota's application of lean to manufacturing. As Ohno said, "All we are doing is looking at the timeline from the moment the customer gives us an order to the point when we collect the cash. And we are reducing that timeline by removing the...wastes" [6]. As a secondary effect, when staff engagement increases, staff are more focused and put in the extra energy on their own; this creates a modest but sustainable increase in effort, which is shown by the modest upward slope of the top "Effort" graph.

4.7 Conclusion

The first step in reducing waste is to recognize it. If we settle for the assumptions of Fig. 4.2, that everything is fine like it is, we will never see the waste that hides in plain sight. As Taiichi Ohno said, "If you are going to do kaizen continuously, you've got to assume that things are a mess" [7].

This chapter has presented the lean equation, a visualization of the relationship between effort, value, and waste. The lean equation is more mindset than math, helping us see the opportunity to direct our energy away from wasteful activities, and this will produce dramatic and sustainable increases in the creation of value. The next chapter will present the 8 Wastes of Knowledge Work to sharpen the ability of knowledge staff to identify the hidden wastes in knowledge work.

References

[1] https://www.lean.org/WhatsLean/.
[2] H. Ford, https://www.industryweek.com/lean-six-sigma/hardest-part-lean-see-waste.
[3] S. Shingo, The Sayings of Shigeo Shingo: Key Strategies for Plant Improvement, Productivity Press, 1987, p 29.
[4] Deming, W.E.Out of the Crisis, Massachusetts Institute of Technology, Center for Advanced Engineering Study, Cambridge, MA, 1986, p 315. Also at: https://quotes.deming.org/authors/W._Edwards_Deming/quote/1538.
[5] S. Shingo, The Sayings of Shigeo Shingo: Key Strategies for Plant Improvement, Productivity Press, 1987, p 6.
[6] T. Ohno, Toyota Production System: Beyond Large-Scale Production, Productivity Press, Portland, OR, 1988.
[7] T. Ohno, Taiichi Ohnos Workplace Management, Special 100th Birthday Edition, McGraw-Hill Education, 2012, 166. Kindle Edition.

CHAPTER 5

DIMINISH: Recognizing the 8 Wastes of Knowledge Work

In order to eliminate waste, you must develop eyes to see waste, and think of how you can eliminate the wastes that you see.

Taiichi Ohno [1]

5.1 Introduction

Chapter 4 presented the idea that lean thinking is a journey of ever-increasing value made possible by relentlessly cutting waste. It made the case that because most of our efforts create waste, modest cuts in waste create large opportunities to increase value. The problem is that those opportunities are unseen because most of the waste we create is hidden. This chapter focuses on the skills needed to recognize the waste that "hides in plain sight" using the "8 Wastes of Knowledge Work," a categorization of wasteful activities. The 8 Wastes of Knowledge Work are intended to create a common language and thus make it easier to identify and communicate about hidden waste.

Only through careful observation and goal orientation can waste be identified. The greatest waste is waste we don't see.

Shigeo Shingo [2]

5.2 The 8 Wastes of lean manufacturing

Recognizing hidden waste, sometimes called "eyes for waste" [1], is a skill that develops over time. We each accept assumptions that normalize our expectations to our experiences, and that normalization allows waste to hide in plain sight. Taiichi Ohno, often recognized as the creator of lean thinking,

Improve
https://doi.org/10.1016/B978-0-12-809519-5.00005-3

saw this in manufacturing. To help people recognize waste, he created what became known as the 7 Wastes [3]:

- *Defect Creation*;
- *Overproduction* (producing more than is needed or before it's needed);
- *Waiting*;
- *Transportation* (of materials);
- *Motion* (of workers);
- *Inventory*; and
- *Extra Processing* (spending resources on something a customer doesn't value).

This list of seven is often augmented with other wastes like *Unused Creativity*, *Stress*, and *Uneven Workflow*, resulting in "The 8 Wastes" or "The 10 Wastes." Ohno's 7 Wastes were never intended to be an exhaustive list [4] or to create a step-by-step method to find hidden waste. The list's value is in creating a common language to help us recognize hidden waste and communicate about it. This list works well in the manufacturing arena and areas close to it. However, it doesn't work well in knowledge work because the factory floor is so different from knowledge domains.

The zero-sum fallacy: A symptom of unrecognized waste

One of my earliest memories of the zero-sum fallacy is as a teenager. Standing in line to have my 1968 Ford Mustang repaired, I noticed a sign behind the register: "You can have it fast, cheap, or done right: pick any two." That mindset, called the zero-sum fallacy, is misguided. It was founded on there being no waste in the shop so that the only way to improve quality was to increase cost or take more time. But this ignores the waste that was most certainly in the shop. Cutting the time employees spend creating waste would have let them spend more time adding value, and this reliably improves the quality, cuts cost, ***and*** *speeds up delivery.*

5.3 Waste diminishes value

The central premise in this book is that waste diminishes value by stealing effort. In this view, the total available effort is constrained so that every hour that goes to waste is not available for creating value. Our starting assumptions are that knowledge staff are good at what they do, and that a lack of competence or dedication is rarely the upper limit to the value a team creates.

Why don't the 7 (or 8) Wastes of manufacturing work well in knowledge work? There are several reasons. The first is that many of the wastes are structured around the factory floor, for example, Inventory, Motion (people moving), and Transportation (goods moving). A common mechanism in lean writing is to create knowledge work equivalents: for example, Inventory is changed from stored goods to open projects. Those analogies certainly have utility, and an occasional equivalence between knowledge work and the factory floor can be illuminating [5–7]. However, lock-step translation of manufacturing wastes into knowledge waste strains the analogy. It can be off-putting to knowledge staff who have no desire to analogize their roles to the factory. Finally, effective analogies illustrate something foreign with something familiar. Since few knowledge staff have deep familiarity with manufacturing waste, these analogies are rarely illuminating for them.

5.4 The 8 Wastes

The 8 Wastes of Knowledge Work are shown in Fig. 5.1. The goal is to balance a set of factors that create waste in knowledge organizations. Some address a wasteful work environment such as "Discord" and "No-Win Contests," while "Solution Blindness" and "Hidden Errors" address how the knowledge staff practice their craft. Some of the 8 Wastes focus on internal mindset; "Inertia to Change" and "Solution Blindness" are wastes related

Fig. 5.1 The 8 Wastes of Knowledge Work.

to how we think. But "Information Friction" and "Inferior Problem Solving" are wastes that arise from a lack of effective knowlege workflows.

> *The hardest part of developing "eyes for waste" in development is that most waste is caused by "doing things right" within the conventional framework.*
>
> ***Allen C. Ware and Durward K. Sobek II [8]***

The next eight chapters will each address one of the 8 Wastes in detail, including tools to reduce that waste. The remainder of this chapter will present a brief overview of each.

5.5 Discord

Discord (Fig. 5.2) captures the waste of the many types of team misalignment including:

- misalignment with team leaders or other team members;
- misalignment to the organization's leadership;
- misalignment to other functions within the organization; and
- misalignment to the customer.

Discord creates waste for at least three reasons:

- Ineffective effort

  In large organizations, one function often takes on work that depends on other functions. For example, Product Marketing may develop a campaign for a product that requires R&D to revise the product to meet a new regulation. If **Discord** prevents the teams communicating with each other, R&D may not support the revision. That

What?	Why?	Examples	Responses
Teams working without coordination or in opposition to each other and/or to the larger organization	• Lack of common purpose • Conflicting priorities • Sense of being excluded, lack of trust	• Lack of passion in knowledge staff • Work waiting on a support group • Bitter or persistent disagreements	1. Engagement Wheel 2. Five steps to high function 3. Socialization and sticky note aggregation

Fig. 5.2 Discord is waste #1.

might prevent the launch so that all the Product Marketing team's effort would become waste.

- Counter-productive effort

  Effort that is expended to oppose other points of view is counter-productive because it tears down the work of others.

- Reduced level effort due to conflict

  Discord is disengaging. Team members who see others in conflict or themselves as shut out of the flow of information are likely to think less of their organization and its leadership, both of which are demotivating.

In Chapter 7, we'll address the **Discord** by increasing engagement with the Engagement Wheel, a simple way to evaluate how change drives engagement in four categories: Inspire, Challenge, Connect, and Protect.

5.6 Information Friction

Information Friction (Fig. 5.3) describes the difficulty of moving information from one part of the organization to another. It occurs any time one part of the organization knows something another part needs but doesn't know. It might be someone in marketing communications who knows that there isn't time to create a scheduled ad campaign, but keeps it quiet to avoid conflict. It could be a team in France that isn't telling the team in Japan something because they don't talk often. It could be two service managers that don't like each other and so hide facts to make the other look uninformed.

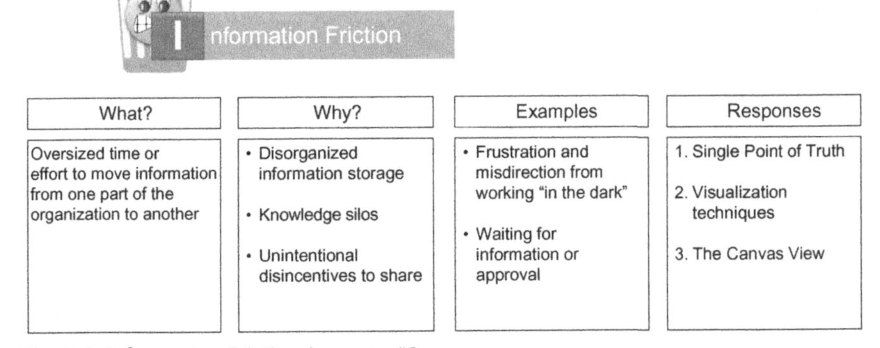

Fig. 5.3 Information Friction is waste #2.

> *Almost all defective projects result from not having the right knowledge in the right place at the right time.*
>
> ***Allen Ward and Durward Sobek [9]***

There are countless reasons for Information Friction: interpersonal conflict, fear of being found out, and physical distance, to name a few. And add knowledge silos: organizations that create one or two people who are the only experts in a narrow area. All of these make it more difficult to flow information to the people that need it. And that creates waste of having to push harder than should be necessary to move the information or, worse, it creates waste when work proceeds without necessary information, which can be devastating to work quality.

In Chapter 8, we will address **Information Friction** with three powerful techniques. We will discuss visualization as an improvement over spoken or written narration. We will use Single Point of Truth (SPoT) to build one well-defined place from which all stakeholders can pull information. And we will rely heavily on the *Canvas View*, a systematic way to tell a story on one page so it can be understood with minimal effort.

5.7 More-is-Better Thinking

More-is-Better Thinking (Fig. 5.4) is working to make progress rather than to accomplish a goal. It includes one team doing their work, even though there's no one available to accept their work product when it's done. It also includes seeing work product is going to be late, then trying a bit

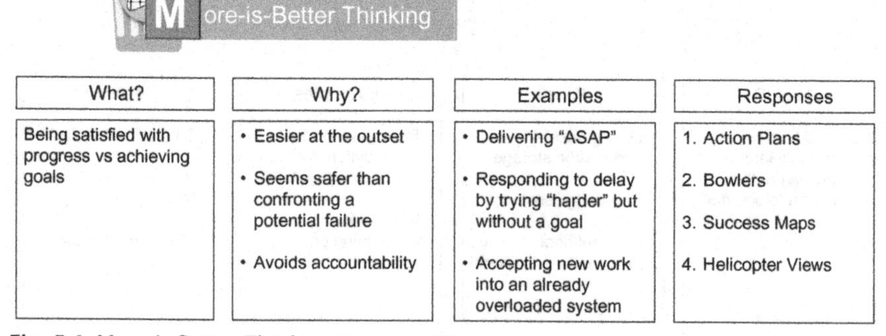

Fig. 5.4 More-is-Better Thinking is waste #3.

harder so it won't be quite as late vs facing that gap and working to deliver on time. It's also hiring the wrong person because they are good enough and we don't have time to get who we really need.

More-is-Better Thinking is any response that throws effort at a problem without having a path to success. Rather than thinking through the options to find a solution that delivers the needed value, **More-is-Better Thinking** grinds through the day doing the "best we can". It is driven by the barrier immediately in front of us. It's easier just to do something even if the overall effort is destined to fail. It's safe because we tried. Of course, all of this is wasteful even if it does consume a great deal of effort.

We will address **More-is-Better Thinking** in Chapter 9 with the *Success Map*, which comprises three artifacts: the Action Plan, the Bowler, and the Test Track. The *Success Map* visualizes a path to success and provides means to track it over time. And we'll use the *Helicopter View* to look at aggregation of multiple individual results.

5.8 Inertia to Change

Inertia to Change (Fig. 5.5) is the avoidance of both real and imagined risks that come with doing something in a new way. For those who are able to just get by in their current work environment, change brings the risk that getting by may become harder, perhaps harder enough that their standing in the organization may be damaged. For those who have actively worked to hide poor results, change brings the risk of being found out. And some people

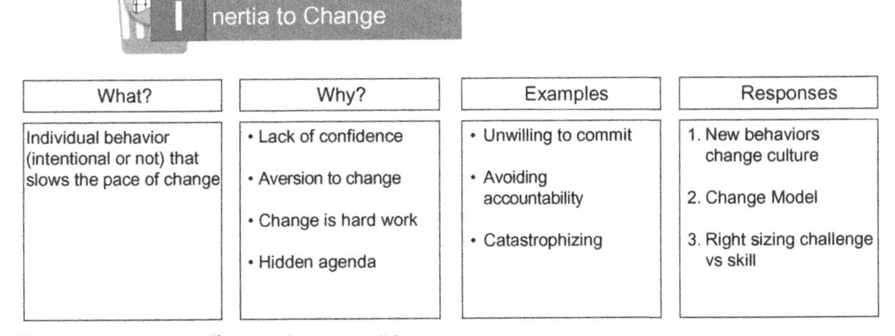

Fig. 5.5 Inertia to Change is waste #4.

have a temperament that is suspicious of change. Whatever the reasons for that aversion, driving change creates tension and pushback in all organizations.

The aversion to change presents in many modes. "We tried that and it didn't work" is a common complaint. Of course, any organization that's been around for a while probably has tried something with a passing resemblance to just about any new initiative. Don't be surprised if some people focus on similarities to past failed initiatives. Some people may be unwilling to try a method they don't fully understand, but the dilemma is they won't fully understand until they try it. Some reservations about change are normal, so you'll need to be sensitive to those who have a reasoned wait-and-see approach before cheering on change vs those who block change and refuse to participate.

We will address **Inertia to Change** in Chapter 10 starting with the Change Model, a well-known explanation for how change is created in an organization. We'll also talk about how to create expectations of ownership and accountability. Finally, we'll discuss growing leaders by balancing new challenges against an individual's skills.

5.9 No-Win Contests

No-Win Contests (Fig. 5.6) occur when a leader places challenges before their teams where there is no realistic possibility to win. A common means to create a **No-Win Contest** is to oversubscribe staff: ask them to do more in a time frame than they can do well. Then the employee has a dilemma: either

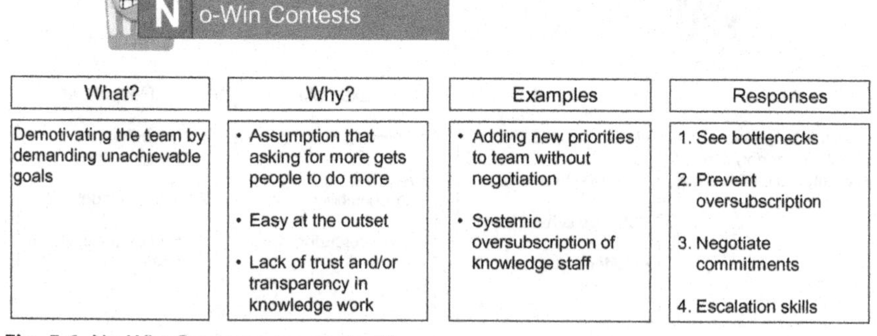

Fig. 5.6 No-Win Contests are waste #5.

do less than asked, in which case they face the immediate ire of their management, or do all that was asked poorly, in which case they face the consequences when the low quality of their work is found out. The latter allows people to avoid confrontation with the boss and so, at least in my experience, is more often chosen. However, both options are disengaging.

Chapter 11 will present several techniques to prevent **No-Win Contests**, starting with understanding and then preventing oversubscription. We will cover how to negotiate with knowledge staff rather than prescribe. Negotiating allows knowledge staff to present concerns before work is agreed to, creating accountability and ownership. Finally, we will discuss how to encourage escalation so that when knowledge staff can't deal with an issue, they quickly raise it to leaders who can.

5.10 Inferior Problem Solving

Inferior Problem Solving (Fig. 5.7) occurs when teams rush from problem definition to solution, never fully comprehending the root cause nor able to ensure effective countermeasures. It is human nature to hear a few symptoms of a problem, connect those symptoms to a known problem, and then start applying solutions. At that point, we stop listening. The problem is, of course, that there are many symptoms that are shared among different problems. When we hear that our IT or Software Development team created an application with too many bugs, we may rush to blame them for incompetence. But maybe those bugs are created because the software is used in applications never envisioned by the developers. One of the most common wastes in problem solving is misunderstanding the problem, in

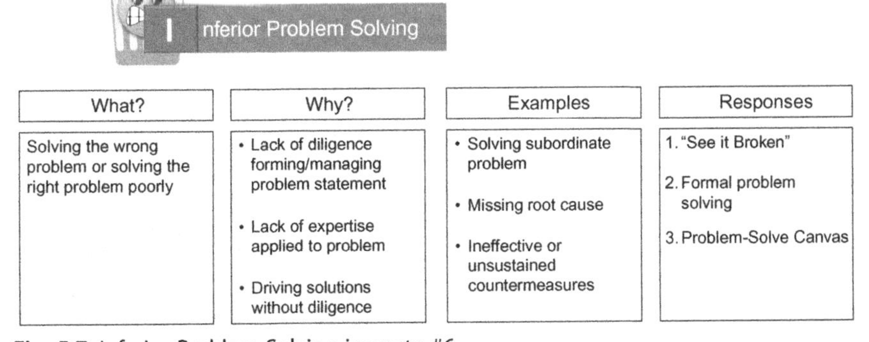

Fig. 5.7 Inferior Problem Solving is waste #6.

which case virtually all your remedies are likely to disappoint. **Inferior Problem Solving** is also revealed when countermeasures are ineffective, which can occur even when root causes are understood. A famously ineffective countermeasure is telling people to "be more careful" or "try harder" when a complex workflow fails.

Chapter 12 will present techniques to avoid **Inferior Problem Solving** such as "See it Broken" where we deeply understand the problem before we attempt to solve it. It proceeds to formal problem solving, a systematic way to move from a problem to effective countermeasures. It also presents a Problem-Solve Canvas view, a single-page problem analysis and solution based on the A3 method developed by Toyota.

5.11 Solution Blindness

Solution Blindness (Fig. 5.8) is the common bias to favor a known solution. **Solution Blindness** is to executing countermeasures what **Inferior Problem Solving** is to planning them. For example, let's say we proceeded through problem solving and emerged with what appeared to be a solid solution. However, as time goes on, we receive feedback from customers that the solution isn't working. **Solution Blindness** describes the tendency to dismiss that information or a failure to take reasonable steps to collect it. Customer complaining? They must be using the product wrong. Operations having problems delivering what we specified? They must need more training. Those who succumb to **Solution Blindness** power through all obstacles with the single-minded focus of finishing what's been started when there was enough information to demand reevaluation.

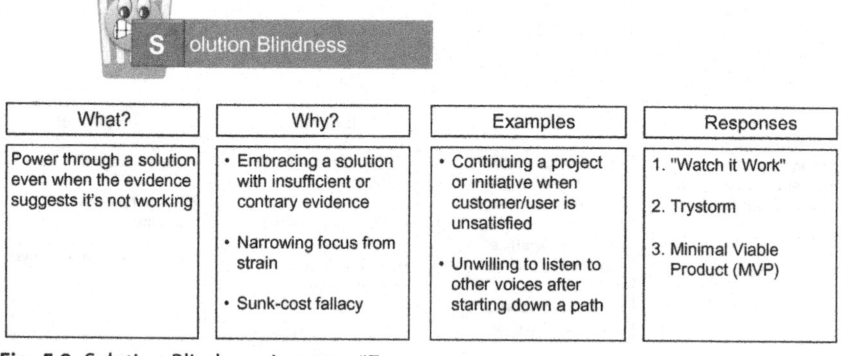

Fig. 5.8 Solution Blindness is waste #7.

In Chapter 13, we will take on **Solution Blindness**. We will discuss the concept of "Watch it Work," the equivalent in **Solution Blindness** to "See it Broken" in **Inferior Problem Solving**. We will also discuss experimentation as a mindset to challenge the uncritical acceptance of our solutions. Finally, we will use the *Test Track*, a unique visualization that is well-suited to tracking how well solutions work in the early stages.

5.12 Hidden Errors

The final of the 8 Wastes is **Hidden Errors** (Fig. 5.9). These are errors that find their way into knowledge work and escape to the customer. In this model, we accept that knowledge work is so complex that some level of errors is unavoidable. If a report requires 500 items and our knowledge staff have to get 99 out of 100 right, this still leaves a handful of errors that might find their way to the customer. In this view, the problem becomes how to detect these errors before they escape.

In Chapter 14, we will address how to detect **Hidden Errors** early, before they "escape" the workflow and cause damage to customers. We'll start by explaining why detecting errors reliably is more effective than chiding knowledge staff. In other words, because we accept an error-free knowledge workflow is unrealistic, we'll focus on creating error-tolerant workflows. We will review the "playbook," a short list of common reactions to detecting **Hidden Errors**. And we will review mistake-proofing, the technique of creating workflows that cannot proceed with errors.

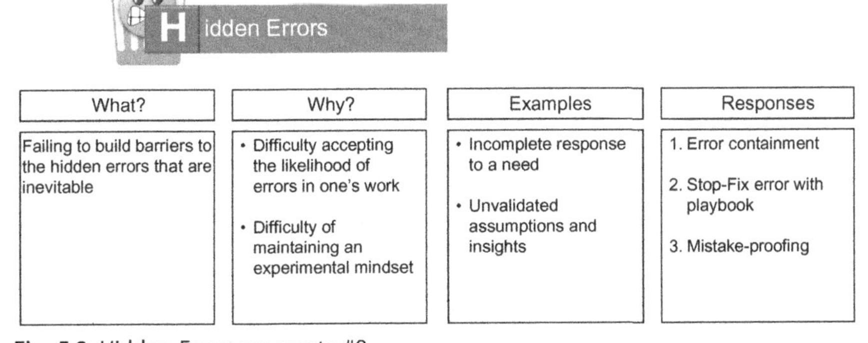

Fig. 5.9 Hidden Errors are waste #8.

5.13 Conclusion

This chapter has presented a first look at the 8 Wastes of Knowledge Work and provided places to look for these wastes. Chapters 7–24 present dozens of waste-cutting tools and techniques from lean thinking, some well-known and some novel. But before we start reviewing individual methods, let's frame a general approach structured along three dimensions: **Simplify**, **Engage**, and **Experiment**.

References

[1] T. Ohno, Taiichi Ohnos Workplace Management, Special 100th Birthday Edition, McGraw-Hill Education, 2012, p. 150. Kindle edition.
[2] S. Shingo, The Sayings of Shigeo Shingo: Key Strategies for Plant Improvement, Productivity Press, 1987, p 19.
[3] T. Ohno, Toyota Production System: Beyond Large-Scale Production, Productivity Press, 1988, p. 129 (English translation).
[4] T. Ohno, Taiichi Ohnos Workplace Management, Special 100th Birthday Edition, McGraw-Hill Education, 2012, p. 175. Kindle edition.
[5] T. McMahon, The Eight Wastes of New Product Development, A Lean Journey, December 3, 2012. Available from: http://www.aleanjourney.com/2012/12/the-eight-wastes-of-new-product.html.
[6] P. Emmons, Get Lean: Avoid 8 Wastes in Development, Adage Technologies, March 29, 2013. Available at: http://www.adagetechnologies.com/en/Get-Lean-Avoid-8-Wastes-in-Development/.
[7] G. Ellis, Project Management for Product Development, Butterworth-Heinemann, 2015, p. 191.
[8] A.C. Ward, D.K. Sobek II, Lean Product and Process Development, second ed., Lean Enterprise Institute, 2014, p. 43.
[9] A.C. Ward, D.K. Sobek II, Lean Product and Process Development, second ed., Lean Enterprise Institute, 2014, p. 31.

CHAPTER 6

Simplify, engage, and experiment

Common sense is not so common and is the highest praise we give to a chain of logical conclusions.

Eliyahu M. Goldratt [1]

6.1 Introduction

In Chapter 5, we discussed *the 8 Wastes of Knowledge Work*, a broad set of sources of waste. In this chapter, we will respond to those wastes. The responses are many, as demonstrated by the right-hand column of each of the eight diagrams in Chapter 5 that detailed waste (Figs. 5.2–5.9), but each can be viewed along three dimensions as shown in Fig. 6.1:

- **Simplify** workflows: Reduce the effort required for a workflow.
- **Engage** staff: Maximize and focus the effort of knowledge staff.
- **Experiment** systematically: Constantly learn, adapting to new information, especially identifying gaps and mistakes in order to prevent their reoccurrence.

6.2 Simplify workflows

Everything should be made as simple as possible, but no simpler.

Albert Einstein [2]

The first dimension of waste reduction is to simplify workflows. The waste associated with overly complex workflows often has concrete measures such as cutting time to do something or reducing the number of people that must be involved. Once these responses are identified, it's usually easy to gain consensus. And results pay off quickly, sometimes within days. Simplification not only reduces effort necessary to get something done, it also increases the engagement of the people doing it. As Ohno said, "Instead of just correcting workers …make the work easier … When you do this, the word

Improve
https://doi.org/10.1016/B978-0-12-809519-5.00006-5

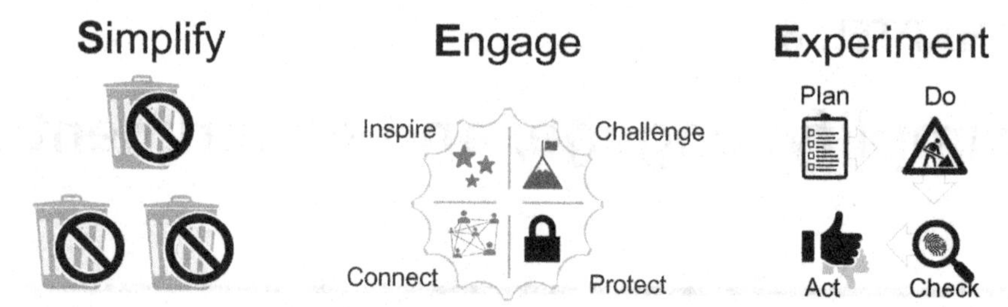

Fig. 6.1 A common sense approach to cutting waste in three dimensions: Simplify, Engage, Experiment.

will spread quickly... Once that happens, people from other areas will also come to you for help..." [3].

There are five main areas we will consider to simplify knowledge workflows (see Fig. 6.2):

1. Eliminate ambiguous calls to action, often called "signals."
2. Shorten long task queues that leave open a large number of partially completed tasks.
3. Simplify overly complex workflows.
4. Speed up knowledge transfer, removing *Information Friction.*
5. Remove unnecessarily varying workflows, where different people work in different ways without clear reasons.

Fig. 6.2 Five common examples of waste that respond well to simplification.

6.2.1 Simplify ambiguous signals

Ambiguous signals waste precious time reacting to problems. In the face of ambiguity, people often do nothing, especially when the response is difficult or challenges the organization. For example, imagine that, on October 1, we discover that a gap in our regulatory testing might cause the FDA to review one of our products. That's something a general manager would want to know quickly, perhaps the day it's discovered. But there are many unintentional disincentives to that pace of escalation—fear of disappointing the boss, not wanting to message bad news, and a misplaced hope that we'll somehow figure something out. I can think of many times when an ambiguous call to action resulted in long delays to me receiving information about unfavorable results.

Three ways to treat ambiguous signals are as follows:

- Stop-Fix alarms (Chapter 14)

  Stop-Fix alarms are unambiguous signals that demand an immediate response, often within a day or two.
- Test Track (Section 9.4)

  The Test Track is a simple visualization that measures early deployment such as the first few times a product or service is delivered to a customer or the initial use of a new workflow. It is a tool that helps the organization gain traction when something new is first deployed.
- Success Map (Section 9.6)

  The Success Map creates a smooth continuum from the Action Plan early on in projects and initiatives, to the Test Track, which measures traction, and finally to the ultimate value the work provides, such as revenue or reduction in work product errors. Success Maps provide a clear path to success with a series of defined points for Stop-Fix alarms; this minimizes the effort for managers and leaders to identify issues, and thus maximizes the time to address those issues.

6.2.2 Simplify knowledge transfer

Information Friction bloats the effort needed to transfer knowledge. **Information Friction** extends beyond the typical case of transfer from knowledge staff one to another to include transfer from knowledge staff to leadership, other functions in the organization, and customers. The problem is well-known; hence, the popularity of daily stand-up meetings, Single Point of Truth (to speed up knowledge transfer within the team), and the lean startup (to speed up knowledge transfer from the customer). Three ways to simplify knowledge transfer are as follows:

- Visualize (Chapter 8)

  Visualization speeds knowledge transfer by converting the knowledge from narration, usually the most tedious means of communication.
- Single Point of Truth or "SPoT" (Section 8.2)

  Creating a Single Point of Truth for the knowledge staff and other stakeholders simplifies the process of collecting needed information. It also draws creativity and expertise from the whole team because it provides a comprehensive view for all.
- Canvas View (Section 8.5)

  The Canvas View tells the whole story in one picture. When a person *sees* (literally) the whole context, understanding the flow of logic is dramatically easier. We will use the Canvas View extensively in this book.

6.2.3 Simplify long task queues

Long task queues cause waste because open tasks have a fixed, though mostly hidden, cost. First, just having tasks open creates intrusive thoughts that make each of us less efficient, a phenomenon sometimes called the Zeigarnik effect. Second, when tasks are open, stakeholders and customers of the task are likely to ask for status updates; when tasks are inactive, status updates are wasteful. If, 2 weeks into a 4-week task, the task is put on hold for 10 weeks, it brings 10 weeks of useless status updates, each of which steals effort.

We can work to shorten task queues with several techniques:

- Cut multitasking (Section 11.2)

  Multitasking is a wasteful behavior that is demanded by workflows with many open tasks. Some level of multitasking is required in most knowledge work—for example, working on one thing while awaiting test results from another. However, more multitasking than is necessary creates waste because of the inefficiencies inherent in multitasking.
- Ruthless Rationalization (Sections 11.3 and 21.3)

  There is only so much work that can be done in a day. To ask for more than that is to oversubscribe, creating a **No-Win Contest** for each oversubscribed person. Rationalizing is a simplifying constraint for prioritization and "Ruthless Rationalization" is the practice of consistently rejecting oversubscription according to the following principle: never accept work that, when properly done, exceeds the team's capacity.
- Just-In-Time prioritization (Chapter 21)

  Just-In-Time prioritization such as the Kanban board draws a sharp line between what's active and what's not, and "what's not" needs no status updates. On-demand prioritization also pushes the decision for resourcing as close to the date work starts as possible, to ensure that the maximum amount of information is used in the decision.

6.2.4 Simplify varying workflows

Producing substantially the same work in different ways—varying by people, by region, or over time—creates waste. The most obvious waste is that surely one of those ways will be superior to the others: it will accomplish the workflow with lower cost or higher quality than the others; in other words, there is one *best* way among the many ways we do things. This simple principle is the key to how lean thinking can deliver quick results—finding a new and better way to do something complex can take a long time, but simply identifying the best way something is already done and deploying it around the organization can be fast indeed. Another type of waste created in varying workflows is **Information Friction**, because when everyone does the same thing a different way, it's harder to share results.

So, we will employ several techniques to reduce unnecessary variation.

- Standardize workflows (Chapter 15)

  Our primary means of reducing variation is creating standard ways to do common tasks such as approvals, reviews, and problem solving. The first step to create a standard is to study the ways the task is accomplished today, devising measures of waste in the workflow, and then selecting the method that has minimal waste. This brings an immediate advantage in that everyone uses the least wasteful method known. But this also leverages the entire organization: any user or stakeholder of the workflow can invest energy to improve it and then that improvement is deployed broadly to cut waste across the organization.

- Mistake-proofing (Section 14.4)

  Mistake-proofing is the design of workflows so that common error modes are either prevented or so obvious that they stop the process until the error is corrected. A simple example is entering a credit card number on a modern commercial website. The error mode of entering fewer than 16 digits is prevented (the "Next" key can be disabled until 16 digits are entered), as is the error of the name on the card not matching the name entered (the sale is stopped until the name is corrected).[a]

[a] This is also called fool-proofing, but Shingo renamed it "mistake-proofing" (*poka-yoke* in Japanese) because of the insulting connotation that the practice is primarily for the benefit of the foolish; in fact, it benefits everyone. Shingo [13], p. 168.

- Checklists and expert rule sets (Chapter 17)

  Checklists and expert rule sets capture the wisdom of the organization. Consider the approval process, which can be improved first with mistake-proofing (for example, the approval template with key questions like "When is approval needed?" and "What budget will this expense be charged to?"). Of course, not all requirements are black and white. For example, the expected answer to a question like "Did you get two quotes?" might be yes, so an explanation might be needed if the answer is no. The template could then provide one or two guidelines on common cases where two quotes are not required. This thinking extends from a simple approval form to the most complex workflows of knowledge work. The more often experts and leaders write their advice down, the easier it is for those that follow them to learn.

6.2.5 Simplify complex workflows

Many workflows in knowledge work are more complex than is necessary. One reason is they are iterated, adding new requirements and not always removing outdated ones. Over time, workflow instructions can balloon. Another reason is that people often hastily bolt together workflows, unaware of how complex practical implementation will be; multiperson approvals are a common example, since getting five or seven people to review something can take weeks if each reviewer lets an email sit in their inbox for a just few days. A third reason is many work instructions are created without consideration for keeping an even "granularity," i.e., that each step in a process should have roughly the same complexity. Three techniques that we will use to simplify complex workflows are as follows:

- Break workflows into multiple steps (Section 15.3.10)

  When steps are overly coarse, they add complexity because each user must interpret for themselves what to do and how to check that it's done well. When we break workflows down to a workable granularity, we bring consistency and simplicity.
- Enabling constraints

  Enabling constraints can be added to simplify workflows and decisions. They work by removing from consideration any alternatives that do not meet the constraint. For example, Stop-Fix alarms are a set of demands that must be satisfied to turn "off" the alarm; there's no need for the team to expend energy on solutions that fail to meet those demands.

- Control granularity (Section 9.3.2)

  Workflows comprise multiple tasks to create work product. Controlling granularity means managing those task definitions so that all tasks are of about the same complexity. When highly detailed tasks are mixed with more general tasks, workflows become bloated.

6.3 Engage the team

When something goes wrong, there seems to be a part of human nature that quickly concludes people are the problem—for example, that a colleague is lazy, incompetent, or careless. We're quick to search for *who* made a mistake rather than *what* the root cause is. By contrast, the assumption in this text is that the great majority of times when something goes wrong, the problem is not in the character or capability of the people, but rather is a result of something structural in the organization. As W. Edwards Deming said, "A bad system will beat a good person every time" [4]. So, when a problem is recognized, start by assuming there's something systemic that caused or at least contributed to the problem. That assumption isn't always right, but it usually is, and even when it isn't, it usually generates good behavior in leadership; that's a lot for a starting assumption to deliver.

6.3.1 Motivation in the workplace

Frederick Herzberg studied motivation in the workplace. He developed the *Motivation-Hygiene* theory (sometimes called the *two-* or *dual-factor theory*), which states that job satisfaction is created by factors that are generally independent of those that cause dissatisfaction [5, 6]. He called the factors that increase job satisfaction *motivators*; they are generally intrinsic, internal to the employee. Motivational factors answer the question, "Is my work meaningful?"

Herzberg called those factors that relate to dissatisfaction *hygienic*; they are extrinsic: for example, monetary rewards, supervisor actions, and company policies. In Herzberg's construction, these items do not cause satisfaction, but if they are missing or done poorly, they do create dissatisfaction [7]. Hygiene answers the question, "Am I treated well at work?" One common misinterpretation of Herzberg's theory is that motivators are so much more important that hygienic factors can be ignored. In fact, in Herzberg's view both were necessary. They have different purposes and those in leadership need to understand both [8].

Monetary reward is not a substitute for intrinsic motivation.

W. Edwards Deming [9]

Salary is an interesting example of a hygienic factor. It's commonly used by companies as a motivator. But money doesn't make people passionate about their jobs [10]. However, if someone learns they are underpaid—for example, when another company offers a substantially higher salary for the same job—it can lead to severe dissatisfaction. For most of us, after we perceive we are fairly compensated, more compensation brings little motivation.

The motivational factors that Herzberg identified were a sense of achievement, recognition, the pleasure of the work itself, a sufficient level of responsibility, personal advancement, and personal growth and learning. There were a larger number of hygienic factors, but they were dominated by company policies and administration, supervision, relationship with supervisor, work conditions, salary, and relationship with peers and subordinates (see Fig. 6.3). Herzberg's seminal work helped redefine how the world looked at engagement, especially separating out intrinsic and extrinsic motivators.

Daniel Pink describes a study of more than 10,000 engineers and scientists working in US companies. The study found that the desire for intellectual challenge was the primary predictor of productivity. Those driven by intrinsic motivators filed more patents than those motivated by compensation [11], even though the two groups expended similar amounts of time and effort. Those intrinsically motivated just accomplished more [12].

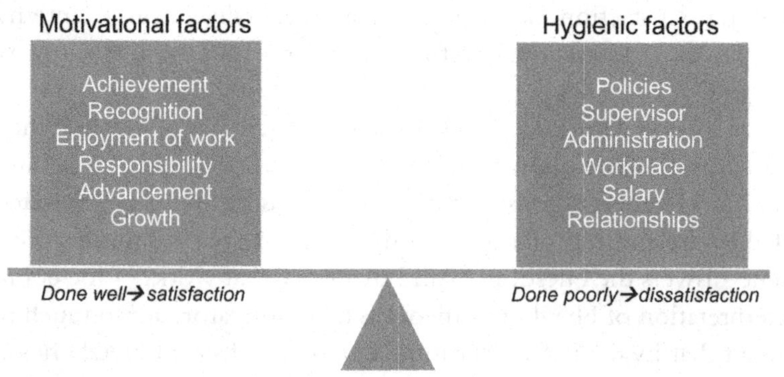

Fig. 6.3 Herzberg's motivational and hygienic factors.

6.3.2 Leading transformation and transaction

All of us need balanced leadership, a combination of:

(1) the ability to get things done (transactional leadership); and

(2) the ability to pick the right things to do (transformational leadership).

My former company president used to tell us, "Do the right things and do things right." He captured in this short phrase the different between transactional and transformational leadership, which Peter Drucker separated into *management* and *leadership*. Drucker is often quoted as saying, "Management is doing things right; leadership is doing the right things."

Unfortunately, a fad of popular writing seems to deprecate management skills while building up "leadership" as if we must pick one or the other; as we'll see, good leaders need both. Certainly, managers without a vision become dull bureaucrats; they will fail to inspire anyone. But transformational leaders who dream big will accomplish little if they don't know how to get things done.

> *The most stupendous improvement plans in the world will be ineffective unless they are translated into practice.*
>
> ***Shingo [13]***

Fig. 6.4 contrasts transactional and transformational leadership. Transactional leadership must handle a large portion of everyday work. When we encounter Problem "A," we react with Response "B." If we are skilled at our jobs, transactional leadership will work most of the time. But no matter how skilled we are, we're going to run into unanticipated problems:

	Transactional leadership	Transformational leadership
How it works	Works within a system. Starts solving by fitting experiences to a known pattern. Asks, "What's the step-by-step?"	Works to change a system. Starts solving by finding experiences that show the old pattern doesn't fit. Asks, "What do we need to change?"
What it does	Minimizes variation of the organization. Expects the entire team to meet a standard. Can be duplicated and sustained. Best at delivering defined results.	Maximizes the capability of each person. Inspires each person to give their best. Requires minimal structure. Best at delivering innovation.

Fig. 6.4 Comparing transactional with transformational leadership.

places where transactional leadership is ineffective. But strong transactional leadership brings great value too; in its absence, every tedious issue will require invention and personal push; nothing will be easy, so little will get done.

Transformational leadership recognizes when a problem doesn't fit the old patterns, when we need to change—literally, transform—to respond well. Transactional leadership ensures the boxes are checked with diligence and efficiency. Transformational leadership asks: why are we checking that box, and what are the boxes we should be checking that we have yet to recognize? Fig. 6.5 demonstrates that neither approach alone works well. A manager without vision attracts no followers—he or she is a dull bureaucrat trapped in the status quo. A visionary who cannot manage leads nowhere. Their big ideas never materialize.

A manager without vision attracts no followers. A visionary who cannot manage leads nowhere.

You will need transformational and transactional skills to lead improvement in knowledge work. Only keen "eyes for waste" can see the bottlenecks that constrain the organization's performance. That's going to take strong transformational thinking. Choose the wrong area to improve, and success will be elusive. Discord and dissension will grow as people sense they are part of a losing team.

You will also need strong transactional leadership. Improving knowledge work demands measurement, management of the details, and the ability to drive solutions when gaps are identified. If you cannot lead transaction, you may find worthwhile areas to improve but doing this won't result in sustainable improvement. Big ideas without results quickly erode enthusiasm.

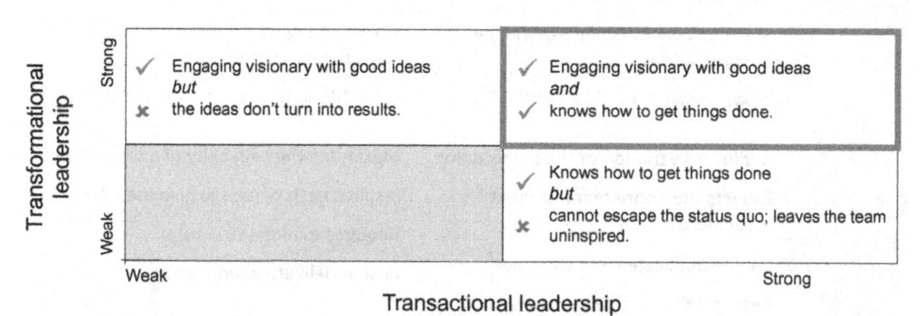

Fig. 6.5 Balanced leadership requires excellence in transformation and transaction.

6.3.3 The Engagement Wheel

The basic elements that help each of us to engage at work are inspiration, connection, challenge, and protection, as shown in the Engagement Wheel of Fig. 6.6.

6.3.3.1 Inspired: We share worthy goals

> *The desire to do something because you find it deeply satisfying and personally challenging inspires the highest levels of creativity, whether it's in the arts, sciences, or business.*
>
> ***Teresa Amabile [14]***

Daniel Pink explains motivation in his book "Drive: The Surprising Truth About What Motivates Us." He describes three types of motivation, which he likens to human "operating systems," that he calls M1.0, M2.0, and M3.0 [15]. M1.0 is basic survival, which dominated motivation for most of human history. M2.0 is a system of rewards and punishment. It replaces the survival mentality of M1.0 with the premise: "Rewarding an activity will get you more of it. Punishing an activity will get you less of it" [16]. He portrays M2.0 as outdated for the workplace and he points out that many companies are yet to recognize that.

M3.0 is based on internal motivators and it is particularly effective for knowledge work. For Pink, there are three main factors: autonomy, mastery, and purpose. Purpose, feeling a part of something larger than yourself, has long been known to be an important part of achieving happiness (Herzberg writes about "a sense of achievement" and Seligman refers to "meaning" [17]).

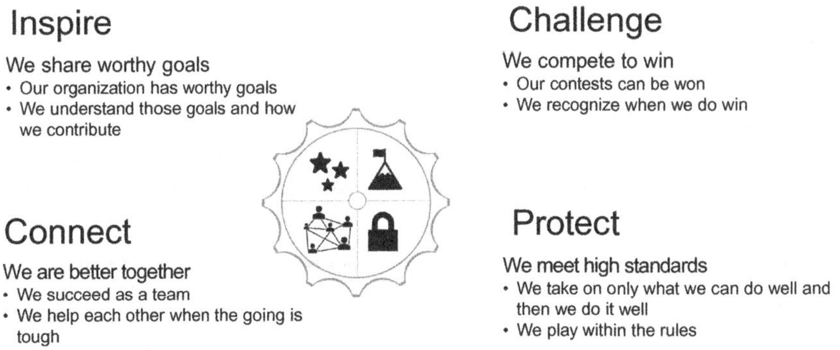

Inspire
We share worthy goals
• Our organization has worthy goals
• We understand those goals and how we contribute

Challenge
We compete to win
• Our contests can be won
• We recognize when we do win

Connect
We are better together
• We succeed as a team
• We help each other when the going is tough

Protect
We meet high standards
• We take on only what we can do well and then we do it well
• We play within the rules

Fig. 6.6 The Engagement Wheel: A balanced view of engagement.

People want to be part of something that makes their community, their country, or the world a better place. They want to sense that their organization has worthy goals. For some, this desire is strong enough that they dedicate their lives to humanitarian causes. For those of us who have careers outside of such causes, those careers provide some portion of meaning: our patients living better, our clients receiving the justice or success they deserve, or our customers and users being better at, safer, and more satisfied doing what they need to do.

6.3.3.2 Connected: We are better together

An article from Teresa Amabile begins with this chilling observation about modern organizations: "creativity gets killed much more often than it gets supported" [18]. She believes creativity requires technical expertise, the ability to think flexibly and with imagination, and motivation. She lists five "levers" that managers can pull to improve the creative environment: magnitude of challenge, degree of freedom, work group structure, amount of encouragement, and organizational support.

Amabile emphasizes team interaction. She believes the team brings many benefits including diverse ways of thinking, having a shared excitement over the team's challenges, being able to get help more easily from peers who understand your tasks, and sharing a sense of respect for each other [19]. She also emphasizes the supervisor's role in recognizing contributions from team members, by greeting new ideas with an open mind, and by valuing ideas for the process that generated them rather than evaluation of the immediate results. This last point is crucial for creating a safe environment—good leaders reward people for doing the right thing based on the information available to them at the time. If you reward only positive outcomes, you'll find your team overly conservative. Wise risk-taking ought to be rewarded, so good leaders will communicate that even when people do the right thing, the outcome cannot always be predicted.

Herzberg was more focused on the connection between a worker and a supervisor: recognition for work done and personal advancement. Google's article in the Harvard Business Review found that their knowledge staff viewed the connection between their career and their supervisor as the most valued aspect of that relationship:

> *Engineers hate being micromanaged on the technical side but love being closely managed on the career side [20].*

Connection is also an entry point to understand engaging effects of high ethics. Stanley McCrystal says that real leaders don't lie, cheat, or steal [21]. In another speech, he asked, should you follow someone who cheats on their taxes? He said he wouldn't because someone who will lie to the government will probably lie to you. People don't connect with those who don't meet their ethical standards.

6.3.3.3 Challenged: We compete to win

Regarding challenge, Amabile says good managers need to match employees with tasks that stretch them enough to make the job interesting, but not so much that they lose "self-efficacy" [22]. This is well-aligned with Herzberg's belief that every job should be a "learning experience." Amabile's experience is that organizations kill the creativity of their associates with fake or impossibly short schedule demands—what we will call here **No-Win Contests**. Mastery, one of Daniel Pink's three elements, is the innate desire to be good at something—it's a powerful motivator for many professionals. Winning a contest we care about is deeply motivating: delivering a better product or service than a competitor, bringing in revenue the organization needs, or receiving an industry award.

6.3.3.4 Protected: We meet high standards

> *It takes 20 years to build a reputation and five minutes to ruin it. If you think about that, you'll do things differently.*
>
> ***Warren Buffett [23]***

Knowledge staff intrinsically want to do a good job, but too often people feel unduly pressured by unreasonable schedules or unrealistic goals. Experts commonly are oversubscribed, which is to say they feel that they lack the time to complete everything they are assigned. Given the option of (1) doing a few things well and not doing the others vs (2) doing all of those things poorly, which losing option should they pick? A common outcome of these **No-Win Contests** (waste #5) is that people sense they are victims in an oversubscribed system, pushed to do work beneath customer expectations and their own standards. Unchecked, this leads to a higher likelihood of mistakes and lower accountability ("It's not my fault...there's not enough time"). People want leaders who inspire in them the courage to take on only the commitments they can deliver and to finish the work to high standards in the face of pressure to do the opposite. This protects the "brand" that the knowledge organization stands for.

The Engagement Wheel also helps us recognize that leading continuous improvement requires that we address the diverse motivations represented in our teams, while avoiding the pitfalls of overemphasizing one area:

- **Inspiration**
  Continuous improvement will reduce routine tasks, freeing mind-share for expertise and creativity
  ...while we get the daily work done that our organization needs.
- **Connection**
  Continuous improvement will bring real consensus and collaboration
  ...while not letting an outsized desire for unanimity make us indecisive.
- **Challenge**
  Continuous improvement creates more wins by bringing clarity and alignment
  ...while not letting the desire to win make us hasty.
- **Protection**
  Continuous improvement will help us be more diligent so that we will protect our brand
  ...and we won't use the "rules" to block progress.

6.4 Experiment to learn

> *We shouldn't fall into the trap of ever believing that at last we see the ultimate light. We are dealing with management science and science definitely doesn't believe in truth, only in validity.*
>
> ***Eliyahu Goldratt [24]***

A mindset of experimentation embraces that what you think today is likely enough to be wrong that it must be diligently verified. It is the most challenging dimension of **simplify-engage-experiment**. "**Simplify**" demands we understand something so we can make it more efficient. "**Engage**" demands we respect others and understand what they need to thrive. But "**Experiment**" demands *business humility*, accepting that there are many important things we misunderstand.

> *You can start change where it feels best. Just don't stop. Don't ever stop. Part of lean development is that you keep learning how to do it better, forever.*
>
> ***Ward and Sobek [25]***

6.4.1 Business humility

The term *business humility* includes accepting there is a lot we don't know, but it's much more. It's accepting there are things you think you know that are wrong—biases, gaps in judgment, and invalid assumptions. The world is too complex to validate every action you take. To survive, you must make assumptions. When I pull my car onto the street, I look left and right, but I don't look up—I *assume* there's nothing dropping out of the sky into my path. When I step into the pasture, I assume my horses aren't vicious and, at dinner, that there's no cyanide in my soup. I must make myriad assumptions to function for just a day. The problem is I make so many assumptions, I'm not conscious of most of them. So, it's not what I know I don't know—for example, quantum mechanics—that gets me in trouble. It's bad assumptions that I may not even recognize I'm making. Business humility requires I accept not only that there's a lot I don't know, but also that there's a lot I think I know that's wrong.

One of the easiest areas to see this behavior is when people rush to solve problems they don't understand. All of us have had the experience of telling someone a problem and them interrupting so fast with their solution that you know they don't understand. They are probably assuming your description fits with something they have seen. This is why such a large part of formal problem solving (Chapter 12) is getting a team to understand the problem deeply before they start solving. And this is why so much waste is hidden. The bad assumptions that cause the most waste are destructive because the people making the assumptions don't know they are bad and, in fact, they often don't even realize they are acting on an unproven assumption.

Business humility also accepts that we can learn something to improve ourselves from every person we come in contact with. The modifier "business" is included because this characteristic is so different from traditional humility. In simple terms, traditional humility is putting others above yourself. Business humility asks you to consider that even those who occupy modest roles have much to teach you. The VP of R&D can learn from the janitor something that makes him or her a better VP of R&D. Ancient wisdom that reflects business humility is often attributed to Confucius:

> *If I walk with two men, each will serve as my teacher. I will pick out the good points of the one and imitate them, and the bad points of the other and correct them in myself.*

In business humility, we remind ourselves that there is a lot we don't know, so we can achieve the mindset most amenable to learning. It's not that we put others above ourselves, but rather we accept that there is something we can learn from them to improve ourselves.

> *Give me the fruitful error any time, full of seeds, bursting with its own correction. You can keep the sterile truth for yourself.*
>
> **Vilfredo Pareto**

6.4.2 30% of what we think is wrong

An experimentation mindset demands first that we recognize that a substantial portion of what we think we know is wrong and so we must look at results. It's not that our thoughts are completely wrong, but they are wrong enough to rob us of success. Taiichi Ohno wrote:

> *A thief may say good things three times out of ten; a regular person may get five things right and five things wrong. Even a wise man probably is right seven times out of ten, but must be wrong three times out of ten... [26]*

The point Ohno made was that even the smartest person is wrong often, let's say 3 of 10 times; the rest of us are wrong even more frequently. This leads to one of the most important concepts in systematic learning, that "30% of what we think is wrong." The quantification is jarring, meant to reveal the common but incorrect assumption that we're rarely wrong. Jordon Peterson, a clinical psychologist and professor at the University of Toronto, is more pessimistic than Ohno: "The probability that...you are going to get it right the first time is zero" [27].

This doesn't mean "wrong" in its binary sense. These are not closed questions such as "What time is sunrise?" or "Is Chicago's latitude below Rome's?" (it is). Those are such simple questions that, with a modicum of diligence, we can be right much more than 70% of the time. But we are wrong much more often in complex scenarios:

- Do we understand the symptoms from all relevant points of view?
- Are we able to understand the barriers to improvement?

Fig. 6.7 "Wrong" is every arrow that misses the target.

- What is the call to action? Is an action even warranted?
- Is our response going to be effective?
- Are all functions in the organization capable of delivering the proposed solution?
- How will we ensure diligence over time?

These are not true/false questions; there are a thousand ways to be wrong. "Wrong" is every arrow that misses an archery target 70 m (~75 yards) away. From this perspective, "70%" is generous for most of us (Fig. 6.7).

30% of what we think is wrong. Not dead wrong, but wrong enough. And wrong in ways we don't imagine.

6.4.3 A falsifiable hypothesis

When you can measure what you are speaking about, and express it in numbers, you know something about it, when you cannot express it in numbers, your knowledge is of a meager and unsatisfactory kind.

Lord Kelvin [28]

A hypothesis demands quantifying our beliefs. It's usually difficult and often uncomfortable, but we do it to communicate. We will call such a hypothesis *falsifiable* to distinguish the common use of the word to describe a hunch or an opinion. When we go beyond the simple insight of "if we do this then we'll get that" to quantifying what we'll measure and how we will decide on

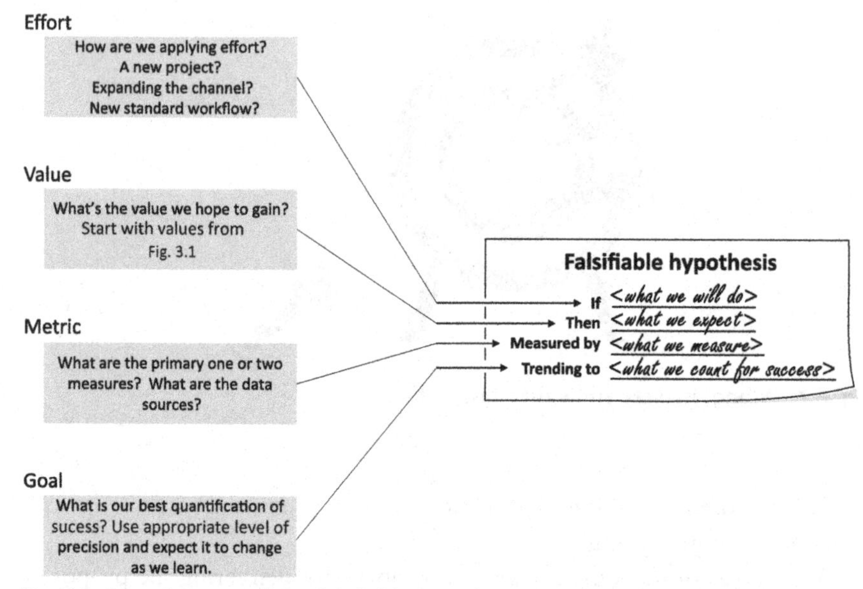

Fig. 6.8 Four elements of the falsifiable hypothesis.

success, we introduce the very real possibility that the data will contradict our insight, and *falsify* the hypothesis. As shown in Fig. 6.8, there are four elements to a falsifiable hypothesis [30].

> *I believe that all qualities we are interested in can be quantified. That means you can put a number on them…this is not about mathematics. This is not about measurement. It's not about statistics.* ***It's about people communicating with each other…***
>
> ***Tom Gilb [29]***

6.4.3.1 Effort: What we will plan and do

What precisely do we intend to do? Are we starting a new project? Adding a new service? Expanding application engineering regionally? Creating new standard work? Offering a product from another supplier? We want to capture in a few words the effort we plan to exert, quantified to a level appropriate for the discussion—elapsed time, people-hours, and expense are preferred. At this point in conversations, I often get the response, "I don't know." That's true—knowledge staff rarely "know" all the data needed to make a sure decision. If there were no unknowns, it wouldn't be knowledge work. So, quantify what you do know and state your assumptions so that it's clear what you don't know. If you judge that the cost of taking action is between $10k and $25k, say that; it has a great deal more information than "I don't know" or "It depends." There's no need to say anything you don't believe or can't quantify with higher precision than expertise and experience support. Just write down what you know.

6.4.3.2 Value: What we expect

Similarly, quantify the benefit. Increase revenue by about $400k next year? That beats "increase revenue," which might refer to $100 or $10M. Are you gaining new clients? How many, roughly? Or upselling the ones you already have, and by how much, roughly? Quantify to communicate; communicate more fully by capturing the unknowns as well as you know them.

6.4.3.3 Metric: What we measure

List the objective measures. In knowledge work, there is always ambiguity—embrace it. For example, you might collect the number of customers won each week vs the number lost because the service was wanting in the customer's eyes. That's subjective, so remove what ambiguity you can—for example, with a template where a sales person who experienced a customer's rejection of a product can write a sentence or two on which customer said this, when they said it, and precisely what they said. Some level of ambiguity necessarily will always remain because you're relying on an expert opinion from the salesperson, but you will get more reliable results as you remove the unnecessary ambiguity.

6.4.3.4 Goal: What we count for success

The final element defines success: what do we need to see in the data to validate the hypothesis? This step requires expert judgment. If we have a performance goal of 5, is 5 required? Is 4.99 sufficient? I often see people straining to achieve unnecessary precision while losing sight of the value of the falsifiable hypothesis. It's fine to say "about 4." This is far superior to stating nothing and having some think "2" is fine while others want "5." Again, experts must often reach beyond their comfort zone and quantify as well as they can today, leaving room to update the goals as they learn in the experiment.

The falsifiable hypothesis is so powerful because it puts our assumptions—even those we are unaware of—to the test. It demands business humility because the act of writing how your hypothesis may be proven false brings the very real possibility that you'll have to admit you've been wrong. But that is the core of learning. People who think they are never wrong don't learn.

6.4.4 Plan-Do-Check-Act (PDCA)

The previous sections described *what* we will do to create an experiment. This section describes *how* we will do it. Our structure will be a beguilingly simple management method: Plan-Do-Check-Act, often known as the Shewhart cycle from Walter Shewhart, who performed a series of industrial engineering experiments for Bell Laboratories in the 1920s. This method was used and improved upon by W. Edwards Deming and others to create the PDCA method, though Deming found the word *Check* to be perfunctory and so replaced it with *Study*. Thus, Plan-Do-Study-Act (PDSA) is the Deming cycle [31], though there is confusion on this, which is quickly seen with a web search of "Deming cycle," for example, Refs. [32, 33]. As stated on the website of the Deming Institute, Dr. Deming's "focus was on predicting the results of an improvement effort, studying the actual results, and comparing them to possibly revise the theory" [34]. This text will use PDCA since it's more common, but embracing Deming's concern that the *Check* must not be a perfunctory test for success or failure, that data and methods used to collect it must be studied before concluding that a hypothesis was validated or falsified.

The PDCA cycle is shown in Fig. 6.9. It amounts to a simple task management method for experiments. The first two steps of Fig. 6.9 are ordinary task management: plan the effort and then do the work. What separates PDCA and the experimental mindset it promotes from simple task

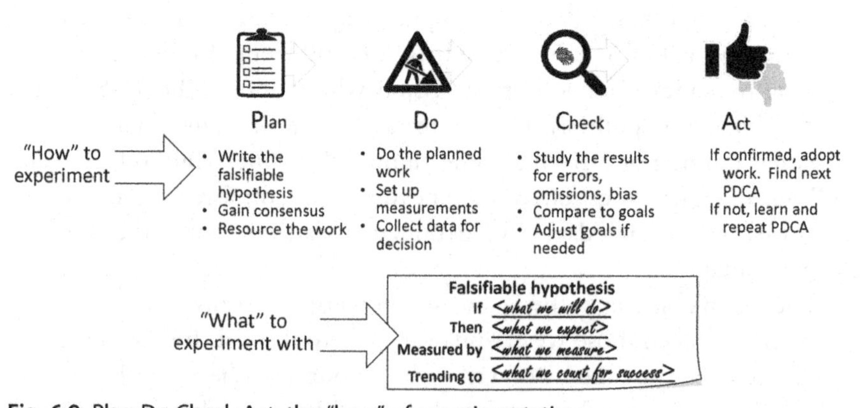

Fig. 6.9 Plan-Do-Check-Act, the "how" of experimentation.

management is that PDCA begins by planning to check results; by contrast, normal behavior starts by assuming that what we do will probably work so there's little need to check anything at the end. The *check* systemically challenges initial assumptions. The final step, *act*, calls out a clear decision to accept or reject a hypothesis. *Act* typically occurs at a point in time and requires some sort of consensus, perhaps just a small group or a supervisor's agreement.

Whether our call to action is wrong just 30% (as derived from Ohno) or 0% (from Peterson), the point is it's wrong too often to accept without verification. Hence there's a need for *systematic learning*. Systematic learning is first a mindset: we cannot avoid being wrong. No matter how clever our idea, how experienced we are, and how hard we work, there will always be **Hidden Errors** in our thinking. That is, we will miss the center of the target, often by quite a bit. So, significant errors are unavoidable and the only logical response is to *experiment*: first, create a hypothesis using all the domain knowledge, critical thinking, and information available. Then, measure the results of that hypothesis.

> *If it disagrees with experiment, it's wrong. In that simple statement is the key to science. It doesn't make a difference how beautiful your [theory] is. It doesn't make a difference how smart you are.*
>
> ***Richard Feynman [35]***

You might think that experimenting is just common sense, but it's not. For centuries, educated people believed that heavier objects fell faster than lighter ones, that objects had a "natural falling speed" that increased with their weight. But a series of experiments on objects sliding down inclined planes by Galileo disproved what today seems like a simple-minded misconception. We can laugh at our ancestors or we can accept that had we lived in their time, we would likely have accepted the flawed principle just as they did. However logical experimentation may be, it doesn't seem to be natural for any of us; once we accept an explanation for a phenomenon, it's unnatural to challenge it. So here we will aspire to create in our actions a system of measurements with the following logic that defines *systematic learning*:

- However convinced we are that we are "right," we accept that our thinking has **Hidden Errors**.
- To identify those **Hidden Errors**, we will measure results and compare them to a set of goals.

- When results diverge from the goal, the divergence likely reveals **Hidden Errors** so that diligent problem solving and countermeasures are called for until the gap is removed.

Human nature seems to be that we work on an idea diligently until errors are removed and then we apply it. There's an implicit assumption that we will catch all the errors of consequence at the outset. Systematic learning advises us to work diligently to create a first iteration and then build workflows that make the inevitable errors easy to see. As John Shook said,

> *The famous tools of the Toyota Production System are all designed around making it easy to see problems, easy to solve problems, and easy to learn from mistakes. Making it easy to learn from mistakes means changing our attitude toward them [36].*

PDCA is a repeating cycle: do something and then learn. The first cycle of PDCA lifts us up a small amount, but enough so we are ready for the second cycle, and then the third, and on and on. Given time and diligence, we can rise to great heights. There is no end of this journey, no point where our bad assumptions are all corrected. Each step in Fig. 6.10 is the same: it is both the result of learning from the last step and preparation for the next.

With all the benefits of *systematic learning*, you might be asking, what's the downside? The problem is cost. Learning through experimentation takes much more time and expense than getting it right the first time. Why do we not experiment with fire insurance? After all, we could decline fire insurance and measure the results—if my house burns down, I'd learn something. Why not experiment with driving at 100 mph down a neighborhood street? If you are considering five designs for a bridge, why not build the bridge five times and see which design can best support the weight of traffic? Such

Fig. 6.10 Systematic learning: A mindset of climbing over knowledge gaps and invalid assumptions by experimenting.

experiments are foolish because either the cost of experimentation is too high, or the cost of failure is unacceptable, or both. In fact, the reason to acquire domain expertise is to eliminate the needless experimentation that results when people attempt to solve problems well outside their competence. I can recall in my first engineering class my professor saying, "Anyone can do what an engineer does…it will just cost them a lot more." So, accepting that experimentation is too expensive to apply to all areas, how do we select the few areas where we can afford to experiment? Here we turn to Goldratt's Theory of Constraints.

6.4.5 Goldratt's Theory of Constraints

The important and difficult job is never to find the right answers, it is to find the right question.

Peter Drucker [37]

In the 1980s, Eliyahu Goldratt presented the Theory of Constraints (ToC) in his management philosophy book, The Goal. The underlying assumption of ToC is that every goal is limited by one primary constraint, the "bottleneck." So, first identify that constraint and concentrate available effort there. Focus all available energy on the bottleneck: improve it by removing waste and increase resources to pry the bottleneck open bit by bit. The organization focuses on the bottleneck and, over time, the results will be stunning.

Since the strength of the chain is determined by the weakest link, then the first step to improve an organization must be to identify the weakest link.

Eliyahu Goldratt [38]

Fig. 6.11 modifies Fig. 4.10, to show how each substantial reduction of waste comes from removing waste in the bottleneck, one iteration after the next, over months and years. Bear in mind that ToC teaches us that making improvements outside the bottleneck has no measurable effect. This helps to explain why so many attempts to transform knowledge work fail: they are not reliably targeted at the one bottleneck an organization faces. If the improvement doesn't address the bottleneck, its effect may be "good," but it will be unmeasurable. According to the Theory of Constraints, only the bottleneck makes a measurable improvement. Unfortunately, efforts that make things "better" without demonstrating sustained,

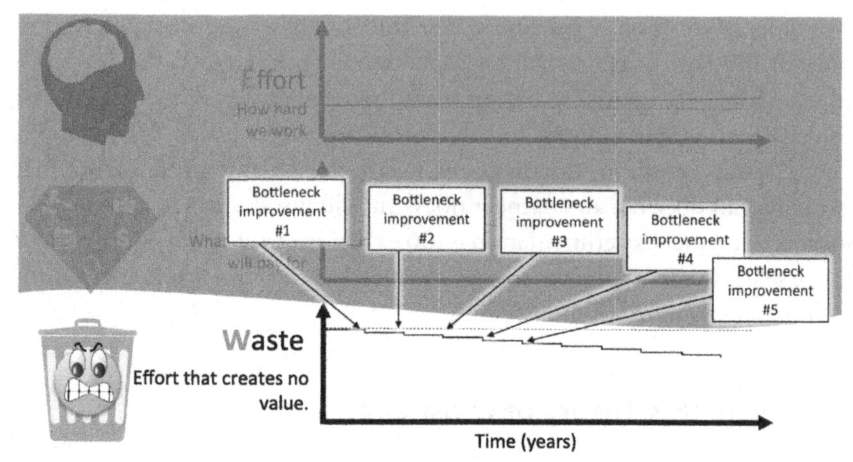

Fig. 6.11 Viewing Fig. 4.10 with the "Theory of Constraints" (ToC).

measured improvement are destined to fail. Continuous improvement demands large, concrete investments and, in the real world, those demand large, concrete returns over time.

6.4.5.1 Reflection: How the bottleneck measures all issues

Reflection is a mechanism Goldratt described that allows the bottleneck constraint to represent all other constraints. The first reaction people often have to the concept of a single bottleneck is that it creates a one-dimensional view of knowledge work. For example, if we say the bottleneck for a project is the schedule (which it normally is), the concern is that the team will abandon all other constraints like quality and ease of use in the narrow-minded pursuit of delivering on time. It's actually quite the opposite. When quality and ease of use are properly reflected in the schedule, then managing the schedule will result in managing all constraints.

As an example, consider how a nonbottleneck quality constraint can reflect into a bottleneck schedule constraint. Let's say 4 months into a new product development, we discover a quality issue that will result in an expected customer return rate of 600 units per year. Further, let's say the normal rate of returns is 100 units per year. So we have a forecasted gap of 500 returns per year. If the bottleneck constraint is the schedule, how do we convert 500 units per year gap to units of calendar time?

1. **Measure the issue directly (the nonbottleneck constraint)**

   Newly discovered quality issue is forecasted to result in about **600 units returned per year**.

2. **Compare measurement to the requirement**

   Returns are normally acceptable at the rate of about 100 units per year; this is an upper limit and we wouldn't likely find anyone happy about it. But, if we judge that a project at that level would be accepted, that becomes a de facto standard.

3. **Estimate the effect on the project schedule to reduce the forecasted return level to meet the requirement**

   Let's say our quality defect proceeds from not having an up-to-date test fixture and that test fixture will take an additional 4 weeks to develop with the existing staff. So, the newly discovered quality gap is reflected to the bottleneck as 4 weeks of schedule delay.

   Notice that the Theory of Constraints does not direct to a simplistic response. It considers all constraints, vigilantly asking: how much schedule will we consume to address this constraint? This is shown in Fig. 6.12. The initial effect of the quality issue reflects to 28 days. In some cases, the schedule delay can be absorbed, but more often it must be countermeasured, at least partially. Let's continue in step 4 assuming we have funds we can apply to mitigate 28-day slip. As you'll see, we can also reflect the cost of mitigation into the schedule.

4. **Countermeasure schedule loss**

   Let's say we have an offer from a consultant to mitigate the slip at a cost of $65k; if we engage the consultant, the team can continue on the project, needing perhaps only a week of team time to implement the consultant's solution. In this case, the gap in the bottleneck constraint—the schedule—is reduced to 7 days. So, the identical issue

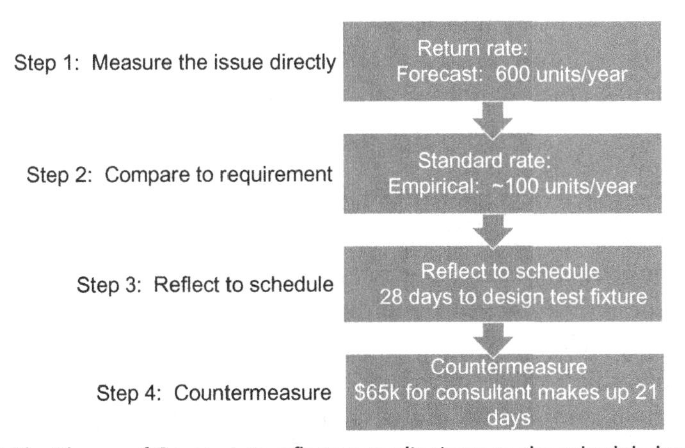

Fig. 6.12 The Theory of Constraints reflects a quality issue to the schedule bottleneck.

may be reflected as 7 days (if we can afford the $65k) or 28 days (if we cannot). Moreover, we have a metric to measure the value of the consultant: for $65k, we can recover 21 days of the schedule. Now we have positioned the organization's leader to ask a simple question: is 21 days of schedule recovery worth $65k? If yes, hire the consultant. Otherwise, absorb the delay. Notice that we never discuss if we are going to address the quality issue: of course we will. The Theory of Constraints allows us to address a broad range of constraints by reflecting their effect back to the bottleneck.

In Section 20.2, we will discuss the Fever Chart, one of Eliyahu Goldratt's most important contributions to project management [39]. The Fever Chart visualizes schedule performance as the bottleneck constraint in project management. It pushes the team every day to ask what issues have been discovered and how we will minimize their effects on the schedule. Though it focuses on the bottleneck constraint, the Fever Chart is not simplistic. It demands rigor in thinking issues through. For example, if a team member sees a quality issue, he or she is encouraged to describe how quality is not meeting a need and, if known, what solution he or she recommends, and then *how much the schedule will be affected*. I've experienced again and again how this type of thinking causes a team to think through issues and take steps to address issues quickly.

6.4.6 The Pareto principle

The well-known Pareto principle sheds more light on the prevalence of bottlenecks. In the 19th century, the Italian economist Vilfredo Pareto identified a distinctly non-*normal* (non-Gaussian) distribution, which has come to be known as the Pareto distribution, commonly known as the 80/20 rule. One example was that, in Italy, 80% of the land was owned by 20% of the people; another was that 20% of the pea pods in Pareto's garden produced 80% of the peas [40]. A modern corollary in the United States is that 1% of the population has 40% of the wealth. The Gaussian or bell-curve distribution describes many phenomena, for example, most physical dimensions within a species: the height or forearm length of a human male is distributed around the mean with roughly the same number of samples below as above (see Fig. 6.13). The Pareto distribution represents something quite different: the "winner takes all" distributions where people compete for things such as wealth or fame.

For example, consider net worth of a US individual. According to MarketWatch in 2018 [41], the average net worth (total assets less liabilities)

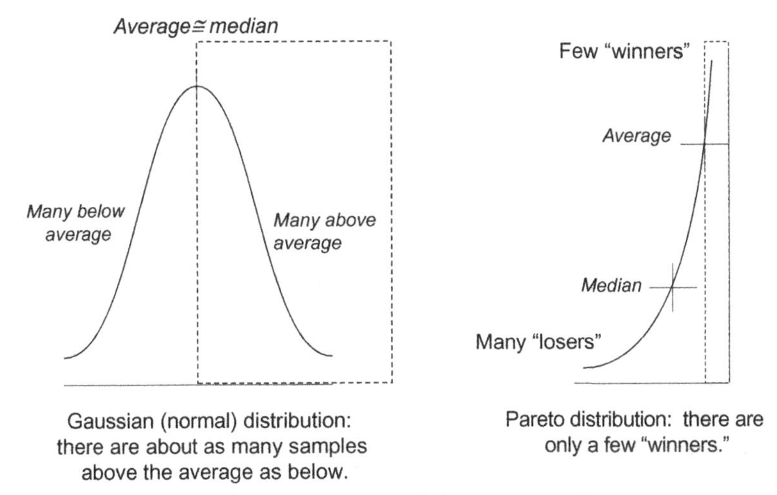

Fig. 6.13 The Pareto distribution represents "winner takes all" contests.

was almost $700k. But the median net worth, the net worth of the person that half the people are under and half are over, is just under $100k. Why is there a 7x disparity? Because Warren Buffett, Jeff Bezos, and a few other "winners" have a disproportionate share, offsetting a large number of people with almost nothing. Net worth is a winner-takes-all contest; it follows the Pareto principle.

The Pareto principle extends well beyond wealth distribution. It describes the vanishingly slim fraction of athletes that play professionally, and within that elite group, there is another Pareto distribution of compensation that goes to a tiny fraction of them. It describes the number of actors that take major roles in movies, the proportion of singers that sing professionally, and the tiny fraction of inventors that patent successful inventions. These are all competitive "winner takes all" contests.

Eliyahu Goldratt's keen observation in ToC lets us see constraints on business performance in the "winner takes all" distribution of Pareto. He goes even further. He explains that the Pareto distribution is for independent causes, and that the effect Pareto discovered is more intense in the business world where most causes are interdependent. For example, if a company has a revenue gap and poor-quality product, it's likely that the quality is partly responsible for the revenue gap. So, the revenue constraint reflects the quality constraint. Goldratt goes on:

> *...most of us really accept the Pareto Principle as the 20/80 rule. In configurations of dependent variables, the Pareto Principle takes the form of 0.1/99.9 rule. Just a fraction of a percent is responsible for almost all the end result [42].*

Thus, ToC teaches that there is one constraint—a bottleneck—that is limiting flow. Focus all your energy on relieving that bottleneck because only improvements to the bottleneck result in measurable results. For example, consider an intellectual property (IP) team at a law firm. Let's say they have a struggling business and want to improve. The first job is to identify the primary goal, which in this case would probably be to increase revenue. Then identify the bottleneck that is limiting revenue. It might be that there are not enough clients, in which case the constraint may be attracting and retaining clients; in this case, they would measure success in client attraction. But if they have an unending source of clients and a systemic gap in capacity to serve those clients, success might be measured as available hours of capacity. Rightly understanding the bottleneck changes how we view success.

The most complex task in continuous improvement is not improving, but rather choosing what to improve. It requires critical thinking to pick low-risk targets and open-mindedness to understand when a preferred solution isn't working. There is neither a reliable rule nor a check-the-box answer; nobody gets it right every time. Eliyahu Goldratt [43] lists knowing "what to change" as one of the three basic abilities of a manager: For the ability to answer three simple questions: "what to change?", "what to change to?", and "how to cause the change?" Basically what we are asking for is the most fundamental abilities one would expect from a manager.

6.4.7 The Hawthorne effect

Finally, as you search for the right place to expend effort, be wary of the Hawthorne effect: the tendency of people who are aware they are subjects of performance experiments to display a temporary improvement. The phenomenon was first recognized in the 1950s by Henry Landsberger. He analyzed a series of manufacturing productivity experiments that had been run decades earlier at the Hawthorne Works, which was part of Western Electric. The most famous experiments are those where worker productivity was analyzed at various levels of lighting. While the research was attempting to discover the relationship of productivity vs lighting levels, they demonstrated an improvement in productivity when lighting levels were changed, whether incrementally brighter or dimmer. Ultimately, it was the change in lighting because when lighting changed it revealed to people that they were being studied, and they strove harder [44]. More recent research has confirmed the effect, but has also discovered that it may have been overstated by Landsberger. Even so, the term is still used in industrial research to describe an unsustainable improvement in productivity, resulting from people who are aware they are subjects in a productivity experiment.

When we create systematic learning, every person in the workflow is part of a continuous series of experiments. We are constantly measuring, comparing, and improving. Perhaps the original Hawthorne effect can be explained in part because people were more engaged during experiments—made part of the process to improve rather than simply workers doing what they were told. The Hawthorne effect adds a precaution to measuring improvement: don't be hasty in accessing improvement from change. Real improvement will sustain over time, but the Hawthorne effect dissipates quickly.

6.5 Conclusion: Simplify, engage, and experiment. Repeat

The techniques in the following chapters present solutions that combine **simplify**, **engage**, and **experiment** to address common problems in knowledge work. Effectively addressing problems that constrain the organization will create sustained, relevant improvements. As shown in Fig. 6.14, over years waste will reduce so that more value will flow to customers. Consider a flooded field where water pools behind a series of bottlenecks. Water flow slows to a drip. As those bottlenecks are opened, the water flows faster and faster. Opening the bottlenecks didn't create water; it merely allowed the water that was there to flow faster. Similarly, reducing waste doesn't create capability in knowledge staff; it just frees the capability that was always there.

Fig. 6.14 Simplify, engage, and experiment to drive waste down year after year.

References

[1] Eliyahu Goldratt, The Goal: A Process of Ongoing Improvement North River Press. Kindle Edition. Location 76.
[2] Attributed to Albert Einstein by Roger Session in How a 'difficult' composer gets that way, The New York Times (January 8, 1950). https://quoteinvestigator.com/2011/05/13/einstein-simple/.
[3] T. Ohno. Taiichi Ohnos Workplace Management. Special 100th Birthday Edition, 2012. McGraw-Hill Education. Kindle Edition, p. 110.
[4] W. Edwards Deming, Deming Four Day seminar in Phoenix, Arizona (via the notes of Mike Stoecklein), cited at https://quotes.deming.org/authors/W._Edwards_Deming/quote/10091, February 1993.
[5] F. Herzberg, B. Mausner, B.B. Snyderman, The Motivation to Work, second ed., John Wiley, New York, 1959. ISBN: 0471373893.
[6] F. Herzberg, One more time: how do you Motivate Employees? Harvard Business Review (Sep-Oct 1987) (originally published in 1968).
[7] J.R. Hackman, G.R. Oldham, Motivation through the design of work: test of a theory, Organ. Behav. Hum. Perform. 16 (2) (1976) 250–279.
[8] F. Herzberg, Jumping for jelly beans, in: The Business Collection, BBC, 1973. Currently Available at: https://www.youtube.com/watch?v=o87s-2YtG4Y.
[9] W. Edwards Deming, Out of the Crisis, (2000) 485.cited at https://blognew.deming.org/w-edwards-deming-quotes/large-list-of-quotes-by-w-edwards-deming/.
[10] T.M. Amabile, How to kill creativity, Harvard Business Review (1998).
[11] H. Sauerman, W. Cohen, What Makes Them Tick? Employee Motives and Firm Innovation, NBER Working Paper No. 14443 (2008).
[12] Pink, Daniel H. Drive (pp. 115-116). Penguin Publishing Group. Kindle Edition.
[13] S. Shingo, The Sayings of Shigeo Shingo: Key Strategies for Plant Improvement, Productivity Press, 1987, P107.
[14] Teresa Amabile as quoted in Pink, Daniel H. Drive (p. 114). Penguin Publishing Group. Kindle Edition.
[15] D.H. Pink, Drive. The Surprising Truth About What Motivates Us, Riverhead Books, 2009, pp. 18–21.
[16] D.H. Pink, Drive. The Surprising Truth About What Motivates Us, 2009, Riverhead Books, p. 34.
[17] M. Seligman, The New Era of Positive Psychology, Ted Talks, 2004.http://www.ted.com/talks/martin_seligman_on_the_state_of_psychology.
[18] Teresa Amabile as quoted in Pink, Daniel H. Drive (p. 110). Penguin Publishing Group. Kindle Edition.
[19] Teresa Amabile as quoted in Pink, Daniel H. Drive (p. 121). Penguin Publishing Group. Kindle Edition.
[20] D.A. Garvin, How Google Sold Its Engineers on Management, HBR, 2013.
[21] S. McCrystal, Listen, Lean… then Lead, TED, 2011.https://www.ted.com/talks/stanley_mcchrystal/transcript?language=en.
[22] Teresa Amabile as quoted in Pink, Daniel H. Drive (p. 117). Penguin Publishing Group. Kindle Edition.
[23] B. Morgan, Warren Buffet as quoted in 101 of the Best Customer Experience Quotes, Forbes (April 3, 2019). Can be viewed at https://www.forbes.com/sites/blakemorgan/2019/04/03/101-of-the-best-customer-experience-quotes/#5805375645fd.
[24] E. Goldratt, What Is This Thing Called Theory of Constraints and how Should It Be Implemented? (1990) p. 124.
[25] A.C. Ware, D.K. Sobek II, Lean Product and Process Development, second ed., Lean Enterprise Institute, 2014, p. 23.

[26] T. Ohno. Taiichi Ohnos Workplace Management. Special 100th Birthday Edition, 2012. McGraw-Hill Education. Kindle Edition, p. 2.
[27] Jordon Peterson. Biblical Series XV: Joseph and the Many-Colored Coat. YouTube. https://www.youtube.com/watch?v=B7V8eZ1BLiI.
[28] W. Thomson, Commonly known as Lord Kelvin, Electrical Units of Measurement, in: Popular Lectures, vol. I, 3 May 1883, p. 73.https://archive.org/stream/popularlecturesa01kelvuoft#page/73/mode/2up.
[29] Tom Gilb. Quantify the un-quantifiable. TEDxTrondheim, https://www.youtube.com/watch?v=kOfK6rSLVTA.
[30] A. Osterwald, Y. Pigneur, G. Bernarda, A. Smith, T. Papadakos, Value Proposition Design. How to Create Products and Services Customers Want. Get Started With..., John Wiley & Sons, Hoboken, NJ, 2014, p. 204.
[31] http://www.apiweb.org/circling-back.pdf.
[32] https://www.isixsigma.com/dictionary/deming-cycle-pdca/.
[33] https://www.balancedscorecard.org/BSC-Basics/Articles-Videos/The-Deming-Cycle.
[34] https://deming.org/explore/p-d-s-a.
[35] Video of Feynman lecture. https://twitter.com/ProfFeynman/status/1010935192901554176.
[36] J. Shook, How NUMMI Changed Its Culture, in: John Shook's eLetters, Lean Enterprise Institute Knowledge Center, September 30, 2009. https://www.lean.org/shook/DisplayObject.cfm?o=1166.
[37] J.H. Dyer, H. Gregersen, C.M. Christensen, From the Innovators DNA, HBR, 2009.
[38] Eliyahu Goldratt, The Goal: A Process of Ongoing Improvement (p. 332). North River Press. Kindle Edition.
[39] G. Ellis, Project Management in Product Development, Butterworth-Heinemann, 2016, pp. 160–162.
[40] S.T. Teich, F.F. Faddoul, Lean management—the journey from Toyota to healthcare, Rambam Maimonides Med. J. 4 (2) (2013) e0007.
[41] MarketWatch.com, What's your net worth, and how do you compare to others?, (Jan 29, 2019)https://www.marketwatch.com/story/whats-your-net-worth-and-how-do-you-compare-to-others-2018-09-24.
[42] E. Goldratt, What Is This Thing Called Theory of Constraints and how Should It Be Implemented? (1990) p. 123.
[43] E.M. Goldratt, The Goal: A Process of Ongoing Improvement, (2014) 337.
[44] S. Robbins, The Path to Critical Thinking, Harvard Business School Working Knowledge, May 30, 2005.https://hbswk.hbs.edu/archive/the-path-to-critical-thinking.

CHAPTER 7

Reduce Waste #1: Discord

None of us is as smart as all of us.

Ken Blanchard

7.1 Introduction

In this chapter, we will discuss **Discord**: waste from people who don't align to the organization as a whole or to individuals within. **Discord** is the first waste because, until this waste is addressed, it's difficult to energize the organization to address other wastes [1].

Discord presents in many ways—for example, when people distance themselves from others in the organization. Perhaps there's tension when two people are blaming or insulting each other in front of others; perhaps colleagues cut away at the reputation of each other in less public ways. Perhaps two departments just don't talk to each other. Silos can rise up between people, between functions, or between regions of a global concern.

When we encounter **Discord**, the starting assumption will be that we have good people trying to do the right thing, but engaged in conflicts they need help to resolve. We'll keep that assumption until the facts prove otherwise, something we'll take up briefly in Section 7.4. Until then, we'll follow the lead of Eliyahu Goldratt:

I smile and start to count...:

1. *People are good.*
2. *Every conflict can be removed.*
3. *Every situation, no matter how complex it initially looks, is exceedingly simple.*
4. *Every situation can be substantially improved;...*
5. *Every person can reach a full life.*
6. *There is always a win-win solution.*

Eliyahu Goldratt [2]

Improve
https://doi.org/10.1016/B978-0-12-809519-5.00007-7

This chapter has two main sections. In Section 7.2, we'll look at four personal styles that gravitate to each of the extremes of the four quadrants of the Engagement Wheel from Section 6.4.3. In Section 7.3, we'll look at five steps we can take to increase team function based in large part on "The Five Dysfunctions of a Team" by Patrick Lencioni.

7.2 The engaging power of diverse thought

In "Pioneers, Drivers, Integrators, and Guardians," the authors explain four archetypical *styles*. Recall that the Engagement Wheel provides four quadrants of engagement: Inspiration, Protection, Connection, and Challenge. Everyone places some value on each of the four quadrants—the loftiest CTO still values high-quality manufacturing and the hardest-driving salesperson knows the value of teamwork. The four archetypes are the extremes, motivated primarily by a single quadrant: Pioneers by Inspiration, Drivers by Challenge, Integrators by Connection, and Guardians by Protection.

The value of understanding the various styles is that it provides a means to discuss, comprehend, and reduce **Discord** as we engage different people in different ways. Johnson Vickberg and Christfort stated: "The four styles give teams a common language for understanding how people work" [3]. Now, let's look at the four archetypical styles.

> *The greatest waste...is failure to use the abilities of people...to learn about their frustrations and about the contributions they are eager to make.*
>
> ***W. Edwards Deming [4]***

7.2.1 The Pioneer: Engaged by Inspiration

Pioneers are creative and spontaneous. They fear risk less than others. They are full of imagination and eager to adapt. They always have a new solution to perfect or a new problem to solve. They naturally gravitate away from process and hate being told "no." Their thinking can be unstructured. They have the biggest ideas, but they can lack the staying power to finish everything they start. As Fig. 7.1 shows, the Pioneer is always thinking of the next thing.

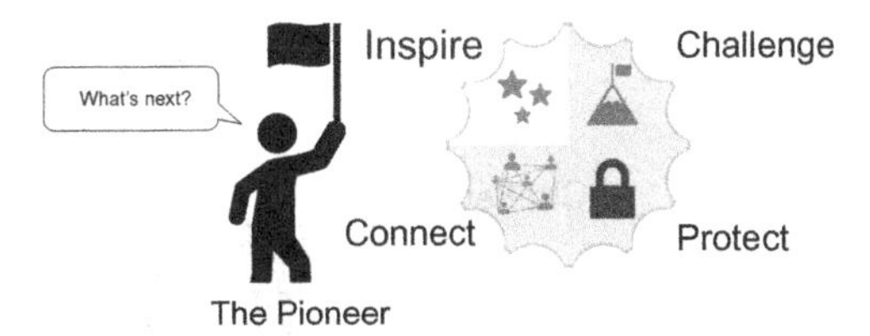

Fig. 7.1 The Pioneer values Inspiration most.

7.2.2 The Integrator: Engaged by Connection

Integrators are collaborative and diplomatic, empathizing with colleagues. They get a lot done through relationships. They avoid confrontation and politics. They are the glue that helps hold teams together, finding consensus and playing down differences. They don't always deal well with conflict and sometimes can be indecisive. As Fig. 7.2 shows, the Integrator values teamwork, knowing that little is accomplished by individuals working alone.

7.2.3 The Driver: Engaged by Challenge

Drivers want worthwhile contests they can win—beating a competitor, delighting a customer, or solving a showstopping operational problem. They are competitive and logical. They can focus intensely in pursuit of a goal. They are direct and decisive. But Drivers can alienate other team members, especially when they sacrifice relationships in pursuit of a goal. They sometimes act with inadequate caution, becoming hyper-focused on winning the prize. As shown in Fig. 7.3, Drivers want to be measured and gain deep satisfaction by winning.

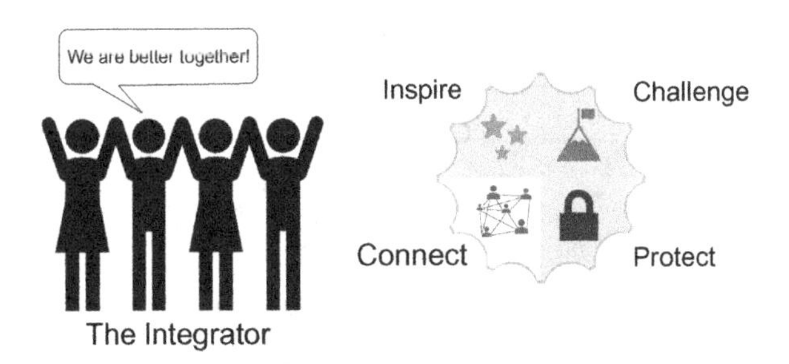

Fig. 7.2 The Integrator values Connection most.

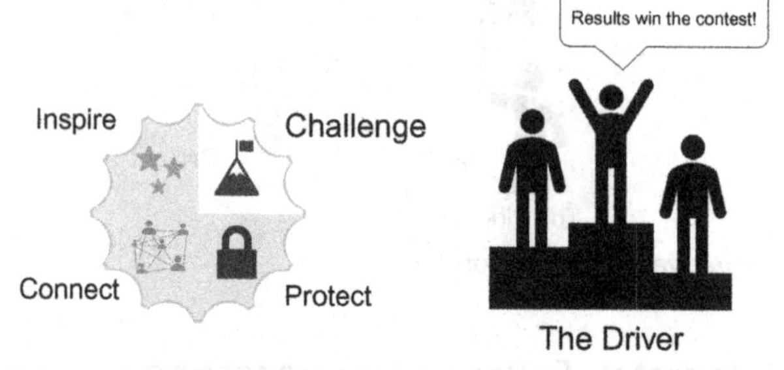

Fig. 7.3 The Driver values Challenge most.

7.2.4 The Guardian: Engaged by Protection

Guardians are highly structured, methodical, and get the details right. They are comfortable playing their part in an organization. They are predictable. Guardians have their feet planted in the practical. The have a natural disdain for time pressure and ambiguity. They are risk averse and can have a difficult time finishing because there's always one more detail that needs attention. As shown in Fig. 7.4, Guardians want to do things right.

7.2.5 Bringing diverse viewpoints together

The power of diverse thinking is that each person brings a needed viewpoint to the team. Drivers and Integrators are two opposites that balance each other. Drivers can bring focus and decisiveness, while Integrators keep the team pulling together. In an inclusive organization, these two archetypes can protect each other from their respective biases. But when out of balance, for example, when the department manager is a Driver with difficulty

Fig. 7.4 The Guardian values Protection most.

listening to Integrators, **Discord** results. Drivers can view Integrators as indecisive people-pleasers, while Integrators can see Drivers as stubborn and concerned more about numbers than people.

Similarly, Guardians and Pioneers are also opposites. Inclusive organizations value the bold ideas Pioneers generate, while also embracing the ways Guardians "Protect the brand" from the shoddy work that comes by pushing ideas out before they are ready. When these two archetypes are out of balance, a protracted battle ensues. The Pioneer derides Guardians as absent of innovative thought and blocking change. Guardians can view Pioneers as disorganized and too willing to compromise standards.

Consider an example of a young leader who wants to make some changes finding themselves in uncomfortable tension with the curmudgeonly veteran who sees too much risk in the leader's ideas. The veteran has lots of reasons why the young leader's ideas need to stay in the bottle, always seeing the risk but never the benefit. Likely, the young leader is driven by challenge and the veteran wants to protect. Most of the time, both have valid points and there is probably a path that can advance most valuable changes while holding risk at an acceptable level. When they realize they both present valid views—that neither is wrong—they better comprehend that diversity creates better decisions.

Leading diverse teams demands sensitivity to all four quadrants. When dealing with any single individual, you can focus on those areas most relevant to that person. If I'm talking to a Pioneer, the discussion will usually focus on innovation and creativity. If I'm talking to a Guardian, the focus will move to the need for high quality, consistency, and complying with regulations. But when the conversation is with a diverse group, which is to say, one where all four archetypes are represented, you should work to address concerns from all quadrants.

As you'd expect, the interpersonal problems that tend to arise when opposite styles come together can put a damper on collaboration. Indeed, 40% of the people we surveyed on the topic said that their opposites were the most challenging to work with.

Johnson Vickberg and Christfort [3]

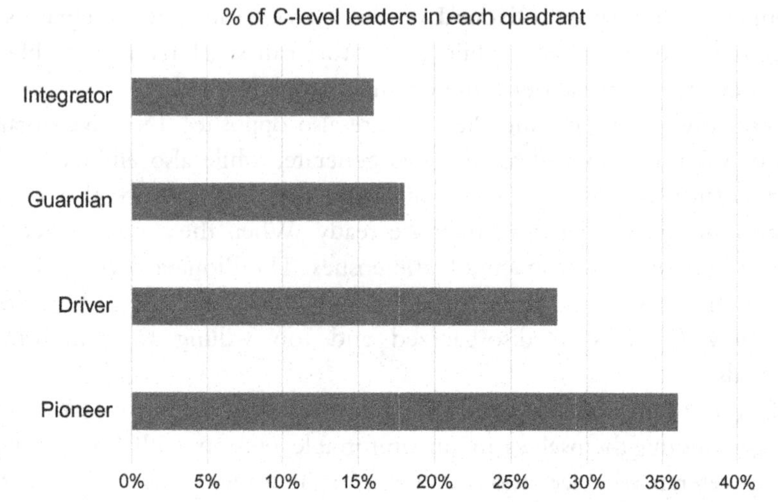

Fig. 7.5 Based on more than 600 surveys of C-level and higher positions, Johnson Vickberg and Christfort found that Pioneers and Drivers are highly represented in leadership roles [3].

While all four styles will be found in an organization of size, they are not evenly distributed. Pioneers and Drivers are more common in C-level positions[a] (see Fig. 7.5).

7.2.6 Engagement Wheel in different regions

> *Respect the culture and customs of every country and region, and contribute to economic and social development through corporate activities in their respective communities.*
>
> ***Toyota Guiding Principle #2 [5]***

Issues of regional diversity can be reflected in the Engagement Wheel. For example, my intuition is that, on average, my Swiss colleagues place a bit more value on connection and protection than we Americans do. When I failed to recognize this difference, it caused **Discord**. I led a regional R&D team in Germany for a few years where I saw something similar. And I had extended relationships with Israeli companies; they seemed to

[a] C-level is the level immediately below general management such as Chief Financial Officer (CFO) and Chief Technical Officer (CTO). It also includes roles without the "C," especially Vice President, like VP Sales.

be more biased toward inspiration and challenge than my US colleagues. This is a small sample set in an unscientific study, but the basic truth is self-evident: the center of the Engagement Wheel for one culture is likely to be shifted in comparison to another culture.

7.2.7 No type is better. All are essential

The model that Johnson Vickberg and Christfort present resonates with me. Most personality tests separate into quadrants that are neutral, neither good nor bad. For example, the Myers Briggs Type Indicator (MBTI, one of the first and still most popular personality test methods) divides people as "Sensing" and "Intuition," a distinction that is neutral. As the Myers & Briggs Foundation website states:

> *All types are equal: The goal of knowing about personality type is to understand and appreciate differences between people. As all types are equal, there is no best type [6].*

It does seem logical that no style is better, but that doesn't imply the types are neutral; in fact, the Pioneer-Integrator-Driver-Guardian model plots its traits in dimensions that are all positive. At the individual level, no personality is better, but at the team level, all should be represented. A team with no Drivers will have a hard time finishing anything. Those with no Guardians are too likely to gamble unwisely and pay later with quality and compliance problems. A team with no Integrators will likely get mired in conflict. Try to do something big without a Pioneer and you're likely to be disappointed by small thinking. All factors must be represented for a team to be balanced.

Communicate Protection and Challenge

For years, I would communicate the need for high-quality work at a high level—for example, at quarterly company meetings and in annual reviews. But when we had an urgent problem such as a late project or a production issue, I focused on the need to move fast. I didn't overtly say quality could be compromised, but I didn't reinforce the need either. When a crisis struck, I talked a lot about Challenge and little about Protection.

In my mind, I struck the right balance in these moments of pressure: I communicated an urgent need and avoided an overt message that quality was second. I learned that many of my team heard something I didn't intend: quality matters at a high level, but in a crisis, it comes second.

Continued

Communicate Protection and Challenge—cont'd

This created **Discord** because I was speaking to the Drivers, but ignoring the Guardians.

I worked to mend this. When I had a difficult message around delivering something at risk, I reinforced our standards of quality and ethics each time I explained the need to deliver, even if that meant covering the topic five times in a week. I believe this communicated a better balance and so increased alignment. One year later, it was rewarding to see the results of our annual survey: reponses to questions that measured our team's view of our organization's dedication to quality went up considerably over the previous year.

While diversity can improve outcomes, it also creates conflict, especially when people do not see the value in the opposing view. In the example above, the veteran may view the new Driver as a person unconcerned about the future of the company, a person who wants to make their mark and then get promoted to their next assignment. The Driver might see the veteran as resistant to change, a symbol of why the organization moves too slowly. There may be truth in both views; it takes business humility to recognize the value in the opposing view.

Despite the havoc such differences can wreak on team performance, opposite styles can balance each other out. Still, that takes time and effort [3].

My own style is probably closest to Driver. If I encounter a challenge that needs focus to improve, that bias usually serves me well. What serves me even better is realizing that sometimes I'm going to see more need for challenge than is really there—perhaps I'm focused on something that doesn't need to be addressed now or even something that is working better than I understand. In these cases, finding consensus with Pioneers, Guardians, and Integrators checks my biases. In the short term, it may generate some friction, but I can point to places where it protected me and my organization from poor decisions; unfortunately, I can also point to some problems that could have been avoided had I listened better.

It's also important to pull your own opposites closer to you, to balance your tendencies as a leader. This is really about generating productive friction. Think Lennon and McCartney, Serena and Venus, the Steves (Jobs and Wozniak) [3].

7.3 Creating highly functional teams

In this section, we will discuss how to improve team function. As in earlier sections, the starting assumption is we have good people trying to do the right thing, but **Discord** is causing them to pull against each other. We will use five steps to remove **Discord**, which are based on "The Five Dysfunctions of a Team" by Patrick Lencioni [7]. Then we'll build up several actions we can take to reduce waste, divided into transactional and transformational leadership skills from Section 6.4.2. Along the way, we'll discuss skills for socialization and group facilitation using a simple Ideate-Aggregate process.

The five steps to move a team from dysfunction to high function are shown in Fig. 7.6:

1. Establish trust

   Build a culture that encourages vulnerability. Own up to the past.
2. Face conflict

   Be transparent and mine the team for conflict. Hear every voice to build true consensus.
3. Negotiate commitment

   Aggregate the views of the team, so everyone is heard and each understands what they are signing up for. Socialize new ideas before moving them forward. Ensure commitments are clear.
4. Nurture accountability

   Build accountability first among team members, then between the team and the organization. Accept that unexpected events will sometimes get in the way of commitments; define the escalation process when that happens.
5. Drive results

   Continuously measure how the team is doing with regard to commitments. React rapidly to performance gaps.

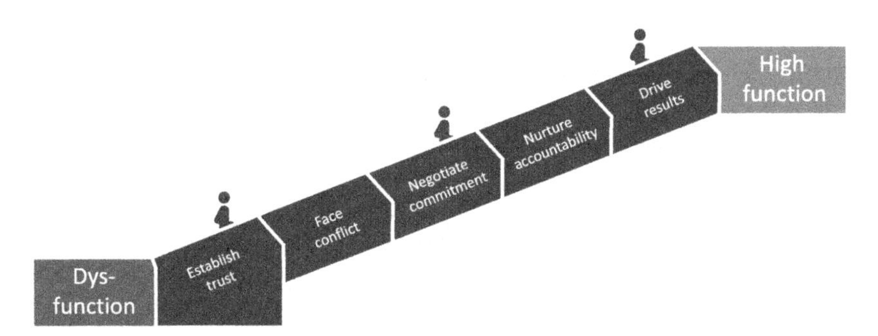

Fig. 7.6 Five steps to move from dysfunction to high function.

7.3.1 Step 1: Establish trust

No quality or characteristic is more important than trust [8].

Patrick Lencioni

Teamwork demands vulnerability. People must trust each other before they will apply themselves fully. If you believe your teammates or organization might squander the fruits of your hard labor, you're going to throttle back the discretionary effort you apply to a new endeavor. What if this project fails to meet expected performance? What if my colleague who is managing FDA compliance does shoddy work? What if the commercial team priced this service so high that no one will buy it? These all demonstrate vulnerability that comes because, if the work fails, even if you do your part, you're still part of the team that failed.

Another part of trust is reconciling events from the past. Every organization has decisions its leaders would do over. It's not possible to get every decision right. The people who were hurt by those decisions will want to know if they can trust future decisions. When leaders misdirect teams, it's unlikely those people will just forget. And why should they? This is why it's often necessary to explain what happened in the past and why it won't happen this time.

A disappointing experience

I can recall early in my career a project that demanded several of us work until very late in the night every night for weeks. Finally, we sent the prototype to our prospective major customer, understanding that we were likely to win a large order. It didn't turn out that way.

We didn't hear much except they were testing the unit. Time passed with no news and eventually I forgot about the project. Four years later I got a note from that customer: there was a box in their lab with my name on it. They were cleaning up their lab and wondered if I wanted it back. That was the prototype from those many late nights. When the box arrived, it was clear it had never been opened.

That certainly injured the trust I had in the group that interfaced with the customer. The incident taught me a lesson: teams doing work should interface directly with customers—never hand that off to someone outside the team. But it also helps me be empathetic to team members probing about past incidents. Rather than dismissing their concerns as grumbling about the past, I try to answer why it will be different this time.

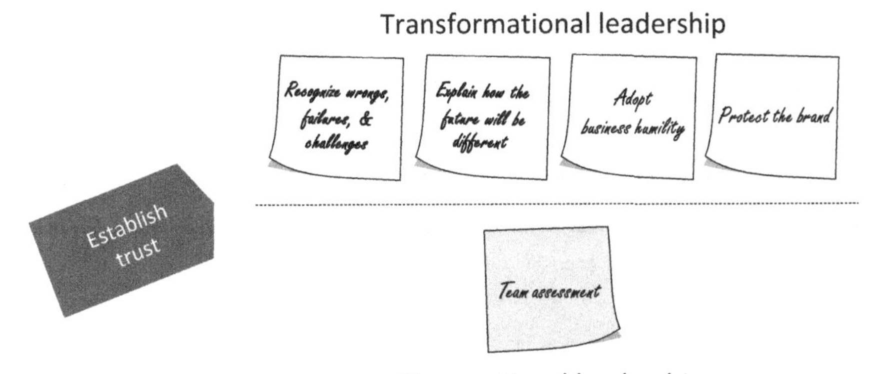

Fig. 7.7 Some concrete steps we can take to establish trust.

Fig. 7.7 shows a few practical steps we can take to establish trust:

- Understand past issues in the organization and explain why those things happened.
- Explain how the future will be different.
- Take on business humility: embrace your own gaps and learn from all those around you.
- Protect the "brand" that the team wants to be known for—things like hard work, high competence, innovative thought, and the highest ethics. That brand is precious.
- Consider doing a team assessment such as a third-party survey or simply have a session run by someone outside the team, perhaps with leaders absent to allow for the most open discussion. Ensure the assessment measures trust in a repeatable way, for example, asking the team what their level of trust is from 0 to 10. Repeat these measures, so you can see progress over time.

Trust is an ambiguous word. You might trust someone's integrity, but be uncertain about their competence. You might trust their competence, but only in their area of expertise. You might trust someone to do their best, but you might not trust that they have a perfect memory. The trust we are describing here is to create safety for those committing to a plan or to a role on a team. I remember once when I said on a conference call that I didn't trust project managers on my team to fill in the data for their projects, by which I meant filling in the data was complex and no one could get it right every time. But that's not what I said. After the call, a colleague phoned me to challenge the wisdom of the words I chose. "When you say you don't trust, it sounds like you're calling them liars." His correction stung, but he was right. I don't think I've used the word "trust" in that way since.

7.3.2 Step 2: Face conflict

> *If team members are never pushing one another outside of their emotional comfort zones during discussions, then it's extremely likely they are not making the best decisions for the organization.*
>
> ***Patrick Lencioni [7, p. 38]***

How does your organization handle conflict? Many people are uncomfortable when conflicting opinions are expressed. They avoid confrontation, fearing that it might cause **Discord**. But usually that's not the case. Knowledge work is difficult, and people dedicate a large portion of their lives to doing it well. Don't be surprised when people are unhappy about a decision or observing inattention to an issue they find important. When you have passionate, engaged people, there will be times those people will get emotional.

When people are passionate, the first step is to ensure their voices are heard. If you sense a team is in disagreement but holding back, don't rush to the next topic—mine for hidden conflict. That conflict may be concealing valuable viewpoints held by quieter members of the team. Good leadership ensures an extroverted Driver isn't running over an introverted Guardian and that a senior Pioneer isn't ignoring input from a junior Integrator.

Resolving hidden conflict is constructive. However, it takes skill to manage the interchanges that might turn destructive—for example, when people attack each other rather than focusing on the problem at hand. Other examples are people complaining about the inevitable or catastrophizing. These are cases of emotional waste, a topic we'll discuss in Section 7.4. So, there are times when good facilitation demands reining in the conversation to protect the more vulnerable team members, especially those who are introverted or junior.

Keep local culture in mind when meetings become confrontational. My experience is that German and Israeli colleagues are more open to expressing their feelings than most of my Swedish colleagues.The passionate exchanges that pass for everyday in Tel Aviv or Düsseldorf would be startling in Stockholm. Facilitators must be cautious about applying their standards in another

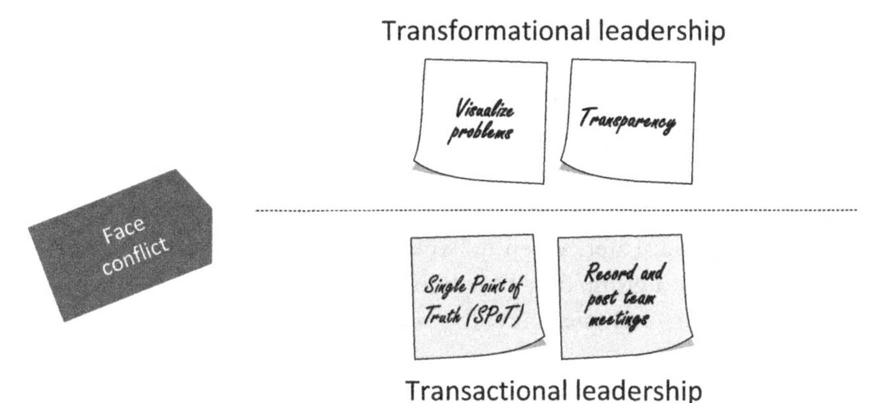

Fig. 7.8 Some concrete steps we can take to face conflict.

region of the world. One technique I've used is to walk through these topics ahead with a local leader who will be present. If the tone seems like it's becoming destructive, I ask the local leader to step in to bring calm.

Fig. 7.8 shows a few practical steps we can take to face conflict:

- Visualize problems

  Representing problems visually rather than through narration allows everyone in the room to absorb the information rapidly. We'll talk about visualization techniques in Chapter 8 and throughout this book.
- Transparency

  Be transparent and demand transparency from all. When the same information flows to all people on the team, it reveals hidden conflict by enabling each person to see the whole story. If someone doesn't agree with a decision, they have the information available to make their case.
- Single Point of Truth (SPoT)

  Build one well-defined place all stakeholders can pull information from—a shared folder, a company intranet site, or physical board. We'll talk about SPoT in Section 8.2.
- Record and post team meetings

  Many people use video conferencing to hold meetings. In those cases, make it a standard practice to record meetings so those who couldn't join can see the meeting.

It's as simple as this. When people don't unload their opinions and feel like they've been listened to, they won't really get on board.

Patrick Lencioni [10]

7.3.3 Step 3: Negotiate commitment

In Step 3, we negotiate commitments with each person taking up the work. This is their opportunity to identify issues and risks they see, to express whether they have capacity, and to clarify their roles. The goal is to create a sort of contract among the team and between the team and the organization. Everyone sees the scope of the problem, the intended solution, and the resources available. Later, when unexpected events threaten the commitment, you'll have a group fully bought in to being successful. You're much less likely to hear "I always knew this wouldn't work" or "I never agreed to this."

Negotiate simple, clear commitments: clear ownership, quantified time allotted, specific resources, and defined deliverables. Capture them in the simplest way possible, preferably a single-page Canvas View. This view will be the foundation for nurturing accountability in the next step. You'll want to return to this commitment over time. Remember that unexpected events will occur; you'll need simplicity and clarity to navigate through those events while maintaining accountability. More on this in the next section.

As much as we strive for simplicity and clarity, we do it knowing that knowledge workflows always have unknowns and some we'll only discover over time—the "unknown unknowns" [9]. So leaders of organizations who depend on knowledge work are faced with a daily dilemma. On the one hand, challenging people with specific goals in an aggressive time frame drives good results. On the other hand, knowledge work is ambiguous so that estimations take considerable expertise; in fact, knowledge staff doing the work will, at least partially, set the time frame they will be measured against. Needling people is a sure way to get overly conservative targets. The answer: negotiate in good faith.

Negotiation requires that all parties see that they are part of an exchange from which all benefit. Managers don't bark demands that staff must accept. Staff don't take for granted that all of the desirable work in their field will come their way. Rather, managers and knowledge staff view work as having value for both parties: leaders will be focused on what value will be delivered to the organization. Staff will want to know how the work adds value as well; they will also want to know the work will be satisfying, build their domain skills, and advance their careers.

As an example, my organization defined a multimillion dollar project that we needed to complete in a little more than a year. Unfortunately, the team estimated a completion time of almost double that. Neither group would budge and we were stuck for months trying to choose from two unacceptable alternatives:

1. Management could prescribe a delivery date the team didn't accept and wouldn't own.
2. Management could accept a delivery date that didn't meet the organization's needs.

We made progress when we realized there was a third alternative: negotiate.

The first step was everyone recognizing that the team yearned to do this project. There were technology developments and opportunities to bring recognition, both of which were engaging. To their credit, the team was not willing just to put down the date that management wanted to get the project. I was sure we could reach a compromise if both sides could give a little. So the technical team spent a week building a plan that compressed the schedule to 14 months—not quite the 12 months initially requested, but well below the 20 months we started at. That was enough to get the project started.

The result was a benchmark: with renewed thinking, the team drove the schedule down almost to the stated need date. They cut time in prototyping by using lower-fidelity hardware that still validated the design. They cut features that, with a more careful eye, were noncritical for the first release. And management held up their end of the negotiation: the project was approved in days, funded for the duration, and resources were left in place until completion. A year later, we had a success story. The project was completed on time, something that managers prescribing an aggressive date rarely experience, and customers were delighted, as indicated by a first year revenue of almost double what had been projected. We found success because we stepped out of a stalemate into a real negotiation where both sides could—and did—win.

Negotiated commitment is not a majority-rule decision. One team member failing to commit can ruin chances for success. Think of a baseball team with eight fully committed players and a second baseman who won't take the field; they would lose every game. It isn't necessary that everyone will agree with every decision—only that each team member is committed to the goal and ready to play their role in creating success.

Waiting for everyone on a team to agree intellectually on a decision is a good recipe for mediocrity, delay, and frustration.

Patrick Lencioni [7, p. 51]

The next two sections present two skills to help teams in their journey to build negotiated consensus: socialization and the Ideate-Aggregate process.

7.3.3.1 Socialization

> *[O]bstinately sticking to your arguments doesn't get you anywhere. You've got to pay attention to human psychology.*
>
> **Shigeo Shingo [11]**

Socialization is the process of gently presenting ideas to an organization, often starting with one or two opinion leaders, then proceeding to small groups, and finally to full teams. There is something in most of us that wants to present our big ideas with a splash. But that's counterproductive in a number of ways:

- It says, "This is my idea," excluding teammates from feeling ownership.
- It minimizes the input of others as the idea is solidifying.
- It risks harsh reactions, for example, when you present your idea to a large group of people and someone challenges you enough to injure your credibility.

Socialization is an efficient way to meld the thoughts of others, which both strengthens the idea itself and reduces friction to the idea in the organization. Socialization starts informally, for example, with a 15-minute phone call or a hallway chat. Each time you present a concept, you'll hear new ideas, not just to the content but also supplying context ("Yes, we have run into this several times over the years" or "Yes, the team in Düsseldorf experimented with that last year...contact Mitch to find out more"). As the idea matures, it will include more history and so become more credible (Fig. 7.9). Also, others can help with format, revealing which parts of the explanation are confusing or off-putting, so you can sharpen the discussion.

Socialization brings harmony. Because objections are raised in smaller, less formal settings, there's less at stake. Compare getting gruff feedback from one of the team's curmudgeons in a quick discussion vs being called out in a meeting of 15 people. Usually, objections are constructive, though occasionally they may be born of hostility or resistance to change. But, as Shingo put it, "No matter what the tone of an objection, seize the essence of what is being said, use it to...develop a better proposal" [12].

- Starts by teaching. Ends in learning.
- Listen and adapt. If you're preparing your response, you're probably not listening.
- Expect iterations.
- Socialization broadly to hear all voices.

Fig. 7.9 Teach to learn: Socialize to sharpen thinking and build consensus.

I came to rely heavily on socialization in my role leading R&D teams. I used it for myself, presenting new ideas to those closest to me first so that if I was offtrack, I'd get a gentle nudge back to reality. I'd work to smoothly expand my audience. All the while, both the quality of the idea and the quality of the presentation of the idea were inching up.

I also recommended socialization for those in my team. If someone came to me with a new idea, I'd normally ask about how their peers felt. If they didn't know, I'd ask them to go find out. Usually, we relied on a visualization based on a Canvas View (see Section 8.5); that was more effective than an open conversation or email.

Nemawashi and Socialization

Nemawashi is Japanese for a form of socialization that in the West we might call consensus building [13]. Jeffrey Liker defines the use at Toyota as "Make decisions slowly by consensus, thoroughly considering all options; implement rapidly" [14]. It captures the process of building consensus in an organization to create alignment [15]. *Nemawashi* literally means to prepare the ground for planting [16], an allusion to the way new ideas are gently introduced into a nurturing environment in order to thrive [17]. Socialization is a way to prepare that ground: introduce ideas in stages and on-board the thoughts of others so those ideas can thrive beyond what any one person can provide for alone. As Shook says, "Approval at the end of the process becomes a...formality. Key approvers and opinion leaders have reviewed the idea and can see their contributions in it" [18].

7.3.3.2 Aggregating ideas from a group

In this section, we'll discuss a simple technique that anyone can use to reduce **Discord** in a meeting. Let's say a group is discussing a problem or creating a new concept and an argument starts. Some are talking over each other and others are rambling; still others are quiet, perhaps fuming or perhaps uncomfortable about jumping in. The group has been talking for 30 minutes and they are no closer to consensus than when they started, and tempers are rising. How can we get an accurate view on what the team is thinking? Here are a few options that don't work well:

1. Keep arguing—that could go on until the day is over.
2. Try to summarize the situation during the meeting—that will probably restart the argument.
3. Break up the meeting and have one person summarize—that will give an inaccurate read, overstating the loud voices and excluding those too uncomfortable to talk.

None of those options is appealing. What we are looking for is a way that is:

- fair: people should be heard whether they are quiet or talkative, congenial or argumentative, eloquent or plain spoken;
- rapid: hear what people are thinking without a rambling narrative; and
- collaborative: allow people to hear others to build new ideas together.

7.3.3.3 The Ideate-Aggregate process

This leads us to a rapid process using sticky notes to collect ideas, present them, and then group those ideas into a handful of "aggregates." This group-consensus process here is called here "Ideate-Aggregate"; it's derived from the K-J Method after Jiro Kawakita. The display it creates is sometimes called an affinity diagram [19]. It works consistently. "One of the most amazing things about the KJ-Method is how well it objectively gets groups to the top priorities. Different groups can analyze the same data and will often come to the same results" [20].

Silent ideation

Start with silent ideation: each person takes a small pad of sticky notes and a felt-tip marker to write down their response to the issue—it might be the right goals, root causes, or potential countermeasures—whatever the core issue is that's causing **Discord**. Spend a few minutes in silence writing down ideas. The ideas must be written in large lettering, so the words can be read from 6 ft. (2 m) away and also to limit the note to the core of the idea. The silence is important [21], allowing each person to think—don't be surprised if you have to gently remind one or two of the more extroverted people that this exercise

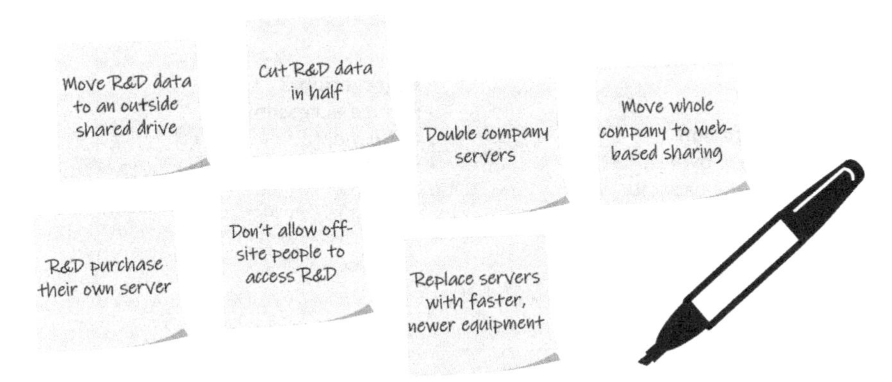

Fig. 7.10 Ideate-Aggregate begins with a silent exercise of each person writing down their ideas.

is *silent*. In 5–10 minutes, each person will have a small stack of sticky notes, each an abbreviated description of one their ideas (see Fig. 7.10).

Structured presentation

Allow each person a short time to explain their ideas, no more than a minute per sticky note. This is structured: people can present the idea, but they may not argue for it. The team can ask questions to clarify, but not to challenge; no other comments are allowed. You are creating an environment that is both comfortable for people who are less talkative and limiting for people who like to chat. Also, at this point, as Shingo said, "judgement is inappropriate. Making judgments about tentative ideas will end up nipping them in the bud." [22]. Expect 5–15 ideas per person; most people present their ideas in a few minutes.

Aggregation

The final step is collaborative; start grouping similar ideas together by moving the notes near each other. Ask everyone to come to the board and move sticky notes around [23]. The leader should hold back at this point and encourage everyone to stand at the board. Don't be surprised if that takes a couple of requests.

The key to aggregation is to group similar solutions; avoid the common error of simply grouping all solutions that address the same element, because this will mix unlike solutions. This is because we can consider different solutions—even opposing solutions—for the same elements of a problem. For example, suppose we're dealing with a problem to increase the profitability of a newly introduced service. We might consider actions to lower price in order to increase volume and thus overall revenue; we might also

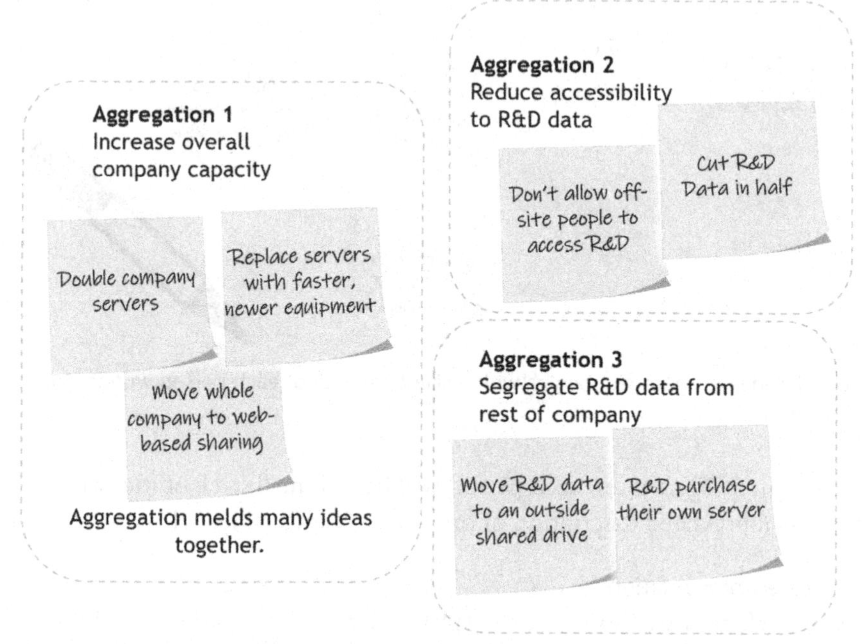

Fig. 7.11 Aggregate as a team: Group similar ideas, then name each aggregate.

consider actions that increase price to increase profit for each unit sold. These are opposing solutions around the same element (pricing), so they need to reside in different aggregates.

All of this takes typically 10 or 20 minutes and most of the ideas will be aggregated into a handful of groups as shown in Fig. 7.11, though there usually are a few outliers. At the end, someone will need to go to the board to move the handful of ideas that didn't get aggregated, and then name the aggregates. Wait until most of the sticky notes have been aggregated before naming the aggregates. This allows unexpected patterns to appear. If you name the aggregates first, you're more likely to subconsciously force ideas into preconceived solutions.

There's a sort of magic when people touch the notes—they are taking ownership. The best is when there are two or three quiet conversations going on together, creating a buzz in the room called "Waigaya" by Honda [24, 25]. Waigaya is onomatopoeic, a word that mimics a sound, like the "woof" of a dog or a "bang" on a drum. Only in this case, it's the sound of many people talking at once, wai-ga-ya-wai-ga-ya-wai-ga-ya. It's something like the English *hubbub*. That sound means people are engaged, explaining, creating, and collaborating. When I hear it, I'm confident we're on the right track.

The lowly sticky note

About a year into one of our largest and ultimately most successful projects, one of my team managers and I were talking. He had a small confession: "When you first insisted we put sticky notes on the wall, I didn't like it. I thought we needed a full-featured software tool to display these ideas. But now I see it works because it helps us understand each other." Dantar Oosterwal tells a similar, though more dramatic, story where an outside lean knowledge expert is explaining a technique based on sticky notes to a design team at Harley-Davidson. "...Product Development was still reluctant to try it. Perhaps it was the low-tech aspect of Post-it® Notes..." [26]. He explains that during the meeting, a chief engineer stormed out of the room. Later, the chief engineer said, "There's no way in hell that I'm going to pay some guy to show me how to stick Post-it Notes® to a wall." These stories illustrate that it's easy to misunderstand the value of group-based visualization techniques, which are naturally egalitarian because everyone can add and move their own sticky notes.

For years, I could not find a "high tech" equivalent for the lowly sticky note, but lately, I've seen an intriguing alternative: the Office 365 version of PowerPoint allows multiple people simultaneously to create and move electronic sticky notes around their screens, which together act as a shared whiteboard. I've used it with people on different continents, and paired with video conferencing on personal devices from an application like Webex or Teams, I have experienced the near-equivalent of people standing around a board talking and moving sticky notes about. While the implementation is complex, the electronic sticky note is still a simple visualization. Nothing automated. Just talented people showing their perspectives and listening to each other to find the wisdom of the team.

Voting

The final step is a simple voting technique: give each person about five to seven votes that they can place in any combination on the aggregates. They should vote for the aggregate they most value for the team to move forward. If they are totally sold on one aggregate, they can put all their votes there. Otherwise, they can apportion them in any way they wish. Hand people markers and ask for votes, placing one tally mark (|) or check for each vote. After each person has voted, you have a simple picture of what the team thinks is important (Fig. 7.12).

The role of this simple voting process is to introduce judgment into the collection of ideas. Shingo said, "Ideas generated by brainstorming are the product of momentary mental flashes. The function of judgment is to choose from among them" [27]. Voting allows you to collect the concepts with the most support for the next round of evaluation without shooting down other ideas. It's a question of what to focus on now so that we learn more. The vote is not "majority rule." Don't simplistically choose the highest vote-getter and ignore the others. However, vote tallies do represent the

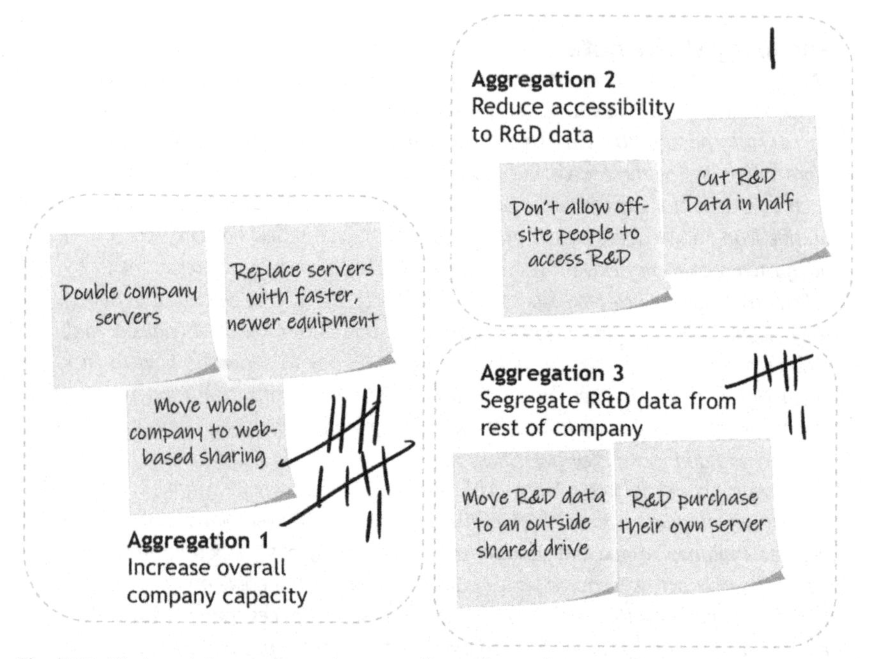

Fig. 7.12 Voting: Ask people to place, say, five tally marks according to where they think focus should be placed.

solutions the group is most passionate about; you don't want to put energy into something with one or two votes, while those with more team support sit idle. On the other hand, if someone is passionate about something only they voted for, they should have the option to pursue it, though they may have to go for it mostly alone. Find ways to let people experiment in those areas where their passion leads them.

The Ideate-Aggregate process can, in 20–30 minutes, end what might have been a long, divisive argument. It builds an egalitarian environment and encourages people to meld their ideas together, rather having the ideas compete as in traditional brainstorming. At the end, the scored aggregates represent the wisdom of the team, and those aggregates are almost always an amalgam of ideas. The fullest aspiration of collaboration is that each person sees their contributions in the aggregates, but no aggregate represents one person's ideas.

Fig. 7.13 shows a few practical steps we can take to create negotiated commitment:

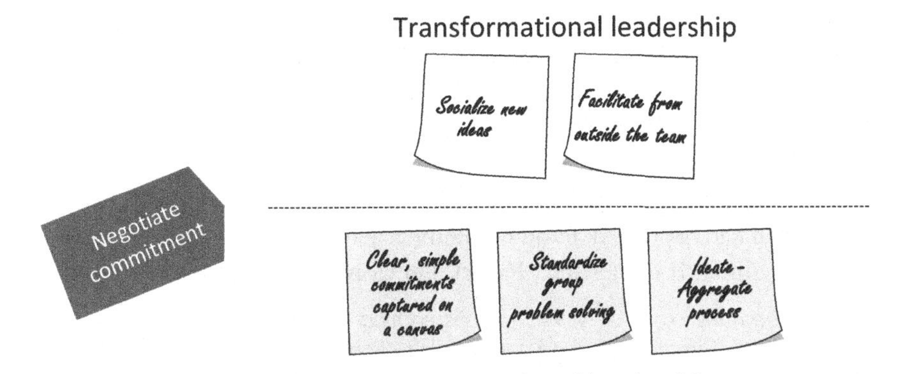

Fig. 7.13 Some concrete steps we can take to reach negotiated commitment.

- Socialize new ideas
  Socialize your ideas to nurture them. Ask your teams to do the same.
- Facilitate from outside the team
  Use an outside facilitator to lead meetings on sensitive topics, perhaps an outside consultant or just a member of another function in your organization.
- Standardize group problem solving
  Ensure diligence and allow all voices through a culture of standard problem solving, as described in Chapter 12.
- Ideate-Aggregate process
  Use the Ideate-Aggregate technique to understand rapidly the wisdom of the team.
- Clear, simple commitments captured on a canvas
  Ensure commitments are simple and clear. Capture them in the simplest way possible, ideally on a single-page canvas.

7.3.4 Step 4: Nurture accountability

Accountability is not simply taking the blame when something goes wrong. It's not a confession. Accountability is about delivering on a commitment. It's responsibility to an outcome, not just a set of tasks.

Peter Bregman [27a]

In the previous section we negotiated a commitment; in this step we will nurture accountability to meet that commitment. Accountability creates emotional responses: it's normal to feel satisfied if you deliver on commitments and to be disappointed with yourself when you don't. Accountability naturally exists between a person and their teammates. It stems from not wanting to let down people you respect. I recall so sharply the look on my boss's face a few years ago when I disappointed him. I committed and missed. I lost sleep that night, recalling the look on his face the instant he registered what had happened. I don't remember what he said; in fact, I'm unsure he said anything. It felt terrible. Patrick Lencioni explains it this way:

> *More than any policy or system, there is nothing like the fear of letting down respected teammates that motivates people to improve their performance.*
>
> ***Patrick Lencioni [7, p. 213]***

Our initial assumption is that the people on our teams want to meet their commitments, but that success demands much more than desire. Leaders can create conditions that nurture accountability. The sections that follow will discuss five or six steps to consider.

7.3.4.1 Kick off the process

When a new commitment is made, kick off the process with the team that will fulfill it. Focus on the committed negotiation captured in Section 7.3.3. Ensure everyone has the opportunity to ask questions and understand their role. Having a simple kickoff, perhaps just 15 or 20 minutes long, makes it clear that the negotiation is complete and the hard work to fulfill the commitment is underway.

7.3.4.2 Negotiate change

Most knowledge work will encounter unexpected events and there will occasionally be a need to modify a commitment. Perhaps a team member has to be moved to another project or perhaps a customer changed their requirements. When this happens, it's as if the original contract changed; return to the team to renegotiate. If the team endures a resource reduction, be willing to accept a fair extension of time. If customers are demanding something new, revise the original commitment to reflect that change. This will allow you to bridge to the new circumstances while still maintaining accountability. Changes made by management fiat injure accountability.

7.3.4.3 Regular review

Your starting hypothesis about a commitment is that the person or group of people who made the commitment are capable of fulfilling it. As with any hypothesis, you may be proven wrong. Protect the organization and the team by testing that hypothesis through weekly or monthly review. Review at a frequent pace so the team has time to identify gaps and make corrections.

7.3.4.4 Define the escalation process

Clear escalation builds accountability because no matter what happens, there is always an action to take. When a person exhausts all the actions they know to take to resolve an issue, they can turn to their leadership. No matter how complex a problem, a person has the obligation to react, even if that reaction is nothing more than asking for help. In Section 11.4, we'll talk about escalation in detail.

7.3.4.5 React to disappointment with equanimity

People are going to miss commitments from time to time. In fact, if a person fulfills every commitment they make, they are probably overly conservative in their commitments. Knowledge work is complex and there isn't time to learn every unknown before we set our path. We need people to take wise risks, and that means some level of miss is to be expected, even when people behave with accountability.

Unfortunately, there will be times when people you work with will miss commitments, and when you look into the details you will find their behavior was at the root. Perhaps they ignored obvious signs of a problem or made unwise choices. Perhaps they saw a miss coming but concealed it or concealed their role in it. When the evidence is clear that a person did not behave accountably, there must be consequences if there is going to be a culture of accountability. When those moments occur, handle the news with equanimity. My supervisor John Marotta told me once about the difference between *reaction* and *response*: your *reaction* is defined by what you do in that moment you recognize someone has not behaved with accountability. You can choose to display anger, ridicule, or disappointment. Or you can choose equanimity and composure. By contrast, your *response* may take weeks or months. React with equanimity knowing that, should the event demand consequences, time is your ally in crafting your response.

A few steps that we can take to nurture accountability are shown in Fig. 7.14.

7.3.5 Step 5: Drive results

Our last step is to drive results, the transactional leadership to finish what we start. There will be a lot of detail on this in Chapters 8 and 9. Some steps we can take to do this are shown in Fig. 7.15:

- Falsifiable hypothesis

  Ensure that goals are set with a falsifiable hypothesis as discussed in Section 6.4.3. If we don't define success, we won't recognize when we are failing.

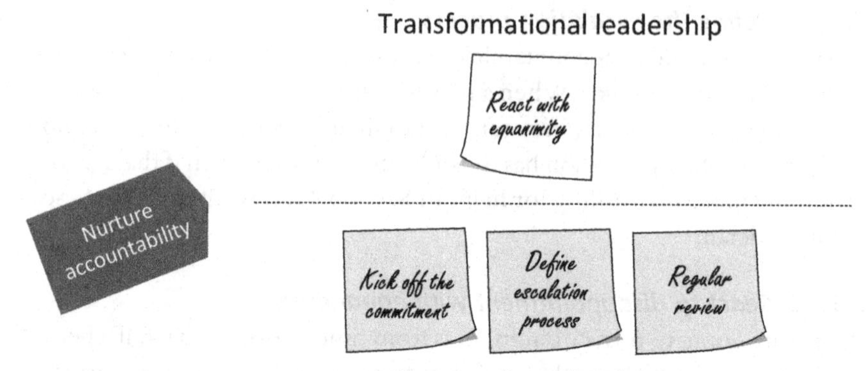

Fig. 7.14 Some concrete steps we can take to nurture accountability.

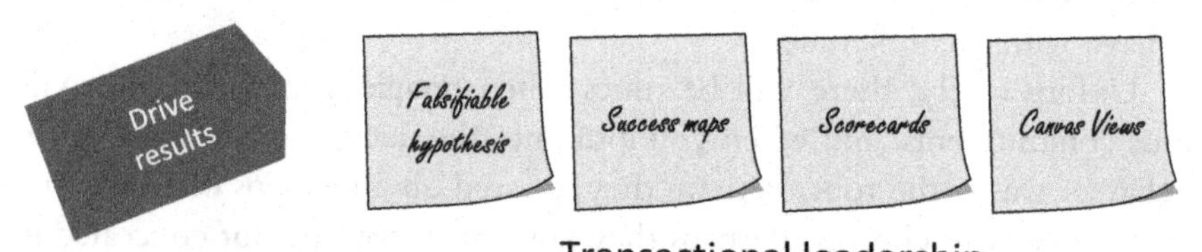

Fig. 7.15 Some concrete steps we can take to create drive to results.

- Success Maps

  Success Maps are a combination of Action Plans and Bowlers, two tools well-known to help teams drive to results, and the Test Track, a tool added between the other two to create a continuous measure of success over the life of knowledge work. Success Maps are covered in Section 9.6.
- Scorecards and dashboards

  Scorecards (Section 8.5.1) and dashboards (Section 8.5.2) are single-page views of a team's results.
- Canvas Views

  The Canvas View is based on Toyota's view that every problem that an organization faces can and should be shown on a single sheet of paper. Canvas Views are presented in Section 8.5.

7.3.6 Measure results from the start

So, how long does all this take? You should see leading indications within a few weeks. People will be less ambiguous in their commitments and more likely to act on gaps in delivery. But it takes months or years to see the full effect of improving team function. Before you get started on improving team function, consider how you will measure your progress. One recommendation is regular surveys of the team, ideally anonymous and administered by a third party. As Judd et al. put it:

> *For decades, having regular employee opinion surveys has been on evidence-based lists of high-performance HR practices. Our internal research at Facebook suggests that...it would be a big mistake to abandon them today [28].*

Your journey will be never-ending because no team reaches a zenith of performance. And people are continuously cycling in and out of organizations. So, old lessons must be retaught over time. And if you don't measure, you risk habituating to new norms. You may bring substantial improvement over a year or two, but by that point many people may have forgotten the starting point.

7.3.7 Take action!

If you're not sure what to do next, consider the recommendation of Fig. 7.16: take a copy of Fig. 7.6 with a small team of leaders and build an affinity diagram to increase function in your organization. Follow the Ideate-Aggregate process of Section 7.3.3.3 and aggregate ideas around the five steps. Make your initial measurement and try a few actions. Then, remeasure regularly to ensure you are improving and sustaining that improvement.

7.4 Ego vs confidence

> *When asked for my absolute best advice ever, I reply, "Stop believing everything you think" [29].*

In "No Ego," author Cy Wakeman contrasts confidence with ego. She sees confidence as positive: having faith in your abilities. But ego she sees as selfish, seeking approval, validation, and a need to always be right. Ego avoids feedback. "In the workplace, the differences between confidence and ego can make or break your career" [30].

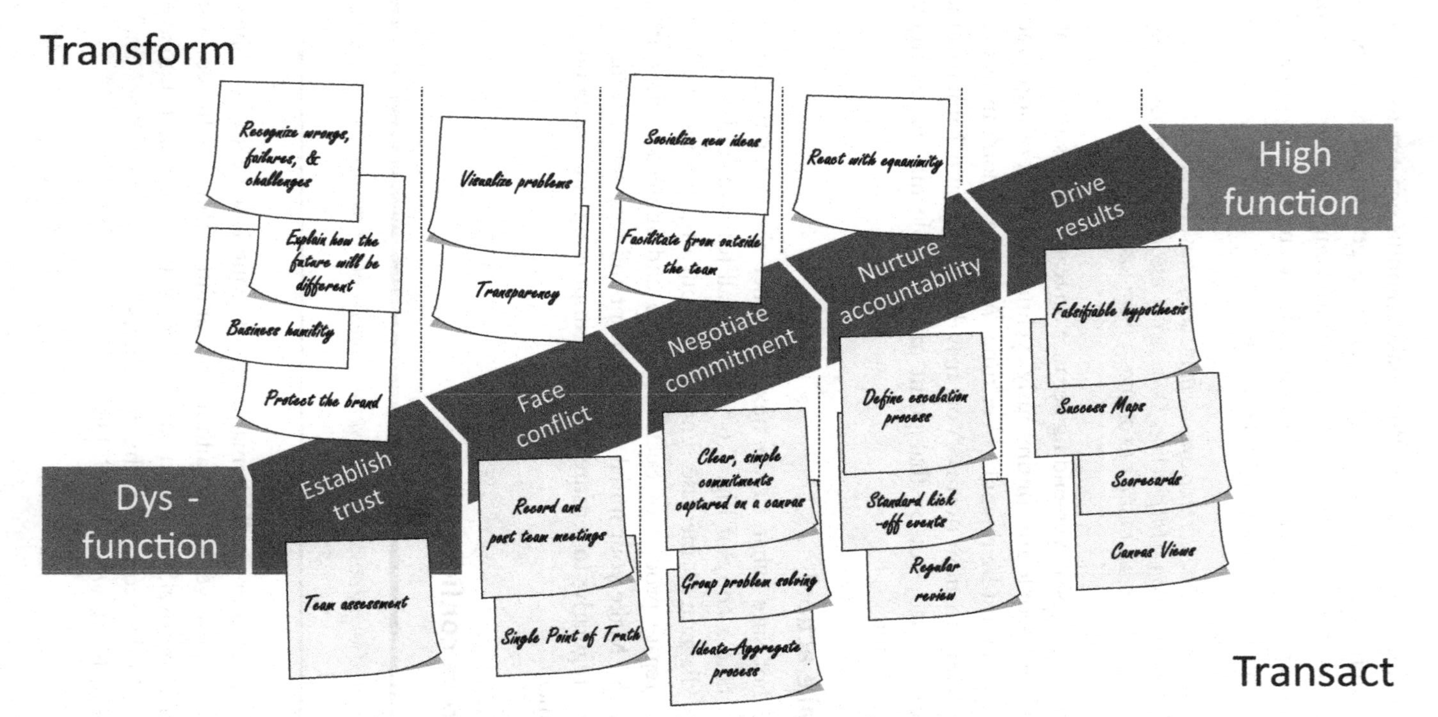

Fig. 7.16 Beginning the process of Ideate-Aggregate to start building higher team function.

She defines emotional waste as proceeding from ego. A few examples of emotional waste are as follows:

- Arguing with the inevitable: wasting time complaining about something where the decision has been made or perhaps there was no real decision to start with.
- Blaming: focusing on how every failure is the result of someone else not being smart or experienced enough, not trying hard enough, or not caring enough.
- Catastrophizing: using fear to undermine an alternative, predicting spectacular failure with vague reasoning. This happens in knowledge work where a domain expert can inflate risk in areas that cannot be readily comprehended by those of less expertise. For example, an expert can catastrophize by just calling a proposed solution a "disaster" rather than thoughtfully comparing it to other alternatives.

Dealing with emotional waste is exhausting because of the time it demands of managers. As Wakeman puts it:

Emotional Waste Is Increasing

The biggest surprise that emerged from the study was that, since the 1990s, time lost in drama at work had increased. The data showed leaders were spending almost 2.5 hours a day in drama that creates emotional waste at work [31].

Her solution when a leader finds emotional waste: move the person out of the team. You've probably noticed in this book that the initial assumption in almost every problem is that "most waste is caused by good people trying to do the right thing." That is usually a good assumption until data shows otherwise. Emotional waste is a problem that manifests over time. By the time you've identified the people generating a disproportionate share of emotional waste, that initial assumption may no longer be appropriate. Depending on the severity, moving someone out of an organization may indeed be the best of the available options.

7.5 Conclusion

The waste of **Discord** worsens every other waste because it makes the team less effective at dealing with problems. If you're a leader—a functional leader, a project manager, or a thought leader—start your journey in knowledge transformation by creating alignment (see Fig. 7.17). Build high team function with the five steps of Section 7.3. Use Ideation-Aggregation and socialization to build consensus, gently melding together the ideas of many.

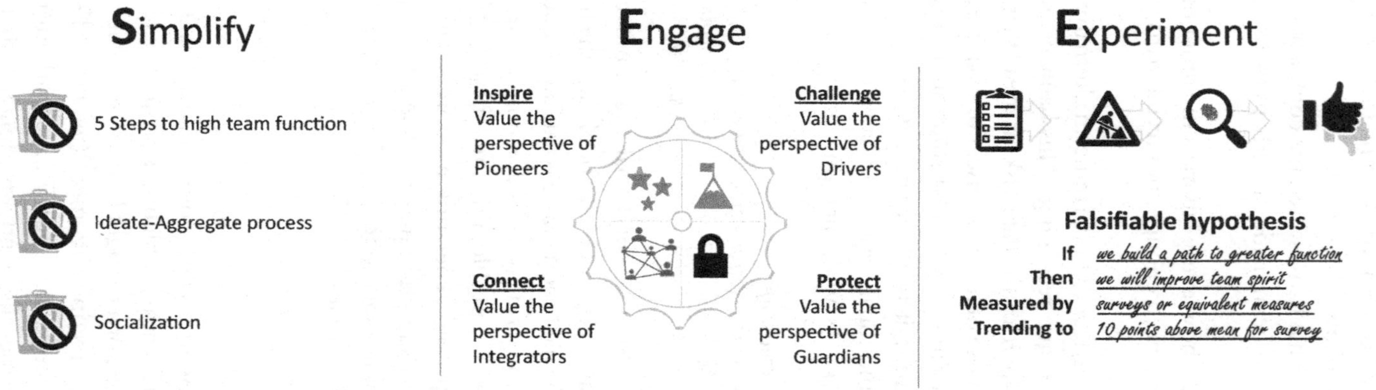

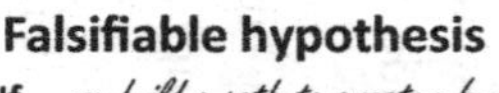

Fig. 7.17 Drive down the waste of **Discord** with techniques that **simplify**, **engage**, and **experiment**.

Socialization applied wisely captures the wisdom of the organization in a single story. The Engagement Wheel in Fig. 7.17 shows how Inspire, Connect, Challenge, and Protect engage our people by valuing the views of Pioneers, Drivers, Integrators, and Guardians. Finally, create a measure to improve: perhaps employee opinion surveys. Reliable improvement demands a falsifiable hypothesis and that demands a measure.

Notice that Fig. 7.17 shows the three dimensions of lean thinking from Chapter 6: **simplify**, **engage**, and **experiment**. A similar diagram ends all the chapters that follow. The left-hand side will show key areas the chapter sought to simplify. The center section applies the Engagement Wheel. The right-hand side shows **"Experiment"** as a falsifiable hypothesis managed with the Plan-do-check-act (PDCA) cycle.

References

[1] K. Blanchard, S. Johnson, The One Minute Manager, William Morrow, May 5, 2015, p.15. Available from:https://www.actionablebooks.com/en-ca/summaries/the-one-minute-manager-builds-high-performing-teams/.
[2] E. Goldratt, The Goal: A Process of Ongoing Improvement, North River Press, Kindle Edition, revised, 2014, p. 360.
[3] S.M. Johnson Vickberg, K. Christfort, Pioneers, drivers, integrators, and guardians, Harvard Business Review (March-April 2017).
[4] W. Edwards Deming, Out of the Crisis, Massachusetts Institute for Technology, Center for Advanced Educational Services, 1982, p. 76.
[5] Guiding Principles at Toyota, Retrieved from, https://global.toyota/en/company/vision-and-philosophy/guiding-principles/, 1992.
[6] https://www.myersbriggs.org/my-mbti-personality-type/mbti-basics/home.htm?bhcp=1.
[7] P. Lencioni, The Five Dysfunctions of a Team: A Leadership Fable, Jossey-Bass, 2002.
[8] P. Lencioni, Overcoming The Five Dysfunctions of a Team: A Field Guide for Leaders, Managers, and Facilitators, Jossey-Bass, 2005, p. 13.
[9] Donald Rumsfeld. https://www.youtube.com/watch?v=GiPe1OiKQuk.
[10] P.M. Lencioni, The Five Dysfunctions of a Team: A Leadership Fable, Wiley, Kindle Edition, 2002, p. 94.
[11] S. Shingo, The Sayings of Shigeo Shingo: Key Strategies for Plant Improvement, Productivity Press, English Translation, 1987, p. 143.
[12] S. Shingo, The Sayings of Shigeo Shingo: Key Strategies for Plant Improvement, Productivity Press, English Translation, 1987, p. 134.
[13] R. Kopp, Defining Nemawashi, Japan Intercultural Consulting, Dec 20, 2012.https://www.japanintercultural.com/en/news/default.aspx?newsid=234.
[14] J. Liker, The Toyota Way, 14 Management Principles From the World's Greatest Manufacturer, McGraw-Hill, 2004, p. 241.
[15] M.D. Fetters, Nemawashi essential for conducting research in Japan, Soc. Sci. Med. 41 (3) (1995) 375–381.
[16] https://www.lean.org/lexicon/nemawashi.
[17] J. Shook, Managing to Learn. Using the A3 Process to Solve Problems, Gain Agreement, Mentor, and Lead, The Lean Enterprise Institute, 2008, p. 69.

[18] Shook, Managing to Learn, 2010, p. 69.
[19] https://templates.office.com/en-us/affinity-diagram-tm06102632.
[20] https://articles.uie.com/kj_technique/.
[21] https://www.nureva.com/blog/business/sticky-notes-the-key-to-gaining-team-input.
[22] S. Shingo, The Sayings of Shigeo Shingo: Key Strategies for Plant Improvement, Productivity Press 93 (1987).
[23] Post-it Collaboration Workshop: David Kilcullen | TEDxSydney 2014, https://www.youtube.com/watch?v=PRXzHpl5yA8.
[24] J. Rothfeder, For Honda, Waigaya Is the Way, Strategy+Business, Aug 1, 2014. https://www.strategy-business.com/article/00269.
[25] Meetings, the Waigaya Way, Management Professor Blog, Robert A. Campbell, https://managementprofessor.wordpress.com/2014/11/05/meetings-the-waigaya-way-f14-7/.
[26] D. Oosterwal, The Lean Machine: How Harley-Davidson Drove Top-Line Growth and Profitability With Revolutionary Lean Product Development, AMACON 220 (2010).
[27] S. Shingo, The Sayings of Shigeo Shingo: Key Strategies for Plant Improvement, Productivity Press, 101.
[27a] P. Bregman, The Right Way to Hold People Accountable, Harvard Business Review, 2016. January 11.
[28] S. Judd, E. O'Rourke, A. Grant, Employee surveys are still one of the best ways to measure engagement, Harvard Business Review (March 14, 2018).
[29] C. Wakeman, No Ego: How Leaders Can Cut the Cost of Workplace Drama, End Entitlement, and Drive Big Results, St. Martin's Press, Kindle Edition, 2017, p. 41.
[30] C. Wakeman, No Ego: How Leaders Can Cut the Cost of Workplace Drama, End Entitlement, and Drive Big Results, St. Martin's Press, Kindle Edition, 2017, p. 27.
[31] C. Wakeman, No Ego: How Leaders Can Cut the Cost of Workplace Drama, End Entitlement, and Drive Big Results, St. Martin's Press, Kindle Edition, 2017, p. 17.

CHAPTER 8

Reduce Waste #2: Information Friction

The greatest value of a picture is when it forces us to notice what we never expected to see.

John W. Tukey [1]

8.1 Introduction

In this chapter, we will discuss **Information Friction**, the waste that arises when information is present somewhere in an organization, but unavailable to or inaccurate for the person who needs it. The lack of needed information causes a host of wastes:

- Bad decisions
- Missed calls to action
- Errors in work product

Information Friction is prevalent in knowledge work. The good news is its remedies generate quick results for modest investment. We will look at three main techniques to cut **Information Friction**:

1. Single Point of Truth (SPoT)
   SPoT creates one place for critical information that all can access at any time. SPoT cuts friction by making the most important data available instantly to key people. It increases accuracy by removing the possibility of differing information across multiple sources.
2. Visualization
   Visualization converts information from narration using tools like graphs, formatting, and tables. It cuts friction because visual information is so much easier to comprehend, retain, and communicate.
3. Canvas View
   Based on the Toyota A3 method, the Canvas View is a technique of telling an entire story in a space constrained to a person's view without

Improve
https://doi.org/10.1016/B978-0-12-809519-5.00008-9

"moving": either physical movement (walking) or virtual (flipping through slides or scrolling across spreadsheets). The Canvas View cuts friction by providing easy access to context and so makes information easier to comprehend and communicate.

8.2 Single Point of Truth (SPoT)

Single Point of Truth (SPoT), also known as Single Source of Truth or Single Version of Truth, is the technique of creating one source of information for all stakeholders [2]. Examples include a shared directory for project data, an intranet site for purchase approvals in process, and a paper schedule posted near the team that uses it. SPoT information can be "pulled" by any team member or stakeholder at any time [3]. There's no need to send emails asking for status or schedule meetings to look at data on someone's PC. SPoT data cuts **Information Friction** for team members and leaders who know what data they need.

SPoT works best to reduce the friction of information that is updated in multiple places. When data is stored in multiple places, it's nearly impossible to keep all those places synchronized over time. As people learn that no data can be trusted, they will invoke the compensating behavior of using a favored source, asking someone they trust. That normally puts them in email queues that may take days. At best, you'll spend 15 or 30 minutes getting information that you should have been able to gather in seconds. At worst, you'll ask several times over a week or more. Eventually, you'll learn to get by without needed data because you cannot sustain the effort to overcome the **Information Friction**; this leads to bad decisions and poor-quality work product.

As shown in Fig. 8.1, creating SPoT data is a journey. At the start, data is *ad hoc*, stored in people's heads or on personal devices, occasionally reviewed in meetings and broadcast by emails. The SPoT journey begins by posting the data in a central place so that it's available to everyone who knows where to find it. The next step may be a physical board, starting with a paper posting that is updated, say, weekly. We can move from there to having a digital source of data that is posted on a monitor the way call centers post average customer wait time. We can move from there to a daily stand-up meeting with a scorecard—electronic or paper—as the centerpiece of the discussion. As you can see, the journey doesn't end with "people can reach it if they want to." There is quite a distance from giving someone a link to click when they choose and fully integrating the information into daily routines.

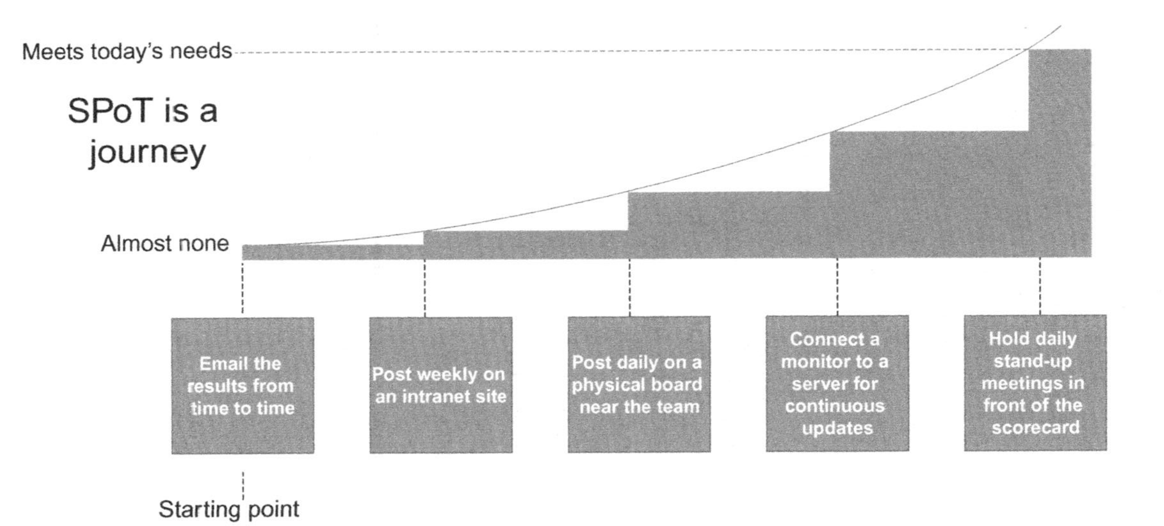

Fig. 8.1 Single Point of Truth (SPoT) increases information accessibility.

8.2.1 SPoT competes with shadow data

SPoT demands constant maintenance to ensure it remains relevant and up to date. It also demands the elimination of shadow data, competing forms of information that, if left to flourish, result in erosion of the relevance of SPoT data. If there are three places where data is stored, it's likely that different team members will have different behaviors: some will update the SPoT location, others the information on their PC, and others will broadcast information in emails or at meetings. As people favor different locations of shadow data, the SPoT location becomes less relevant. Users perceive it's not trustworthy and stop relying on it. There's less and less purpose to update the SPoT data, and the cycle continues until SPoT falls into disuse.

SPoT data competes with shadow data. We can complain about it, we can talk about what people "should" do, or we can recognize the unintended incentives for shadow data and organize ourselves to win the competition. Foremost among those incentives is that shadow data is just easier in the moment—SPoT data requires that I learn the standard format and, if that format is electronic, that I jump through a few hoops to get access. If I'm in a rush, it's easier to update an Excel file on my PC and send an email to a few people.

For SPoT to win against shadow data, the benefits need to draw people to it. There's little point in wagging a finger at people for using shadow data; it's more effective to draw them to SPoT because it meets their needs better. To do this, SPoT must meet several characteristics:

SPoT requirements:

Used regularly by the team
Accurate and up to date
Relevant to the organization
Accessible to all

- Used regularly by the team
  First and by far the most important: it must be used by the team on a regular basis. It should be the center of the team meetings. Otherwise, some team members will likely consider it only marginally relevant. If the team's primary goal is to deliver work product to high quality standards on time, SPoT data should reflect the work product quality against the standard and the delivery time against the schedule.

- Accurate and up to date
  SPoT data must be the most up-to-date, reliable information available; that will pull more people to use it. The first time someone makes a bad decision because they relied on outdated SPoT data may be the last time they use it. On the other hand, if people go down a bad path because they relied on bad shadow data, they will see more benefit in SPoT data.
- Relevant to the organization
  If the team is striving to provide 1500 marketing leads by September 1, relevant information to the organization might be the number of leads acquired to date, the total revenue value of those leads, the rate at which new leads are being acquired, and, of course, a central location for the detail on those leads. Avoid information irrelevant to important goals. A famous quote from Taiichi Ohno, the father of the Toyota Production System, is "the more inventory a company has, the less likely they will have what they need." He meant it for physical inventory, but it's applicable for inventory of SPoT information, especially if we modify the last phrase to "the less likely they will **be able to find** what they need."

The more information stored in SPoT, the less likely we can find what we need.

- Accessible to all
  SPoT must be accessible to all team members and stakeholders. This is a tough requirement because it's natural for the team to create information that is so detailed that those outside the core have difficulty interpreting it. To attract stakeholders, SPoT needs to include relevant summary data. For example, let's say our SPoT data shows all sources of project delay. Perhaps our main entry point could show total delay and whether the project was on track to schedule.

The techniques in this book are all improved with SPoT. For example, we will in later chapters work on formal problem solving. Having SPoT on the formal problem solving results allows more people to contribute to a robust set of solutions and countermeasures. And so it is with all of the metrics, Canvas Views, Success Maps, Kanban boards, and the other dozens of techniques found here. Each provides new information and so each can benefit from elements of SPoT.

8.3 Visualization vs narration

The most time-consuming and difficult way to understand complex ideas is to have to decipher a lengthy report filled with technical descriptions, business jargon, and tables of data. More efficient is the visual approach.

Jeffrey Liker [4]

"Data visualization refers to the technique of communicating complex data or information by converting it into a visual object or graphical representations in order to aid in visual processing and comprehension" [5]. By contrast, *narration* is meant to include both spoken and written prose or, more simply, words. Narration is often the first way we communicate something complex. Most of us know how ineffective narration is in explaining complex ideas. Consider the narration necessary to communicate the need to stop an automobile (Fig. 8.2). You may think of "Stop" as narration, but it is most assuredly not, a fact that became clear to me when I saw stop signs in Germany, France, and Israel, countries that speak languages that don't include the word "Stop." It is the visual attributes—the octangular shape and red color—that communicate the need to stop; the word "STOP" is an embellishment.

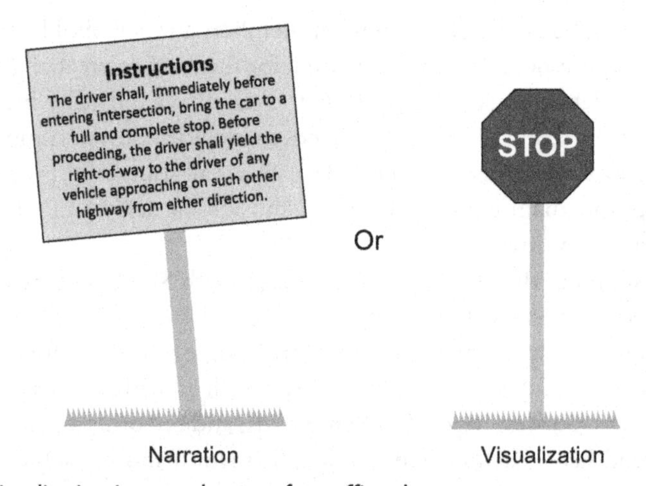

Fig. 8.2 Visualization is second nature for traffic rules.

There is magic in graphs. The profile of a curve reveals in a flash the whole situation—the life history of an epidemic, a panic, or an area of prosperity. The curve informs the mind, awakens the imagination, convinces.

Henry D. Hubbard [6]

Those of us in knowledge work have to admire architecture and construction engineering as domains where visualization has been practiced so widely. Where it seems most domains limit visualization to within the domain (for example, electronic schematics speak mainly to electronic experts), architects and builders communicate with each other and those outside the domain primarily visually. They have a full set of visualization techniques and regularly use the equivalent of "Canvas Views" and "hierarchical visualization" described in the next two sections. That is, they show views that encompass the whole structure and add drawings in descending hierarchy to show greater and greater detail. For example, they can proceed step-by-step from an elevation view of a building, to a single floor layout, to the electrical layout, to a detailed view of the main power distribution for a floor.

Today mainly text and numbers rather than visual formats are used in the transfer of knowledge in organizations. Architects, in contrast, practice the opposite: they transfer knowledge mainly by using complementary visualizations...Further, architects are experts in interfunctional communication because architects transfer knowledge to individuals from different backgrounds, such as engineers, workers, authorities, and customers [7].

Those of us in other domains can only aspire to move so much of our communication to visualization. Visualization reduces Information Friction by cutting at least three wastes of narration, as shown in Fig. 8.3.

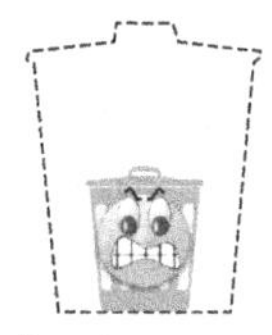

Visualization cuts waste from narration:
- Cuts focus on progress vs results
- Cuts partially thought-out ideas
- Cuts Hidden Errors

Fig. 8.3 Visualization cuts the waste rampant in narration.

8.3.1 The waste of focusing on progress over results

Narration encourages the speaker to focus on progress in at least two ways, both of which can create long and rambling discussion with little value:

- to laud what was done well, allowing the recent past to dominate the discussion; and
- to excuse gaps in performance, focusing on why the team didn't do as well as hoped, sometimes even before the listeners understand precisely what the anticipated gaps are.

Here's an example:

> (Supervisor): "Gabriella, is Project X on time for the September launch?"
>
> (Gabriella): "We were able to finish an important test that has been holding us back. And Crystal is back after several weeks out. And our UL consultant finally got back to us after 3 weeks of ignoring our request..." (and on and on).
>
> (Supervisor): "Makes sense. But is the project on time?"
>
> (Gabriella): "We have run into some problems. Like I said, UL was delayed and speeding it up is proving difficult. And Hans from Finance won't approve our expense for $20k so we have not been able to start on the CE testing..." (and on and on).
>
> (Supervisor): "Right. But is the project tracking to launch on time?"
>
> (Gabriella): "We definitely have run into some delays. Tolani, the Product Manager, keeps changing the specification, so we may have to redo some work..." (and so on).

Notice that the supervisor asks each time, "Will we make the launch date?" which is asking if we are expecting to deliver the results we committed to. And each time, Gabriella diverts attention to accomplishments in the past, areas of concern, and comments about people outside the team creating problems. In my years leading knowledge staff, I found this type of response to be almost universal. I've concluded it's not by chance—Gabriella has probably concluded that the project is going to be late, though she may not have looked in enough detail to answer quantified questions. It's likely she is not even viewing the problem in such a data-centric manner. Likely she feels disappointment for the results, and the diversion from the supervisor's questions results from a natural aversion to confronting this issue; it's usually not a premeditated intent to mislead. But however innocent Gabriella's motivation might be, it brings waste, both the waste of discussing something relatively unimportant when a serious issue needs attention and the waste of losing mindshare (e.g., people become distracted with email [8] or meeting time expires) and people who genuinely need to know remain ignorant of the gravity of the situation. Gabriella's answer could have avoided these flaws by starting with an unambiguous answer: "yes" or "no," which could then be followed with detail. But that doesn't seem to happen very often without coaching.

Shingo talked about this: each time he'd ask about a problem, people would give long narratives, explaining every twist and turn in a long process. Finally, the speaker would reveal the result at the end: either it was solved or not. To use a term from my former General Manager, Kevin Layne, they told a "mystery novel" when he wanted the answer first. Shingo said that people will not listen carefully if they think the problem is solved [9]. I'll go further: they don't listen carefully unless they know something important isn't working. To understand the problem more, both Shingo and Kevin asked a series of why-questions. Think of this as a sort of information "pull system": let the reviewer *pull* information in small pieces to build the context. Unfortunately, the natural tendency seems to be to *push* information at people in large, meandering batches.

Visualization reveals the tendency to focus on progress and excuses. And it can be avoided, for example, when an organization provides clear guidelines or even a template ahead of the meeting to steer people to focus on results. For the example above, the supervisor might ask for a block diagram showing each part of the project necessary for the launch in calendar form (e.g., the Visual Action Plan of Fig. 20.8), displaying the likely completion date of each task. Then, before asking the first question, the supervisor will see if the project is off and, if it is, by how much. This can allow the discussion to shift from "why we might be late" to "what we'll do to catch up." Visual artifacts like this slash wasteful narration.

8.3.2 The waste of presenting partially thought-out ideas

Narration seems to encourage presentation of partially thought-out ideas. As soon as the new concept comes to mind, depending on how extroverted a person is, he or she can share fresh ideas to gather input from others. This can be a positive tendency in informal discussions including teasing out new ideas and building consensus through socialization (Section 7.3.3.1), but presenting poorly thought-out ideas to a large number of people is wasteful.

Visualization, on the other hand, inherently shows context. For example, a "how it works" diagram or a "flow sequence" diagram conveys information on many levels at once. These types of visualization show gaps to the presenter often before the first presentation, so that they can be addressed ahead. And for those gaps that do get through the presenter's first draft, people in meetings usually feel more comfortable asking questions when they are confident that they understand the context.

8.3.3 The waste of Hidden Errors

Good visualization also exposes **Hidden Errors**. Where the paragraphs above dealt with gaps—areas where we didn't think through something—**Hidden Errors** are areas we did think through but still drew an incorrect conclusion. Visualization conveys ideas faster and more fully, so **Hidden Errors** are likely to be revealed to colleagues.

8.3.4 Comparing narration and visualization

Table 8.1 compares narration, both spoken and written, to visualization. Visualization is superior in a number of ways. It focuses presenters and listeners alike, pushing out clutter. While both take a significant amount of time to prepare, visualization is much easier to consume and so demands less time from the organization overall, especially when the audience is large. Finally, visualization is easier to sustain. Long narrations—lengthy presentations and reports—are rarely updated. Visualizations are much shorter and so can be revised with modest effort.

8.4 Visualization tools

Visualization tools are the individual and collected elements that communicate without prose or, at least, more than prose can alone. These tools range from text formatting to graphical display of data and how-to diagrams. Don't mistake text in a slide deck for effective visualization. Slide decks often use the smallest sliver of visualization and, when that's the case, they are just an alternative form of narration. True visualization takes time—time to create diagrams, time to find icons, and time to display data in meaningful graphic forms. It's much more than bulleted text in landscape orientation. Visualization tells stories that can be absorbed with an order-of-magnitude less effort than narration.

This section will briefly review five types of visualization tools (see Fig. 8.4). This list is certainly not comprehensive; it is here to provide a simple way you and your organization can discuss what "visual" means. I've lost count of the number of times people asked me, "What does visual mean?" I believe this arises in large part because people confuse reformatting

Table 8.1 Comparing two common forms of narration with visualization.

	 Spoken narration	 Written narration	Visualization
Focus	Speakers often wander to less important topics	Superfluous information easily creeps into lengthy documents	Improves focus because clutter in visualization is easy to see
Time	**Large** preparation time and **Large** consumption time		**Large** preparation time but **Small** consumption time
Efficiency	Different people absorb at different speeds. People accept partial understanding, often without being conscious of it		Engages visual processing of the human mind, which can consume information astoundingly fast
Sustainability	Difficult because material usually must be presented again after major updates	Difficult to update long narration	Easy to keep up to date

narration from a text document to a slide deck with effective visualization. This list demonstrates that formatting is just the first and lowest level of visualization.

- Formatting text
- Widgets
- Graphical data
- Diagrams
- Collections of the above

Visualization can change an organization

In one of my recent assignments, there were large amounts of waste from excessive narration. People would draft long specifications and white papers, and then convene meetings to review. At the end of sometimes hours, we'd break up; I'd leave confused about what we agreed to or what was learned. At first, I thought I was the only person that didn't comprehend the presentation. I learned over time that almost no one was absorbing much information, but most were uncomfortable saying so. I was new, so it was easier for me to admit I wasn't following. Don't underestimate the waste caused by unintentional incentives to keep silent, especially the desire to avoid looking unintelligent.

We instituted a near-universal requirement for visualization, founded at first on the Canvas View (Section 8.5), initially in R&D and then throughout the organization. In many review meetings, we eliminated slide decks—single-page problem solves replaced 35-slide presentations. The results were outstanding. People spent more time collaborating ahead. That's no surprise—it's difficult to collaborate on a 25-page white paper. With more collaboration, ideas arrived at meetings more polished. And everyone comprehended the decision point better, especially those of us in leadership with less domain knowledge. The ability to convey complex technical information faster left more time for meaty discussions, which resulted in faster and better decisions. This improvement was reflected in the engagement of key knowledge staff who were able to get decisions critical to their work. Over the course of just a few years, we remade our meetings by cutting the **Information Friction** of narration.

I came across an almost identical story from Paul Williams in his book "Visual Project Management" where he recounted the story of introducing a project "one-pager": "Leadership satisfaction with the process improved. However, the main positive impact in the process changeover was realized in the preparation time and effort" [10].

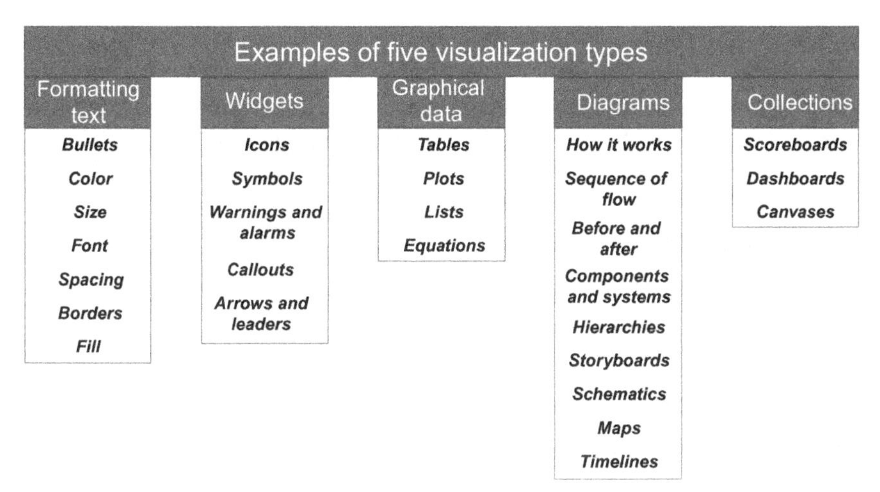

Fig. 8.4 Five types of visualization tools.

8.4.1 Formatting text

Formatting text is one of the simplest ways to start the transition from narration to visualization: modify the text to augment information. It includes bulleting text, varying font for emphasis (e.g., italics or bold), size, color, and outlining. One highly accessible tool for knowledge work is Microsoft Excel conditional formatting, where nonprogrammers can change these elements automatically based on results of calculations. Just the power of red and green autoformatting brings life to a set of data.

8.4.2 Widgets

Widgets is the moniker for small graphics like icons, symbols, callouts, and leaders. Widgets are elements that can draw attention to and explain areas of the visualization. Icons are broadly available and easy to use. For example, Microsoft Office products, a couple of years back, added hundreds of "icons," many of which are used in this book. Using icons can be as simple as the top row of Fig. 8.3 to help clarify the columns (the speaker-at-podium is a Microsoft Office "Icon").

8.4.3 Graphical data

Graphical data comes in a large variety, but generally falls into a few categories of tables, charts, and equations. Of course, within these categories there is immense variation, especially with what must be thousands of types of charts. In this text, we'll emphasize a couple that are less well-known that have high utility for continuous improvement: histograms (Section 22.6.4) and waterfall/double waterfalls (Chapter 23). Again, Excel is a strong tool for nonprogrammers to bring data from many sources and convert it to a range of standard charts and an even larger number of add-ins types.

8.4.4 Diagrams

Diagrams here are schematics that tell a story from how something works to how systems are connected to timelines and schedules.

8.4.5 Collections

Collections combine the examples above. They are examples of *complementary visualization* [11]. Common examples are the dashboard and scorecard discussed in Section 8.5.1, which combine a handful of tables and charts into a single visualization.

8.4.6 Graphical excellence

The goal of these visual tools is to help people tell their story more clearly and accurately. Of course, these tools can be used to oversimplify a problem or otherwise poorly represent an issue. The demands for clear and honest representations are perhaps higher with visualization because so much of what good visualization does is direct attention to critical areas. Simply omitting or visually downplaying a problem can be misleading. Accordingly, those who seek excellent visualization skills need to redouble their efforts to avoid misleading visualization. "Make it ugly," a former business unit president said as he wisely advised me to use visualization to highlight problems so that they could not be overlooked.

> *Graphical excellence consists of complex ideas communicated with clarity, precision, and efficiency...and graphical excellence requires telling the truth about the data.*
>
> ***Edward Tufte [12]***

8.5 The Canvas View

The Canvas View is a technique of telling an entire story on one page. By requiring a full story, every reader has the context. Transferring knowledge requires each person to reconstruct the knowledge as they absorb information [13], and having context is a necessary part of that reconstruction.

The A3 [a type of Canvas View] should tell a story that anyone can understand...The reports don't merely state a goal or define a problem in a static or isolated manner... An A3 shares a complete story. There is a beginning, a middle, and an end.

John Shook [14]

The Canvas View requires that the author provide context around insights, conclusions, and plans. This helps every reader understand the issue better; moreover, the canvas format lays bare the author's thinking first to himself or herself and so encourages authors to think more carefully through the issue before the presentation. It also invites feedback—collaborating on a single slide, where (1) the author has trimmed away all unnecessary information, (2) told a coherent story, and then (3) put it on a sheet that doesn't move during the conversation; it is an order of magnitude easier to follow than a 60-sheet slide deck.

The whole time he talked about the report, I was imagining a large book-like document. Suddenly it dawned on me that he was talking about an 11x17 sheet of paper and how he was going to put the entire budget on that one sheet.

Jeffrey Liker [15]

The Canvas View derives from Toyota's A3 process, a single-page view for problem solving. Many knowledge staff will initially see the single page as too constraining, too simple for the complexity of their problem or their domain. But the weight of evidence is against them. Every story can be told on a single sheet of paper. If more detail is required, a second A3 can drill down to the next layer and provide another canvas, which builds "hierarchical visualization," which is discussed later in this chapter. But most of the time a single page works well. The reason is that even complex problems can be described, analyzed, and solved in a confined space. In fact, Toyota determined many years ago that an A3 could hold every issue they faced:

> *Toyota's insight many years ago was that every issue an organization faces can and should be captured on a single sheet of paper. This enables everyone touching the issue to see though the same lens.*
>
> ***John Shook [16]***

So, what do we mean specifically by "one page?" In "Managing to Learn," Shook focuses on the A3, a single sheet of paper that is roughly 11″ × 17″

meant to be held at arm's length. That is an example of a canvas, but the term *canvas* is broader:

A canvas tells a story in a space constrained to a person's view without movement, neither physical (walking or turning pages) nor virtual (flipping slides or scrolling on a page).

For an A3, the view is meant to be held at arm's length, roughly 0.5 m (18″), so the font size would need to be large enough that most people could read it at that distance, no smaller than about 10 pt. font. But the canvas size can increase as the distance to the viewer increases. If the canvas is a poster, that is, on a wall 2 m (6′) away so that multiple people can view it simultaneously, the lettering probably needs to be 20 mm high (about ¾″). As Fig 8.5 shows, there's not a strict rule for font size, but the typical person needs to be able to view it without walking—that is, without having to step in to read detail and step back to see the whole.

In this book, we'll recognize three main variants of the canvas view, each of which is shown in Fig. 8.6. The first (leftmost) is the A3 view espoused by Toyota. The second is the "digital" equivalent, a monitor, which has roughly the same dimensions. The third is the poster-sized canvas, a wall-mount view that must be large enough to be viewed by the local team.

There is no specific physical limit to the poster; it could be 10 × the size of Fig. 8.6C, but then people would have to step back 10 × as far and font size would need to be 10 × larger. It's still a canvas view as long as people can

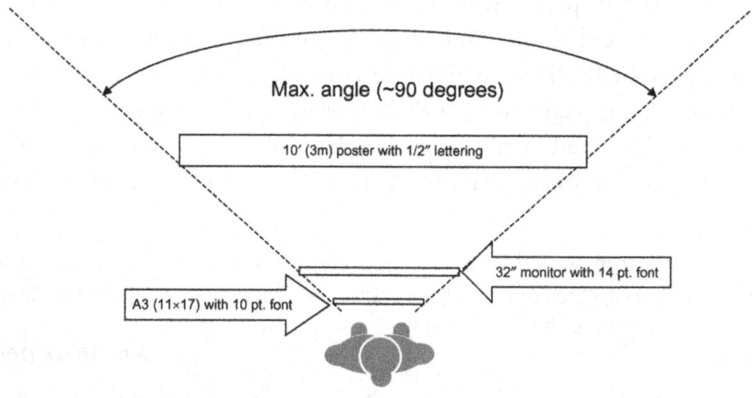

Fig. 8.5 The Canvas View is a combination of fitting information in a viewable angle and large enough detail to be distinguished at the intended distance.

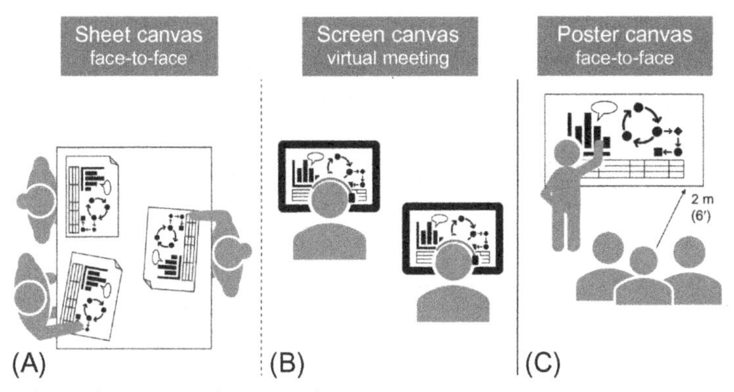

Fig. 8.6 Three alternatives that provide a *Canvas View*.

absorb the information without having to move. As shown in Fig. 8.6, the Sheet canvas and Screen canvas can be mixed in a single meeting. For example, if a meeting is partly in a conference room and partly with people attending remotely, you can print the *sheet* version for people in the room and show the *screen* version for remote attendees.

The rules on movement are crucial because it is the single view that empowers people to understand the information rapidly. A 10m (~30 ft) long poster in a hallway that requires a person to walk is not one view—it's many snippets that must be reassembled in memory after the walking is finished. If the font is too small to read at 2 m, a minimal distance for even a small meeting, then people must step forward to read and then step back and "reassemble"; that's not one view. I've watched people absorb a canvas, watched their eyes dart left and right, up and down, as they digest information, mentally asking and then answering questions as they build the context up. This leads to the four "do nots" of a canvas: do not use tiny fonts, do not require walking, do not require flipping slides or pages, and do not require scrolling. All of these constitute motion between views, indicating you've missed the mark (Fig. 8.7).

Having a canvas is important, but it isn't sufficient to guarantee the story is visual; for example, a story told wholly in text on an A3 may be a canvas view, but it isn't visual.[a] When people work to tell a story in a canvas, they typically start with text. Then as space fills, they are forced to compress some of the story. They must eject less important information to focus on what's

[a] For example, John Shook's book "Managing to Learn" has five full-size A3 examples tucked in the back cover pocket, all of which tell their stories visually.

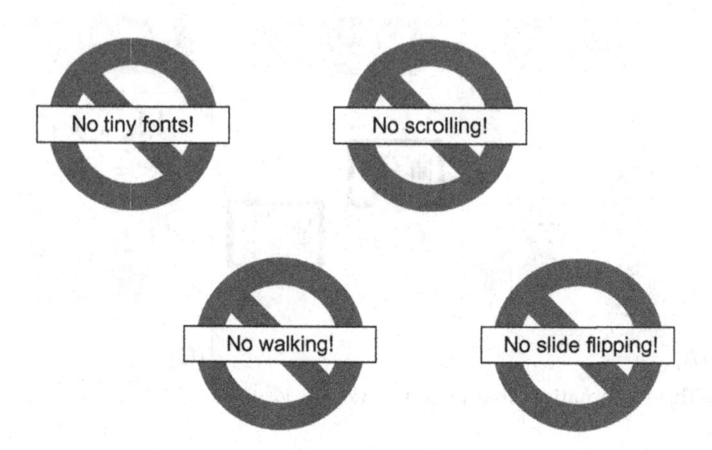

Fig. 8.7 Four cardinal "do nots" for a Canvas View.

necessary to tell the story. All of this unburdens the reader who must otherwise fight through less relevant narration. This naturally leads to more and more visualization as the canvas is refined.

The Canvas View cuts **Information Friction** by providing easy access to context and so makes information easier to comprehend. It guides presenters to focus on what's critical and remove everything else. It also reduces friction to information fed back from colleagues to the presenter since it's easy to knit new thoughts into the canvas.

8.5.1 Dashboards

The elephant looks different from the side.

Richard Dugan

The dashboard is "a collection of performance data, represented in graphical format, to provide an 'at-a-glance,' real-time snapshot of the health of a project, process or work units" [17]. It is an important variant of a Canvas View adapted to providing a comprehensive snapshot using high-density graphics representing varying points of view. The benefits of a dashboard include the following [18]:

1. Simplifies complex measurement
2. Identifies outliers and negative trends
3. Shows connections between data

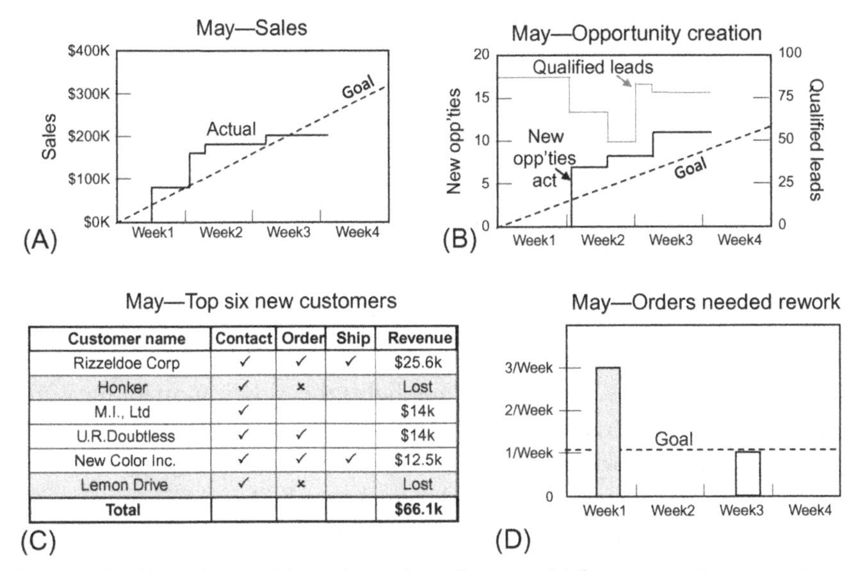

Customer name	Contact	Order	Ship	Revenue
Rizzeldoe Corp	✓	✓	✓	$25.6k
Honker	✓	×		Lost
M.I., Ltd	✓			$14k
U.R.Doubtless	✓	✓		$14k
New Color Inc.	✓	✓	✓	$12.5k
Lemon Drive	✓	×		Lost
Total				**$66.1k**

Fig. 8.8 Dashboards combine viewpoints from multiple perspectives to give a comprehensive Canvas View.

4. Displays alignment of strategy to performance goals
5. Saves time on report creation

Fig. 8.8 is a sample dashboard that might be used in a monthly sales review. There are four perspectives: (a) monthly sales, (b) new opportunities during the month, (c) activity with largest new customers, and (d) a quality measure: how many orders needed to be reworked after being placed.

In their article, "The Balanced Scorecard" from 1992, Kaplan and Norton [19] provide a structured way to build a dashboard for any issue using four perspectives, each with its own goals and measures:

- A Financial Perspective providing goals and measures that would be most meaningful to shareholders
- A Customer Perspective that shows how customers view us
- An Innovation and Learning Perspective that guides the organization to improve
- An Internal Perspective to show the most important challenges we must win

For those developing new dashboards, Kaplan and Norton provide a good starting point. Also, Williams gives half a dozen helpful examples of dashboards [20], and Dennis provides a large number [21].

8.5.2 Scorecards

Scorecards are similar to dashboards: both are a single view showing a snapshot for progress or performance from multiple perspectives. The use of the terms in business varies. Here, dashboards will be used for the whole organization and scorecards will be for the team doing the work. For example, if a sales team meets every morning to review daily progress looking at last yesterday's orders and leads, that would be a scorecard; the monthly summary of sales performance for leadership review (e.g., Fig. 8.8) would be a dashboard. The scorecard is managed primarily by and for the team; by contrast, the dashboard is provided by the team for others to review.

Creating a simple scorecard that highlights the most important measures is a step that can bring rapid improvement in team execution speed and quality. This is roughly the equivalent of a dashboard on a car. Imagine driving with no dashboard. Suppose the car's speed was displayed not on a speedometer, but was just logged in the car's memory, and accessible hours later via a website. This concept seems ridiculous for driving, but knowledge work processes are often designed this way, with key data hidden away in a shared folder or intranet website. If the team's #1 goal is to sell $400k in new products this month, they will immediately benefit from having the daily sales displayed prominently and reviewed against that goal, similar to the way a driver continuously monitors a speedometer.

A scorecard is a type of SPoT and, as such, requires the following:

- Display only what information the process users need. Consider an automobile dashboard: how many parameters of car operation could be displayed vs how many are there? Every dashboard is an example of designers picking the critical items to support the driver and leaving everything else out. Typically, a process scorecard will have three to five items—enough to guide most users most of the time.
- The information should be trustworthy: built from reliable data and updated regularly to minimize the number of information defects. Trustworthy information isn't free: it takes time and energy from a "scorecard owner."
- The information is actionable. When a speedometer shows the speed is excessive, the driver immediately knows to lift the accelerator pedal. Similarly, information on a scorecard should create clear calls to action.
- The information is accessible. Put the information in a place everyone can reach it, such as an internal website. Better yet, if the team is located together, display it on a physical board mounted where the team sees it several times a day. Even better again, hold short daily meetings where the team reviews the scorecard and decides what action to take.

8.6 Hierarchical visualization

A hierarchical visualization describes a means to tell a story that needs more than one canvas. It is "hierarchical" in that there is always a top-level canvas of the whole story with the ability to drill down to a lower level. This can proceed to multiple levels such as a canvas that presents every product a company produces, which rests upon a number of canvases of that company's products by market, which could then drill down another level to an individual canvas for each product.

Hierarchical visualization competes with linear visualization, for example, a long slide show. In linear visualization, you must "move" through many slides to find the one you need. And that assumes there is one you do need, which is often not the case—a fact that may take quite a bit of time to determine. A series of highly detailed views does not replace a true hierarchical representation any more than a series of 100,000 street-level maps (my rough math for the number required for the United States) would replace a system that lets me drill from a view of the United States to any street in five or six steps.[b]

Fig. 8.9 shows a two-tier hierarchical view. In this case, let's assume there is a key customer furious because of multiple problems from a component we provide. In this case, the organization leader will probably want a top-level canvas of how we are dealing with the customer. That canvas will probably have customer-facing information such as how we handle communication, the salespeople calling on the customer, the total revenue at risk, current status on the active issues, and so on. At this level, there is no need for a great deal of technical details. But the top-level canvas will probably point to views one level down for the design and operations teams working on the individual problems. Such a view allows a single high-level commercial view that is not overloaded with technical detail, but that is supported by detailed technical canvases focused on problem solving. So, the president of the unit would have visibility in the problem solving including the option to review any or all of them. But his or her primary interest would be at the high level.

I find the hierarchical view is an important option to support, but not for the reason I first thought. I had imagined we'd have many canvas sets structured two and three tiers deep on a regular basis. That was not the case. In fact, for every 25 times someone has started a canvas with a complaint about

[b] A practical example of hierarchy is finding my home on Google Maps. If I start with a map of the entire northern hemisphere, it takes just five or six "zooms" and about 1 minute.

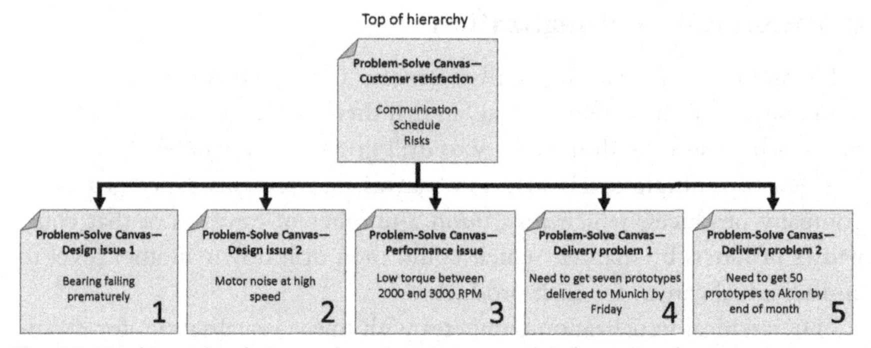

Fig. 8.9 Use hierarchical views when there's too much information for a single canvas.

the size constraint, perhaps one has ended up needing anything more than a single page. It seems Toyota's insight that one sheet can hold any issue was correct. Nevertheless, I continue to support the hierarchical view because it provides a ready solution for those who feel they cannot get their story on a single page. The option, though rarely used, quickly ends conflict about the concept of canvases and increased engagement for people learning how to use the tool.

8.7 Conclusion

This chapter has presented three techniques to cut the waste of **Information Friction**. Fig. 8.10 shows how visualization **simplifies**, **engages**, and **experiments**. We started with Single Point of Truth, a set of techniques around how to make information accessible to every stakeholder, and then moved to visualization tools and the Canvas View, two methods to make information easier to digest. We will use these three methods throughout the remainder of this book. These methods are engaging: they build a shared vision, eliminate tedium, and flow information to the full team for strong connection. Goals are sharply stated to help create a winnable challenge. And accurate information is broadly available, allowing a diverse set of experts to validate it. Finally, scoreboards and dashboards create an experiment in value generation; if our work falls short of expectations, these techniques reveal this quickly so we can learn and improve.

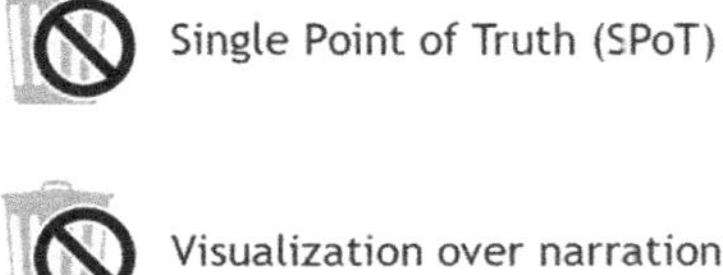

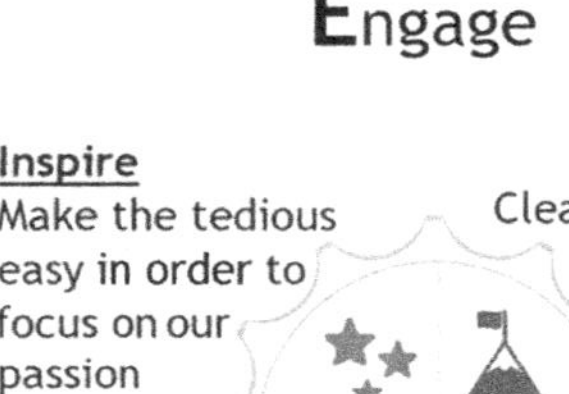

Fig. 8.10 Visualization techniques simplify, engage, and encourage experimentation.

References

[1] J.W. Tukey, Exploratory Data Analysis, Pearson Modern Classic, Pearson, 1977.
[2] https://www.forbes.com/sites/brentdykes/2018/01/10/single-version-of-truth-why-your-company-must-speak-the-same-data-language/#26a261031ab3.
[3] P.R. Williams, Visual Project Management, Think for a Change Publishing, 2015, 5.
[4] J. Liker, The Toyota Way. 14 Management Principles From the World's Greatest Manufacturer, McGraw-Hill, 2004, 244.
[5] P.R. Williams, Visual Project Management, Think for a Change Publishing, 2015, 8.
[6] W. Brinton, Graphic Presentation, Introduction by Henry D. Hubbard. Also Available at:http://www.archive.org/stream/graphicpresentat00brinrich#page/2/mode/1up, 1939.
[7] R.A. Burkhard, The Use of Complementary Visual Representations for the Transfer of Knowledge. A Model, a Framework, and Four New Approach, A dissertationSwiss Federal Institute of Technology Zurich, 2005.
[8] P.R. Williams, Visual Project Management, Think for a Change Publishing, 2015, xv.
[9] S. Shingo, The Sayings of Shigeo Shingo: Key Strategies for Plant Improvement, English translationProductivity Press, 1987, 139.
[10] P.R. Williams, Visual Project Management, Think for a Change Publishing, 2015, xvii.
[11] R.A. Burkhard, The Use of Complementary Visual Representations for the Transfer of Knowledge. A Model, a Framework, and Four New Approaches, A dissertationSwiss Federal Institute of Technology Zurich, 2005.
[12] Edward R. Tufte. The Visual Display of Quantitative Information second ed., Graphics Press. Cheshire, CT. p. 51.
[13] R.A. Burkhard, The Use of Complementary Visual Representations for the Transfer of Knowledge. A Model, a Framework, and Four New Approaches, A dissertationSwiss Federal Institute of Technology Zurich, 2005.
[14] J. Shook, Managing to Learn. Using the A3 Process to Solve Problems, Gain Agreement, Mentor, and Lead, The Lean Enterprise Institute, 2008, p. 16.
[15] J. Liker, The Toyota Way. 14 Management Principles From the World's Greatest Manufacturer, McGraw-Hill, 2004, 157.
[16] J. Shook, Managing to Learn. Using the A3 Process to Solve Problems, Gain Agreement, Mentor, and Lead, The Lean Enterprise Institute, 2008, 7.
[17] P.R. Williams, Visual Project Management, Think for a Change Publishing, 2015, 69.
[18] P.R. Williams, Visual Project Management, Think for a Change Publishing, 2015, 72.
[19] R.S. Kaplan, D.P. Norton, The Balanced Scorecard—Measures that Drive Performance, HBR, 1992 From the January–February 1992 Issue.
[20] P.R. Williams, Visual Project Management, Think for a Change Publishing, 2015, pp. 73–78.
[21] P. Dennis, Getting the Right Things Done: A Leader's Guide to Planning and Execution, Lean Enterprise Institute, 2006, 56f.

CHAPTER 9

Reduce Waste #3: More-is-Better Thinking

It is not enough to do your best; you must know what to do, and then do your best.
W. Edwards Deming [1]

9.1 Introduction

In this chapter, we will take on **More-is-Better Thinking**, the waste from a mindset of grinding away at a task without a plan to succeed. It is hard work but lazy thinking, mistaking being busy for creating value. We will start by addressing the difference between work and artifacts of work. We will then take a detailed look at three artifacts to measure success: the Action Plan, the Test Track, and the Bowler. These three artifacts measure knowledge work product at different stages of maturity. Together, they provide a comprehensive measure of success; we combine the three into what here is called a "Success Map," a tool that can rapidly cut **More-is-Better Thinking** in a knowledge organization. We'll end with a discussion of two different perspectives of knowledge work: the *ground view*, the view of the team doing the work, and the *helicopter view*, an aggregation of many instances of work, for example, a snapshot of a dozen active initiatives.

Table 9.1 compares **More-is-Better Thinking** to its counterpart, Goal Orientation. In the first case, consider two very different reactions to a customer complaint: do we (1) try harder or (2) determine what a customer is demanding and then meet that demand? Or compare working harder to avoid software defects to finding relevant measures of software quality and creating methods to meet them? Or, in the last row, do we work harder because sales are disappointing, or set sales goals needed to succeed and then do what's required? **More-is-Better Thinking** meets new demands with a grimace and a promise to "do our best." But just trying harder usually amounts

Improve
https://doi.org/10.1016/B978-0-12-809519-5.00009-0

Table 9.1 Comparing More-is-Better Thinking to Goal Orientation.

Example	More is Better	Goal Orientation
Customer problem	☒ We need to work harder because our customer is unhappy	☑ What do we need to do to make the customer happy?
Software quality	☒ We need to work harder to reduce our software defects	☑ What are our software quality goals? Are we meeting them?
Sales	☒ We need to work harder so that we can sell more	☑ What are monthly targets? Are we meeting them? Why not?

to surrendering to the circumstances. After all, aren't people already trying hard? And why would newfound knowledge of a looming failure cause anyone to try harder, let alone cause them to sustain that extra effort for long enough to make a real difference? Goal-oriented thinking meets each challenge with two steps: define what's needed to win and then do what's required.

Goal-oriented thinking doesn't deny reality by pretending we can never lose, but it does refuse to waste effort on lost causes. It's a mental posture that recognizes that if there's no winning strategy, there is no reason to think we'll win. If we believe we'll win, sometimes we win and sometimes we lose; if we believe we'll lose, we lose almost every time. So, determine what's needed to win and decide to do it, or decide winning isn't likely enough to spend more effort.

9.2 Artifacts vs work

The first step to avoid **More-is-Better Thinking** is understanding the difference between "work" and the "artifacts of work." Artifacts are imperfect evidence of work. A ubiquitous artifact of learning in modern education is the exam. Students apply themselves for weeks at a time listening to lectures, reading, and solving sample problems. This is the *work* of learning. But in about 1 in 10 class sessions, each student provides a key *artifact* of their learning: an exam. We need an artifact because learning cannot be observed directly. And, for all the complaints about the inaccuracy of testing, it's one of the best artifacts available to our education system. As shown in Fig. 9.1, work is hidden but artifacts, for all their imperfections, can tell us most of what we need to know about the work.

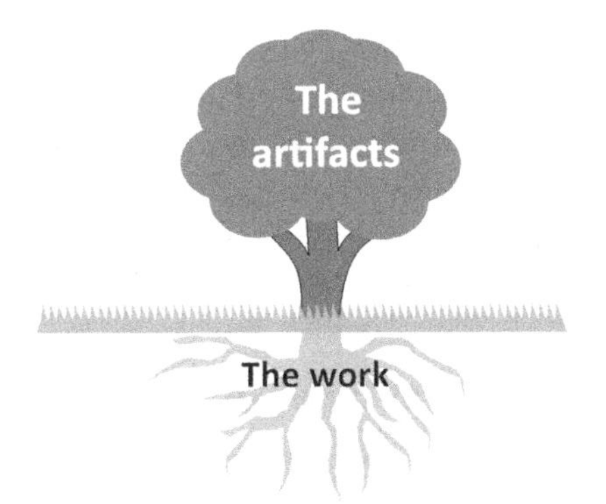

Fig. 9.1 The work is below the surface; we can see only the artifacts.

9.2.1 Action Plan: Start with artifacts of change

Tasks in different parts of an initiative require different artifacts. Early in an initiative, the team needs to bring change: a new procedure, a new distribution model, or new products to launch. A flurry of activity in different directions is needed: we might need to coordinate a task with Finance to collect a new metric or with Operations to prepare to support a new offering. Here the Action Plan is ideal: the only uniform measure is the dates by when tasks are finished and clarity around ownership. It gives a structured *when* and a *who*, allowing the *what* to have wide variation. For example, let's say we want to open a distribution channel in Brazil. Such an initiative would start with a number of disparate tasks such as identifying Brazilian distributors, modifying information systems to allow sales in Brazil, and translating user documents into Portuguese. An effective visualization for change is the Action Plan, which we will review in Section 9.3.

9.2.2 Bowler: End with artifacts of value

At the end of an initiative, the demand is to deliver value to the organization: more revenue, for example, or fewer defects. The *what* is much more constrained here than it was in the Action Plan; in fact, it's probably an aggregation of one or two of the types of value defined in Section 3.6:

- Fitness for purpose (quality and completeness)
- Profitability (higher revenue and lower cost)
- On-schedule delivery
- Innovation (differentiation from competitors and alternative methods)
- Protection (compliance with regulations and industry standards, both formal and informal)

For the example to develop a Brazilian distribution plan, the key artifact would probably be revenue generated in Brazil. We may sell thousands of units, but there is one primary measure: the sum of revenue. Where the Action Plan allows the flexibility to gather unrelated tasks, value relies on visualizations that focus on one or two aggregated measures. Here we will use the Bowler, which is presented in Section 9.5.

9.2.3 Test Track: Transition by measuring artifacts of traction

The artifact to bridge between *change* and *value* is *traction*: the first examples of the change delivering value. While the Action Plan and Bowlers are common, they leave a gap because there is usually a lot of time between when *change* of the Action Plan is complete and when *value* flows evenly enough to rely on the Bowler. It's common to measure traction in defined cases: a pilot run in manufacturing or an early prototype in product development. But in this chapter, we will create a systematic approach so that we measure traction consistently for all types of knowledge work.

For our example of distributing in Brazil, the end of *change* would approach when the distribution channel is set up: the website is ready, the products are in the inventory, all needed documents are translated, and the first set of distributors has been selected and trained. But the flow of sales will be just a trickle. It may be months before orders come in smoothly enough to be well-represented by the sum of sales in a Bowler. That period between *change* and *value*, *traction*, is measured by the Test Track in Section 9.4.

9.2.4 Flexibility vs aggregation

Each of the artifacts of the previous three sections has its place in the life of an initiative or project. At the start, high flexibility is required; at the end, aggregation such as summing revenue is needed. As shown in Fig. 9.2, over time the relevant measures of initiatives span from *change* to *traction* to *value*:

- Early: Create change

  At the start of an initiative, there is a great deal of *change*: new needs, new methods, new process, new design, and new people. There are many disparate tasks that must be managed during periods of high change. Here the Action Plan fits best because of its flexibility.
- Middle: Get traction

  As the project or initiative proceeds, the areas of *change* stabilize. We now need *traction*, using the new solutions and carefully watching the

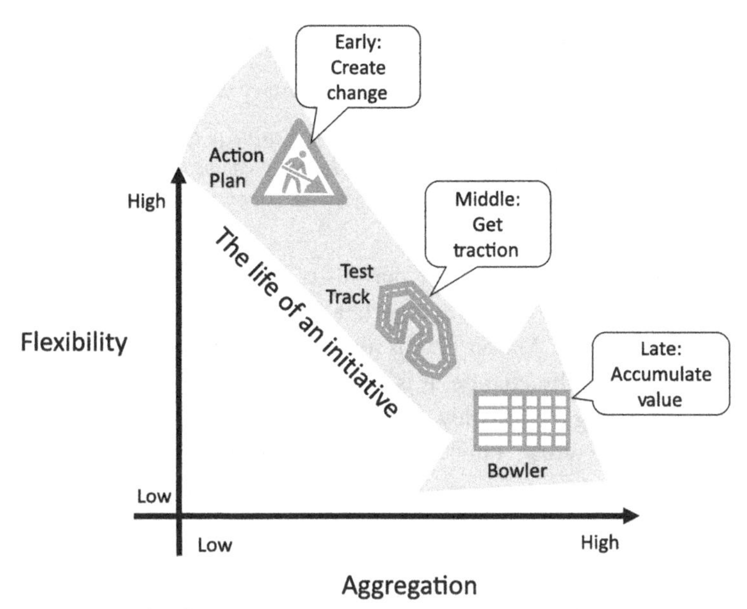

Fig. 9.2 Over the life of an initiative or project, the Test Track fits between the highly flexible Action Plan and the highly aggregable Bowler.

first instances to ensure they are working. We try our new website by closely watching seven key customers using it for the first time. We may carefully review every order received from a new distributor for their first 10 weeks. Flexibility is less important because we don't have a disparate list of changes. But we are not ready for full aggregation—we must watch each instance work.

- Late: Accumulate value

  Toward the end of the initiative, the focus shifts to building standard work with sustainable results. Here many events are summed into a few metrics that correlate to *value*: sum all the revenue from many sales, count the user downloads across many regions, or add up all the patent applications submitted to the patent office. The demand for flexibility is low because many results are being combined into just two or three measures, but the need to accumulate results over time is high. Here the Bowler charts fit well.

The next three sections will review the three artifacts of Fig. 9.2 in detail. Following that, we will discuss the "Success Map," a combination of the three artifacts that allows a single view that tracks results for the life of an initiative or project.

9.3 Measure change with the Action Plan

An Action Plan is defined by Business Dictionary as:

> *A sequence ... that must be performed well, for a strategy to succeed. An action plan has three major elements (1) Specific tasks: what will be done and by whom. (2) Time horizon: when will it be done. (3) Resource allocation: what specific funds are available for specific activities [2].*

The Action Plan is best suited for periods of high change: there are many things to do and people working together is the only way to get them done well and on time. The Action Plan is much more than a list of things to do. At the beginning of the change, it is a detailed way to communicate where we are going; during the journey, it tells us where we are and if we are meeting the goals of the work. The "we" is crucial. Action Plans are an informal contract between the organization leadership, the plan owner, the people who are responsible for or "own" the tasks, and the people who take on the tasks. The Action Plan owner may write a first draft of an Action Plan, just as a home buyer may write a contract to purchase a house. But that draft means little until all the task owners have accepted the plan. This *negotiation* builds team ownership, an element foundational to the success of any complex plan (see Section 7.3.1 for a detailed discussion on negotiation).

The basic Action Plan of Fig. 9.3 cuts the waste of **More-is-Better Thinking** by defining success for each task in the plan, tracking four elements necessary to be able to recognize success:

- Result

  Measurable results, a tangible deliverable of this task. Always results; never effort. For example, if one action is to have a meeting to select a software vendor, the action could be characterized as "Select vendor at meeting," a result, not the *effort* of "Discuss vendor selection."

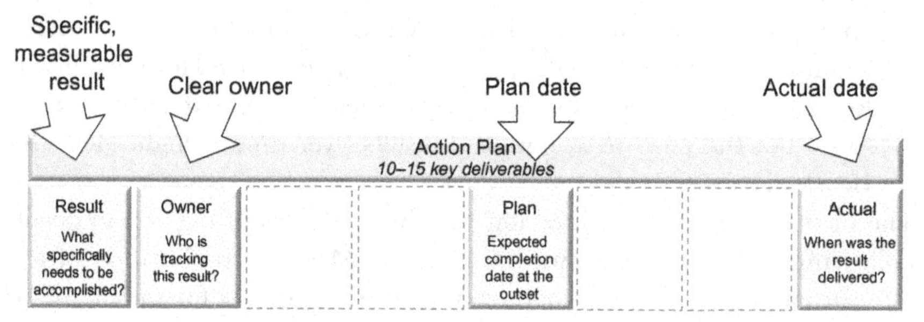

Fig. 9.3 The basic Action Plan.

- Owner

  The single owner of each deliverable. Of course, many actions will have multiple people involved. But one and only one person should be responsible for communicating and managing the action. Occasionally, it will happen that two people will be working on a task equally; make the hard choice and pick one to own the action and be responsible to interface to others working on the action.

Divide responsibility and no one is responsible.

W. Edwards Deming [3]

- Plan

  The plan is the date the result is needed as understood at the outset. The plan date does not change during the life of the Action Plan.
- Actual

  Initially blank, the actual date is filled in with the date the result is realized.

9.3.1 The Action Plan vs the To-Do list

The Action Plan is the alternative to the "To-Do" list. The To-Do list states requirements with so much ambiguity that it is normally ineffective for coordinating a team. The Action Plan with its focus on results, unambiguous ownership, and defined series of dates makes the plan clear at the outset and, after that, provides a visualization of execution effectiveness.

A sample To-Do list is shown in Fig. 9.4. "Improve" in the first bullet is ambiguous—improve by how much, e.g., 5% or 150%? And how is improvement measured? Bullet two has ambiguity in ownership: "Marketing" is unclear. Bullet three has an ambiguous date. What is mid-year? July 1? Bullet four has unmanaged granularity, because in a single step we are doing something large (signing on four distributors) and something small (updating one discount schedule). All tasks on an Action Plan should be at roughly the same complexity.

9.3.2 Managing task granularity

Task granularity quantifies the size of tasks within an initiative or project. It can be measured by the number of people-hours necessary to complete it, total time, or external expense. Managing granularity is the skill of keeping

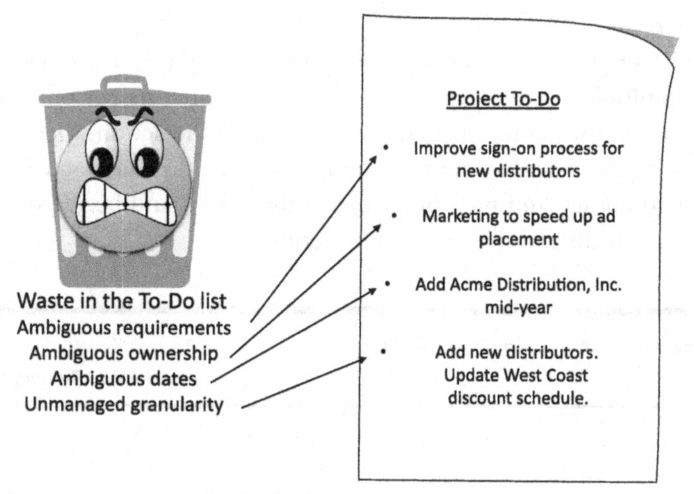

Fig. 9.4 The Action Plan vs the To-Do list.

all tasks in an Action Plan right-sized. Three common mistakes are (1) defining tasks too finely, (2) defining tasks too coarsely, and (3) failing to manage granularity so that large and small tasks are mixed together in the same Action Plan.

9.3.2.1 Finding the right-size granularity

An Action Plan fulfills an overall goal; it is used when that goal is too complex to accomplish in a single step. The goal is a mountain the team needs to climb; the Action Plan maps the path the team will follow. But, like a map, the plan should fit in a single view. For an Action Plan, that means the list of actions should be limited to perhaps a couple dozen, the amount that can reasonably be placed on a single piece of paper.

Too-fine granularity can result when the project or initiative leaders overestimate the Action Plan's ability to manage small details. This leads to action lists of hundreds of tasks spanning multiple pages; the path to the goal is lost in the details. It's something like asking for directions to a nearby grocery store, expecting a handful of instructions like: left on Spruce St., right on Main, and the store is about a mile on the left. Instead, you get 500 steps of minute detail:

1. Walk out front door.
2. Walk to car.

3. Open door.
4. Sit in driver's seat.
...
498. Pull into open parking spot.
499. Turn ignition off.
500. Remove key from ignition and place in pocket.

The first way to avoid overly fine granularity is to limit the number of actions that will be tracked in the plan. A project simple enough to be managed with an Action Plan can be broken down into about 25 tasks so that the team and the leadership can maintain an overview of the work. At the other end of the spectrum, granularity can also be overly coarse. Let's say we have a 3-month project; if a single task were, say, 2 months, that task would be almost as high a mountain to climb as the 3-month project. This doesn't work well because people often overestimate task completion progress; it's common for a task owner to think a task is "80%" done when it's actually less than half complete. This is from the Pareto principle, which often is characterized as "20% of the effort gets 80% of the results." Its dark corollary explains why people almost always overestimate progress: "when 80% of the results are delivered, the work is only 20% done."

Unrealistic optimism probably derives from the Dunning-Kruger effect, the well-known phenomenon where people who know very little about a topic think they know the topic well. Without even realizing it, we seem to estimate the complexity of a topic against what we comprehend within that topic. When we know very little, this can lead to a gross underestimate of the complexity of the topic, and this leads a person to think the little they know is most of what there is to know.

Perhaps a derivative effect happens when people do work. They complete a portion of work and, not fully comprehending how much more there is to do, conclude they are nearly done. Only when they start validating their work do they understand what's required. For example, a website programmer with the task of supporting a new product for e-commerce may sense she is 80% done with her work when the product appears on the website and she can enter a mock order. But really, 80% of the work remains: testing from multiple regions, evaluation of compliance with regulations, translation, feedback from potential customers, and ability to order multiple variants of the product. The newer someone is to an area, the more they are likely to fall prey to this effect.

An effective way to combat the negative effects of overoptimism is to constrain individual task granularity to no more than 10% or 15% of the total project or initiative. For example, in a 3-month project, no task should be no longer than a week or so; in this way, overestimating progress on any single task can misstate actual progress by no more than a few days.

This leads to two simple rules to manage granularity: limit the maximum number of tasks to about 25 and don't allow any tasks to be longer than one-tenth of the total project. The natural question is: what happens if, when we break down tasks to 25 or so, the tasks are still too large to be managed on a daily or weekly cadence? In other words, what if those 25 are each a mountain too high to climb? There are at least two choices. One recommendation is to go to a Visual Action Plan, which is discussed starting in Chapter 19. Here a single page can hold 50 or 100 tasks—2 × to 4 × the density of a traditional Action Plan. The second is to add a level of hierarchy; this is the topic of the next section.

Two simple rules to manage granularity of an Action Plan:

1. No more than 20–25 tasks
2. No single task takes longer than 10%–15% of the project/initiative length

---

9.3.2.2 Multiple levels of granularity

When projects are too complex to manage in a single level of granularity, we can still avoid an overly long Action Plan by using multiple levels of granularity, as shown in Fig. 9.5. Here we can start with an 18-month project and break it down in about 25 major deliverables; this is our coarsest granularity. Those deliverables can be segregated into project phases that are 2–4 months each. Those phases can then be broken into about 25 tasks that are 1–2 weeks long; this is medium granularity for this project. Those can be gathered into sprints of 2 weeks, and each of those weeks can be divided down again into daily tasks.

The primary advantage of this approach is *horizon-variable granularity*. For Fig. 9.5, we have an overview of the goals of the 18-month project, a ~10 × more detailed view of the next phase (2 months) and a ~10 × more detailed view of the next 2 weeks. Compare this to an Action Plan hundreds of lines long with *flat granularity* where daily we must deal with the detail of the entire large project. In the second case, we are planning high detail in the far future which will surely be wrong enough to need rework. All the time

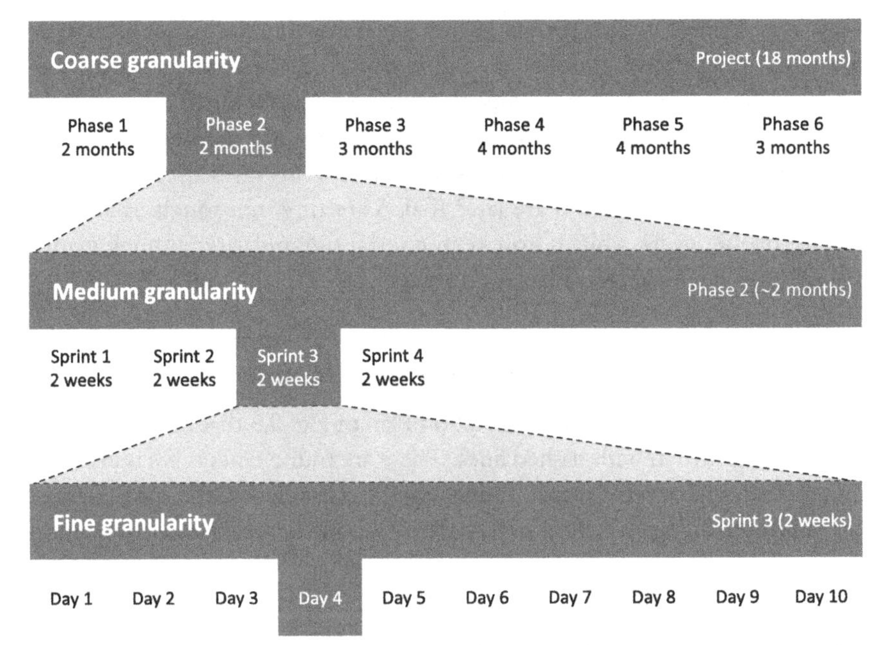

Fig. 9.5 Hierarchical view of tasks: Three levels of granularity are common in project planning.

we spend planning something that has to be planned again is waste. Horizon-variable granularity focuses the team on the right among details for the long view, for the view of the next week or two, and for views in between.

Three tiers of granularity are necessary only for complex projects. The ordinary improvement initiative or small project can almost always be managed with one level of granularity; just about everything else can be done with two levels. Where multiple levels are required, bring the team together at each step of the coarser levels of granularity. For the example of Fig. 9.5, you might have a major team meeting at the start of each phase (coarsest level) to divide that phase down to ~25 major steps. Further, there might be a 60- or 90-minute video conference with the full team at the start of each sprint to lay out ~25 tasks. For projects of this complexity, consider using the more advanced techniques like the Visual Action Plan (Chapter 19) and the Visual Action Plan with Buffer (Chapter 20); these techniques allow a single-page view like Fig. 20.9.

Finally, project leaders need to keep granularity even. Each task in a task list should be about the same complexity. Think of coarse granularity like the floors of a building, as shown in Fig. 9.6. Fine granularity could then be seen as the steps between floors. Avoid mixing coarse and fine granularity in the same Action Plan. This avoids the common problem in Action Plans where, partway through the project, a few actions are incomplete but it's unclear whether the plan is on track. If you mix 10 tiny tasks among 20 larger ones, what does it mean if 5 are late? If all 5 are tiny, not much. If all 5 are large, probably quite a bit. If you keep constant granularity, a quick glance will tell you if the Action Plan is on track.

9.3.3 Adding forecast date to the basic Action Plan

You may have noticed in the basic Action Plan of Fig. 9.3 that there were four empty blocks shown with dashed lines. These are four elements we will add to further reduce **More-is-Better Thinking**. The first element to add is forecast date, as shown in Fig. 9.7. The forecast date is the likely date of task completion

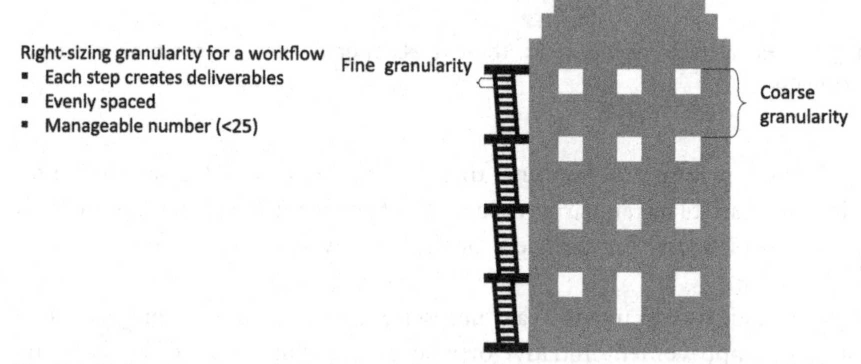

Fig. 9.6 Right-sizing granularity is one of the most complex project planning skills.

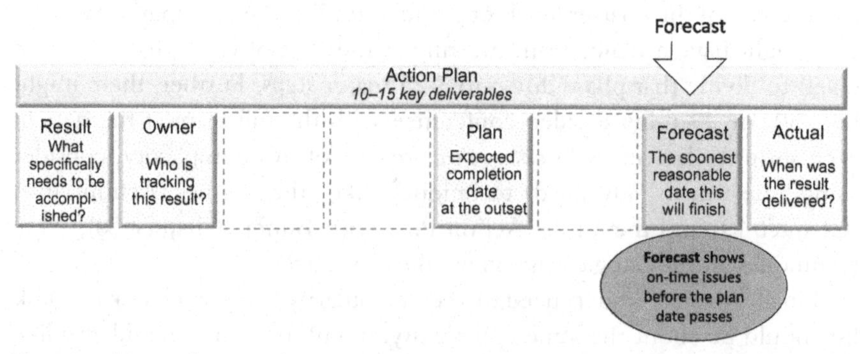

Fig. 9.7 Adding a forecast date to the Action Plan helps resolve hidden delay issues.

based on what we know today. While the plan date never changes, the forecast can change weekly or even daily.

9.3.3.1 The milestone delay problem

The forecast date is added to avoid the *milestone delay problem*, the natural tendency for people to report a task is late only after the plan date passes, even when the delay was known well ahead or could have been predicted easily. It's not that people are scheming or being deceitful; it's more that addressing the looming delay is easier to put off for another day. The problem is that this day never comes.

Let's consider an example. Suppose that we are building an Action Plan to open a sales office in London. One of the key deliverables is to sign a lease on the office property by January 1, a date that looks reasonable at the outset. Let's say Christian has taken ownership of that and, on November 1, we plan for the result by January 1. Unfortunately, our first choice in locations is full and it's taking longer to develop an alternate location than was allowed for. On December 1, Christian knows that the January 1 date is unlikely; in fact, he doesn't see a path to close before January 15. But he hasn't told the rest of the team because he is working as hard as he can—another form of **More-is-Better Thinking**.

Milestone delay problem: the tendency to report a delay only after a milestone has passed, even if the delay is known or could easily have been forecast well before that.

So, we have one of the most common problems in keeping a team aligned to a goal: the current plan is not an accurate reflection of reality, but not everyone knows that. It's quite natural for Christian to keep that information to himself until the plan date approaches or even passes. But by that point, there's nothing that can be done to countermeasure the problem. You might be thinking, why doesn't Christian just tell the team? Clearly, that would be the most effective reaction: pulling everyone together to address schedule gaps as soon as they are detected. But it doesn't happen often. You might ask, isn't this as simple as Christian just doing the right thing? Wouldn't it work to admonish Christian to flag a delay as soon as it is recognized? Probably not.

We must accept there are real disincentives for people in Christian's position to share the full story, the primary ones being that Christian may feel he's let his team or his boss down or that, by revealing the gap too quickly, he's communicating that he is giving up. Another disincentive to revealing a gap is the concern about added workload such as having to explain what's happened, why, and what we're doing about it. Another issue is that starting a conversation about a newly identified gap isn't always easy; someone with a great deal of self-confidence might be willing to expose promptly problems they likely contributed to, but not everyone will be in that position. Another reason is that people may fear the consequences of revealing their part in a performance gap. These are some of the reasons why simply telling everyone to be "transparent" is usually insufficient to stop **More-is-Better Thinking**.

Instead of telling employees to ignore the disincentives we know exist (and then being surprised when not everyone does that), let's reduce those disincentives. The first step is adding forecast, a place for Christian to record a delay in a factual way and without communicating that he's given up. The second is having regular team stand-up meetings, a place where people are more comfortable communicating forecast delay.[a] The problem with forecast date is that once the forecast date is beyond the plan date, the task is "late"; often there's nothing that can be done to remedy the delay, even when it doesn't affect the success of the project. So, no matter how hard you work, you cannot eliminate this problem; therefore, it becomes a false call to action: it shows "red" on the Action Plan, but cannot be remedied. This is where we add the next element to our Action Plan: the replan date.

9.3.4 Adding replan date

The second element to add to the basic Action Plan of Fig. 9.3 is the replan date, as shown in Fig. 9.8. Once the forecast date exceeds the plan date, the task is late—some sort of alarm should be signaled. Perhaps the whole action turns red. The first course of action is to work to pull the forecast back in, but that's often impractical because the person reporting the delay has probably done what they can before highlighting the delay. For our example with Christian and the contract for office space in London, assuming Christian is competent (a good starting assumption), it's unlikely there's anything so obvious that another team member will mention something that can pull the task back. Removing delay from a single task is a highly constrained problem. A much less constrained problem is: how can we meet project expectations being informed about this delay? This broader problem includes remedying the delay as the first step, but is augmented with options

[a] Another method we'll discuss in Chapter 20 is adding buffer so the team can absorb modest delay without impacting commitments and thus further reduce the aversion to recognizing delay early.

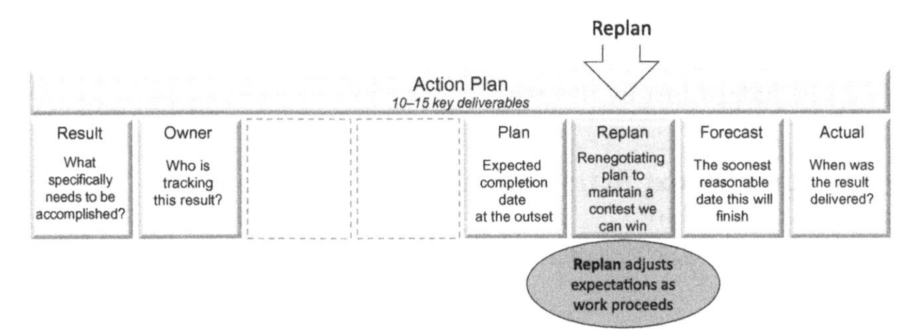

Fig. 9.8 Adding a replan date allows expectations to be adjusted so there is always a contest we can win.

to "re-plan" other project tasks so that the overall goals are met and calls to action ("alarms") are turned off. One of the cardinal rules of Action Plans is that plan dates don't change because they represent the initial commitments; if the team relaxes those, the stakeholders may sense the team is not living up to earlier commitments.

The replan date solves the dilemma of choosing between two bad options: the overly rigid plan date (firing an "alarm" that cannot be turned off) and teams changing plan dates (giving the sense of not respecting commitments). In our example where Christian cannot get the new lease until the middle of January, the replan date could reflect that issue. However, the team would then need to review the remainder of the Action Plan to ensure no other dates are affected. It's likely a few tasks might need to be replanned as well, but as long as the primary goals of the Action Plan are met, replanning a few individual tasks is probably going to be acceptable to stakeholders. Having the "Plan" and "Replan" dates side-by-side demonstrates that the team is committed to delivering results and also shows where they have been agile enough to make minor modifications in the Action Plan for success in the face of unexpected events.

Unfortunately, sometimes there is no way to accommodate the delay and maintain the Action Plan goals. This will be reflected in the replanning process when the end of the Action Plan is delayed as an effect of the task delay. In such a case, the replan becomes the primary visualization for renegotiating with leadership. After the team creates a recommended replan, they would seek approval from Action Plan stakeholders. The replan doesn't remove the stigma of failing to deliver results, but it allows the recognition of unrecoverable delay. It allows the team to turn the alarms off, bringing order to daily meetings that otherwise can become chaotic when the Action Plan goes permanently "red" partway into the project. So, the team's daily work is regulated by the replan, but the ultimate results will always be judged first by the plan.

9.3.5 Adding Next Action and Next [Action] Date to the Action Plan

The final feature we will add to the basic Action Plan requires two elements (see Fig. 9.9):

- Next Action

    A breakdown of a long task, extracting the next action of finer granularity
- Next [Action] Date ("Next Date")

    The date by when the next action is planned for completion

Next Action and Next [Action] Date allow a nimble way to add a tier of detail to one task without the complexity of a full level of granularity, as shown in Fig. 9.5. They provide a way to deal with the occasional task that violates the 10% max. rule of Section 9.3.2.1. Recall that each task is for a deliverable and we seek to divide the Action Plan so that each task deliverable requires no more than about 10% of the project total time. Sometimes there are deliverables that require effort over a longer time, but there isn't an obvious choice of an intermediate deliverable.

Next Action creates smooth flow in these long tasks—tasks that may require many team meetings to complete. At each meeting, the team can divide off the next action needed to keep the deliverable on track. Next Action is particularly useful for tasks that take a long time but require small effort any given week, often just enough to monitor and ensure that another

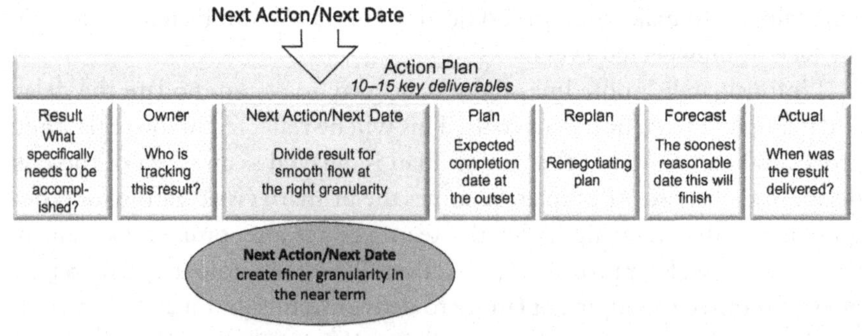

Fig. 9.9 Next Action and Next Date allow finer granularity in the near term without bloating the Action Plan.

team is fulfilling the need. At each team meeting, there can be a short discussion about the overall task result and then a minute or two dedicated to what needs to happen this week. This occurs each week with a simple question: what do we need to do this week to maintain timely delivery of the full "Expected result" as shown in Fig. 9.10?

For example, let's say we need a review from the legal department, which will require 4 weeks in an 8-week project, well over the 10% upper limit. But there is no identifiable deliverable until the review is complete. Here is where Next Action can fill a gap: capture the effort needed this week to keep the 4-week deliverable on track. For example:

Start date: May 1
Task: Legal review (May 28)
on May 1: Next Action: confirm commitment from Megan in Legal (May 5)
on May 8: Next Action: meet with Megan for preliminary review (May 12)
on May 15: Next Action: book May 28 meeting with Megan/project team (May 19)
on May 22: Next Action: (none—Legal review should be complete May 28)

As you can see, each week, the team works out what should happen in the next week to maintain the May 28 date. Next Action/Next Date allows the team to create a partial-level granularity just where needed. Because only the

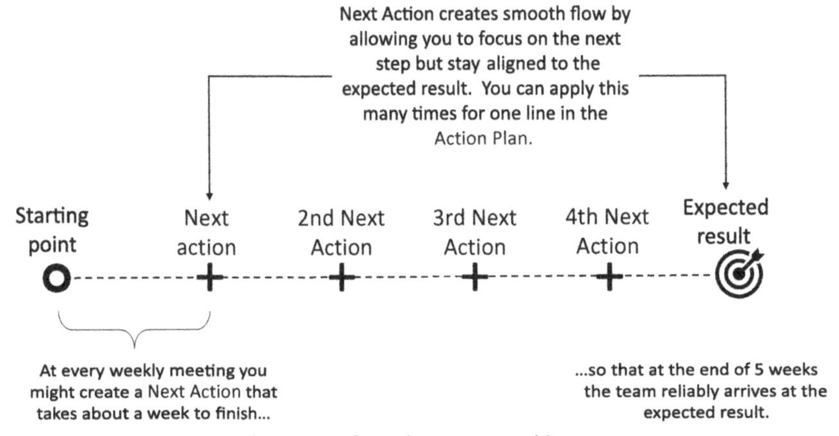

Fig. 9.10 Next Action is dynamic, often changing weekly.

Next Action and the full expected result is discussed in any one week, it's a lightweight technique to add detail.

9.3.6 Review cadence

Rule 9: "I review progress toward my goals on a regular basis" [4].

It seems to be natural to create an Action Plan and then push it off to a deserted folder on a PC. Perhaps a month later, on reviewing old emails or getting an unexpected request from the boss, it gets cracked open for the first time. Inevitably, some things didn't happen. Now the contract the Action Plan may have represented a month ago is lost as people have deployed to other work. There's a choice of two bad options: resurrect the Action Plan and work through the gaps or start over. Neither works well, so a better solution is to avoid that problem with high-cadence reviews from the start.

The Action Plan should be reviewed regularly because whatever was thought at the beginning of the work will be in some degree wrong. No matter how smart people are or how hardworking, there will always be unexpected events. One or two weeks into the Action Plan, progress will be slightly off; another week, off a bit more; at the end of a month or two, progress will be so far off the plan is no longer valuable. It is through the small corrections of frequent short meetings that an Action Plan can navigate to the goal.

The cadence of the meetings should match the actions. If we're monitoring a metric like quality that is measured monthly, probably our actions will conform to that cadence. If we have an irritated customer calling in daily, the dynamics of the actions we will need to take demand we meet every day or so. It is the flow of the actions that set the cadence of review. That said, if there is a bias in human nature, it's probably to a lower cadence than would be most effective. So, if in doubt, set the cadence on the high side; if that's too frequent, you'll know it in the first few meetings and it's easy to slow down the frequency of meetings.

Reviews should include all the task owners—the people who are responsible in the team for getting tasks done. In other words, it should include everyone who is listed as an owner. That doesn't mean everyone who works on a task has to attend. If Pilar is the owner of testing and she's working with Isabel and Tony in the lab to finish a test, probably only Pilar needs to come to the meeting.

High-cadence meetings must be short. For example, you could put the Action Plan on a screen, and work through all the actions that are open or that should open before the next meeting. Ask three questions: what did we get done, what are we getting done before the next meeting, and what barriers do we need help with? Barriers are areas outside the direct control of the task owner, such as getting a needed resource or approving an expenditure. Target 10–15 minutes for a daily review meeting and perhaps double that if you're meeting weekly. It's crucial you keep high-cadence reviews short. They will bloat as people discuss issues in more detail than is necessary for the full team. I've watched this in daily meetings that were targeted to 15 minutes but extend to 75 or 90 minutes every day! This will destroy the value of the meeting—some people will stop joining so that there will be no quorum and thus no ability to react to serious issues. Besides, those long conversations usually involve just a few people—it's wasteful to keep the rest of the team from their work. It's not easy to lead quick meetings; you'll find yourself saying often, "That's outside the scope of a daily meeting" or interrupting an argument with "Can you two get together on that later today and let us know tomorrow what you found?"

The review meeting is usually the place the Action Plan is modified. The Action Plan is a contract between the initiative leader and all the task owners. If the plan is changed without their being aware, they may cease to accept ownership. So, limit making changes to the Action Plan such as closing tasks, replanning, and adding new tasks, to times when the team is together.

9.3.7 Example of an Action Plan

Table 9.2 shows a sample Action Plan with eight tasks, the first two of which use the Next Action feature. This is a project to improve the approval process that includes creating virtual, recorded meetings, adding an approval checklist[b] for applicants to understand requirements better, and root causing problems in approvals that don't go well—for example, those that took too long to approve. The virtual meeting tasks were delayed by an unexpected IT issue (hence both are replanned), but the initiative is still on track as indicated by the later tasks remaining on schedule; hence, there is no need to replan the later steps.

9.3.8 Summary of Action Plan

The Action Plan cuts the waste of **More-is-Better Thinking** by maintaining a definition of success with measured deliverables, clear ownership, and specifying due dates. Managing granularity creates a visualization that is

[b] See Chapter 17 for more discussion of checklists.

Table 9.2 Augmenting the Action Plan with "Next Action" allows flexibility as the project proceeds.

Expected result	Next Action	Next Date	Owner	Plan	Replan	Forecast	Actual
Virtual/recorded meetings							
SPoT for recorded meetings	IT meeting	15-Feb	GM	1-Jan	1-Mar	1-Mar	
Standardize off-line approval	Trial: record one approval	20-Jan	GM	15-Jan	30-Jan	30-Jan	
Checklist							
Checklist signed off by approvers			ALN	1-Jan	31-Jan	31-Jan	
Checklist piloted on one approval			GM	1-Feb		7-Feb	
Checklist standardized for EU			RWC	1-Mar		1-Mar	
Root cause individual cases							
Admin. assistant to track approvals			HMC	1-Jan		1-Jan	
Set up monthly review			MMC	1-Feb		1-Feb	
Report on first 3 months			RE	1-Apr		1-Apr	

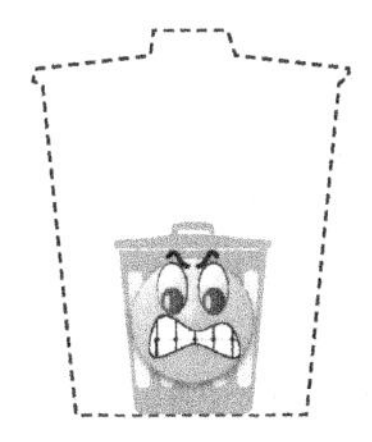

Fig. 9.11 Action Plans cut the waste of "More-is-Better Thinking."

intuitive: it allows the team and the stakeholders to see at a glance if the work is on track. Tracking forecast brings faster recognition of impending issues. The replan date allows the team to recognize past delay and create unambiguous alarms even when the project has slipped from the original plan. It is an artifact well-structured for periods of high change, such as installing new processes or creating a new service or channel, and can work for simpler product or service development (Fig. 9.11).

9.4 Measure traction with the Test Track

The second artifact of this chapter is the Test Track. While the Action Plan and Bowler are in wide use, the Test Track is not. It is meant to fill in the space after change is complete but before value flows reliably enough to measure directly (see Fig. 9.12). Traction is the characteristic of a project that the early results are working: the first few customers are delighted, the expectations of the first 5–10 users of an internal IT improvement were met, or a new review process went well. These are examples of "watch it work," the diligent analysis that the change instituted by the project is working. In most companies, traction is often monitored informally or not at all; if we rely on the Bowler to count revenue, it may be months before we learn about gaps we could have recognized in weeks.

Projects and initiatives commonly follow a path like Fig. 9.13. They begin with *change*: "See it Broken" brings deep understanding of the problem, which leads a process of creating, developing, and solving. They then proceed to a period of *traction*, the period of first prototypes and trials to "watch it work," observing early use cases diligently. After that, the work

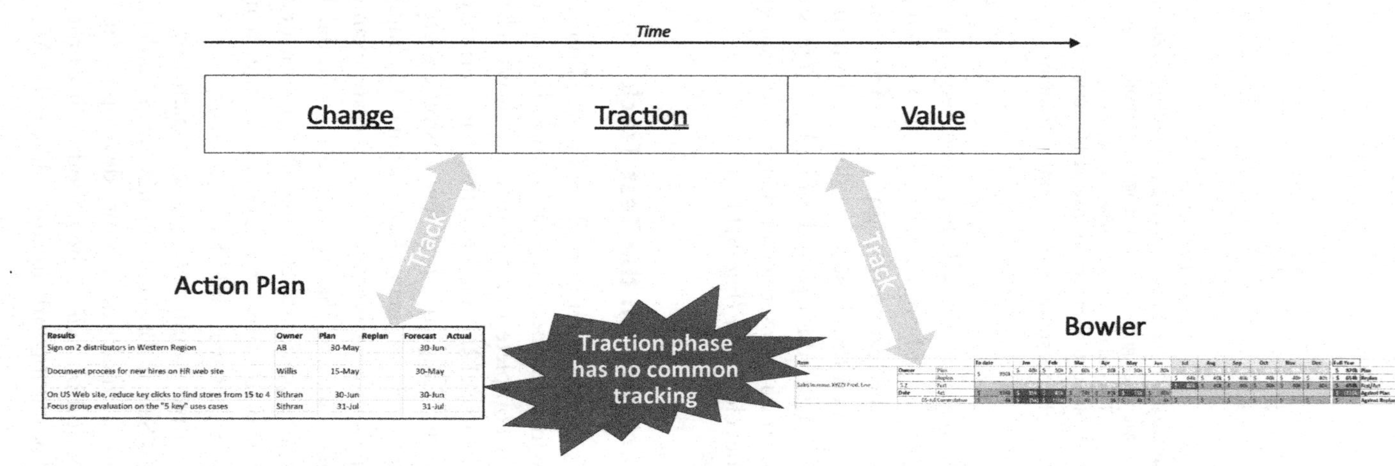

Fig. 9.12 Traction has no common measurement visualization.

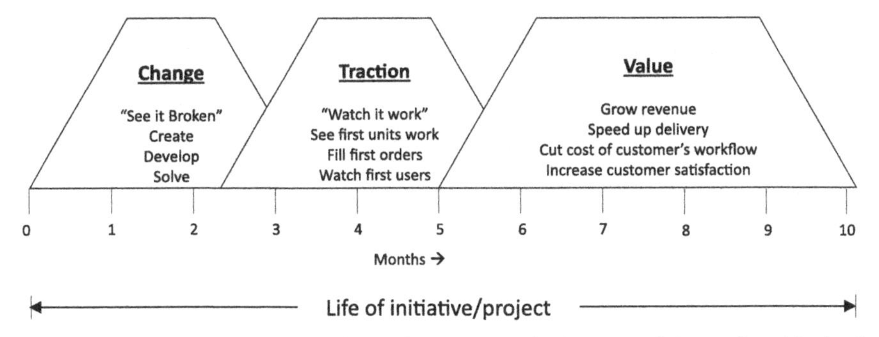

Fig. 9.13 "Traction," monitoring the early successes, fits between "change" and "value."

moves to *value* where metrics can be applied to aggregated data to show benefit to the organization. The Test Track is a kind of insurance that's needed because *change* is rarely perfect and Bowlers reveal that too slowly.

The Test Track is insurance. It's needed because change is rarely perfect and Bowlers reveal that too slowly.

The Test Track has two dimensions as shown in Fig. 9.14:

- "Laps," the stages necessary to graduate to success, one for each column
- "Entrants," the number of cases being monitored, one for each row

Laps are stages of maturity toward success. For example, let's say we have a new service we want to offer in our law firm: trademark support. We'd begin with change: selecting a leader for trademarking activities, putting a team in place, creating standard fees, and so on. If we somehow knew the change was "perfect," complete and valid in every detail, we could just start offering the service and counting the new revenue. But we know change is rarely perfect. Our website might be confusing. We might leave out an important service or offer it at uncompetitive rates. We might lack the necessary infrastructure to deliver trademarks at the pace new clients demand. Or a hundred other things.

The Test Track is a succinct way to measure if our initial cases are meeting expectations. A glance at Fig. 9.14 shows we have a clear plan for the first client, lost a couple of weeks in getting the agreement signed, but success with that client is still expected by the original plan date. This is completed for only one "entrant"; a Test Track will have several clients to gain a more robust measurement derived from broader customer representation.

	Laps				
Entrants	Client identified	Client agreement signed	Client request form done	Application submitted to USPTO	Trademark notification received
Client 1	Plan: May 1 Replan: Fcst: May 1 Act: May 1	Plan: Jun 1 Replan: Jun 15 Fcst: Jun 15 Act:	Plan: Jun 30 Replan: Fcst: Jun 30 Act:	Plan: July 20 Replan: Fcst: Jul 20 Act:	Plan: Sep 20 Replan: Fcst: Sep 20 Act:
Client 2					

Fig. 9.14 Test Track: Measure maturity to success "laps" (the columns) for each race entrant (the rows).

The expected number of entrants is probably between 3 and 10, depending on how diverse the customer base is and how complex it is to run a set of "laps."

To create the laps, we start at the right-hand side—what's the indicator of success for each case? For our trademark service, we might monitor the experiences of three or five customers. That would end perhaps with customers receiving notification of their trademark. Then we would work our way backwards, asking, "What lap comes before that?" For this case, a set of five laps might be:

Lap 5: Notification of trademark
Lap 4: Trademark application submitted
Lap 3: Trademark request template completed by client
Lap 2: Client agreement for trademark complete
Lap 1: Client identified

The Test Track shows "traction." If customers aren't coming to us (first lap), maybe our marketing isn't a good fit or perhaps we're offering the service at uncompetitive rates. If it takes longer than expected to fill in the request (third lap), maybe our template is confusing or incomplete. If it takes too long to create the application (fourth lap), perhaps we lack capacity in our legal team. Or if the trademark office is slow awarding the mark (fifth lap), maybe we have given customers unrealistic expectations or maybe our applications have errors that are delaying the process. The presence of these or many other root causes can be detected quickly with the Test Track. If we went directly to the Bowler, it might take months before significant revenue gaps were recognized because revenue is such a lagging indicator.

The Lean Startup and the Test Track
In 2011 Eric Ries wrote "The Lean Startup: How Today's Entrepreneurs Use Continuous Innovation to Create Radically Successful Businesses." Ries described how to engage customers early in product development, taking early prototypes and inducing customers to use them in real-world conditions. It was one of the rare cases where a book on lean knowledge earned a position on the New York Times Bestseller list. Ries's profound understanding of how often initial assumptions are wrong and how quickly early customer feedback reveals that has changed product development. The Test Track can be a simple visualization to monitor the results of Ries's lean startup method, with laps tracking customer "entrant" responses to different prototypes.

9.5 Measure value with the Bowler

The third artifact of this chapter is value: the revenue we generate, the cost we cut, and the increased the volume of work product we flow to customers. The Bowler is a common visualization of how value is delivered. Time is divided into increments and each column holds data for one increment. In the simple cases, the increments are equal, for example, where each column reflects the data for 1 month in a year-long 12-column Bowler.

9.5.1 Key Performance Indicator (KPI) metrics

A Key Performance Indicator is something that can be counted and compared; it provides evidence of the degree to which an objective is being attained over a specified time [5].

The first discussion about Bowlers is what data they will represent. Just as there seems to be a natural inclination in Action Plans to include excessive detail, the sort of thinking that creates 200-line Action Plans, people often want to collect large amounts of data in long Bowlers. Managing work to 10 or 20 Bowler metrics ignores the basic truth of Goldratt's Theory of Constraints (Section 6.4.5), that there is one bottleneck: focus on that bottleneck to improve performance, because the bottleneck reflects almost all of the issues.

We will focus on Key Performance Indicators (KPIs), the two or three metrics that measure the bottleneck. For example, if I'm driving on a 2-hour trip that I must complete by 5:00 p.m. and time is tight, my bottleneck metric is probably the GPS estimated arrival time. I'll probably watch a couple of other metrics like the amount of fuel (because an empty gas tank renders my GPS estimation invalid) and the speedometer. So, there are three gauges a driver may watch, but the trip is primarily managed to the GPS estimate because it estimates most directly the likelihood of arriving on schedule.

The bottleneck metrics should directly align to value. Five examples are shown for the five facets of value in Fig. 9.15. If a company has determined that there are too many compliance violations, they may start a compliance test program to reduce that number. On the other hand, an excessive number of customer complaints is an issue of fitness for purpose. Table 9.3 gives examples of bottleneck and supporting metrics for several project types.

Keep metrics concrete to drive action. Complex metrics such as net present value (NPV), which accumulates revenue including cost of capital over time, have an initial appeal of creating a measure that takes more factors into account. But they come at a cost of obscuring the call to action. For example, a revenue gap in a simple sum is clearer; add the cost of capital and now perhaps we can spend a month trying to find the right cost of capital and doing nothing to increase revenue. The primary purpose of metrics is to drive action, creating formulas that are accurate, but obscure action is precisely the wrong thing. This is a common weakness in metrics. Taiichi Ohno wrote the same point about "The Blind Spot in Mathematical Calculations" in cost calculation: "Costs do not exist to be calculated; costs exist to be reduced" [6].

When defining metrics, avoid inventing a new measure and, where you must invent, invent as little as possible. Defining metrics is a process of discovery: discover what the organization uses today and squeeze out variation to create a single best way to measure it. For example, if some of your organization reports customer complaints as a percentage and others as a number,

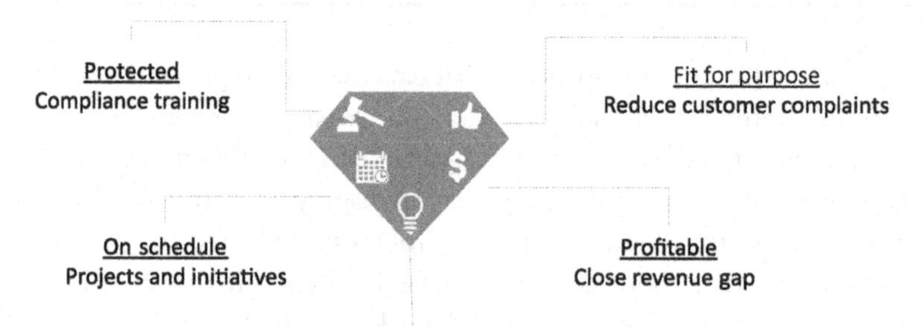

Fig. 9.15 Examples of knowledge work and how bottlenecks would align to the five facets of value (from Fig. 3.1).

Table 9.3 Example project types with metrics.

Type of knowledge work	Bottleneck	Value	Supporting metrics
Reduce travel cost	Travel cost	Profitability	Travel costs for noncustomer-related trips
Medium and large project (Chapters 19 and 20)	Days to schedule completion	On-schedule delivery	Days forecast late
Small projects (Chapter 21)	Projects completed per month	On-schedule delivery	
Reduce errors reported by clients	# Client-reported errors	Fitness for purpose	
Close revenue gap (Chapter 23)	Revenue gap	Profitability	Bookings, # new customers

use the metric discussion to pick one, not because one is necessarily better than the other, but because having a single way is better than having two ways. But don't use the creation of a new metric as a reason to add new complexity. Good metrics reflect the values of the organization; they don't create them. Finally, all things being equal, choose the alternative that is most concrete, avoiding percentages, differences, and ratios.

9.5.2 The basic Bowler

A basic Bowler is shown in Fig. 9.16. It tracks a KPI called "Sales Increase, XYZZY Product Line." Similar to the Action Plan, a good Bowler tracks a measurable KPI with a clear owner. An owner must [7]:

1. be able to take actions that influence the KPI;
2. contribute to the value that the KPI is driving;
3. be fully bought into the KPI;
4. deeply understand the data sources that contribute to the KPI; and
5. keep the KPI up to date.

The example Bowler shows data in two rows: one (above) for plan and the other for actual; each column of data represents one month. The Bowler of Fig. 9.16 adds a third row, a measure of cumulative gap to plan. It also adds a "to date" column (left of January) to sum all past data; in this case, January to June. Bowlers need explicit thresholds defined. Here we choose the simplest threshold with two colors: red (dark gray in the print version) when unfavorable

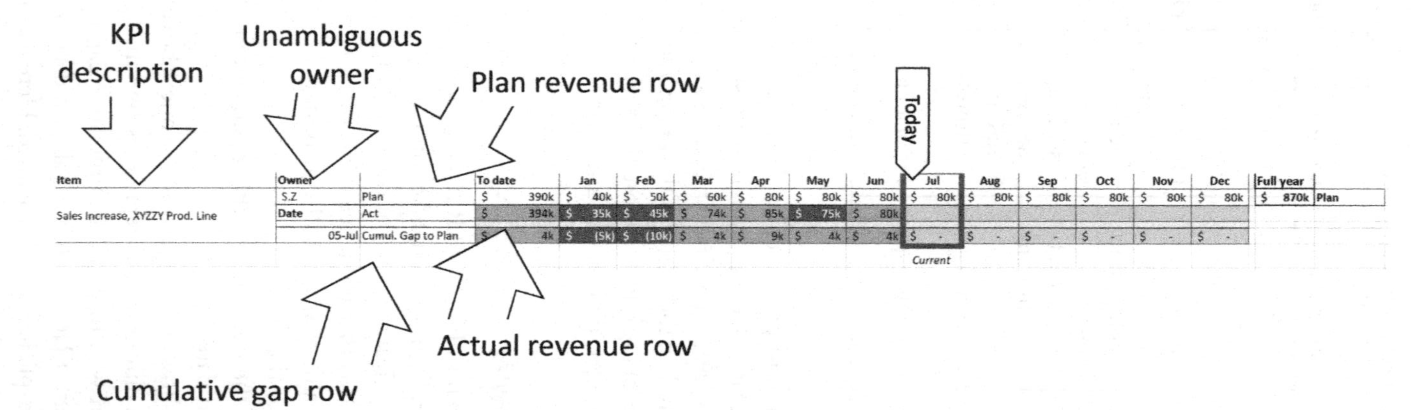

Fig. 9.16 Basic Bowler.

and green (gray in the print version) otherwise. Another common threshold method uses three colors: setting a margin for green (gray in the print version) (say, within 5% of target), red (dark gray in the print version) when the KPI is well below acceptable (say, 20%), and yellow (light gray in the print version) in between [8].

As with the other artifacts, the absence of a forecast row subjects readers to the *milestone delay problem* of Section 9.3.3. For example, let's take the case where a major customer has decided to exit the industry that uses the XYZZY product line. Further, perhaps we have some other bad news indicating that another customer has left XYZZY, choosing a competitor instead. As it turns out, neither event has much effect until July. The Bowler shows no problem because there is no forecast. Thus, we will not see an identified problem in the Bowler until the events pass. Of course, there will likely be conversations on these types of looming problems, but narration that contradicts visualization is weak; better to have the Bowler reflect known issues so that narration and visualization are aligned. This is the topic of the next section.

9.5.3 Adding the forecast row

Adding the forecast row to the Bowler provides an opportunity to represent known issues before the effects are reflected in past measurements, avoiding the *milestone delay problem*. The forecast line in Fig. 9.17 reflects the imminent loss of the two customers that went unseen in the basic Bowler of Fig. 9.16. It also allows a new entry on the right: the full year forecast/actual, which sums the actual data from past months and forecast data from future months. This is often the single best summation to indicate if the Bowler metric is on track. In our example, the new forecast row (far right) shows we are off track, forecasting a $216k shortfall.

9.5.4 Adding replan row

As with the Action Plan and Test Track, the forecast row can create an unwinnable contest. In our example, while we would doubtless take steps to improve XYZZY revenue, the likelihood of fully recovering from two lost major customers in a few months is low. This is because of the long cycles in gaining (or winning back) customers. It may be that we need to make significant changes in the product, such as adding features or changing the manufacturing model to be able to offer the products at a lower price while staying profitable. Nothing in the Bowler says we should not take those sorts of steps, but the forecast row should be transparent, showing when those actions are unlikely to bring full resolution. The value of the Bowler is that it informs the

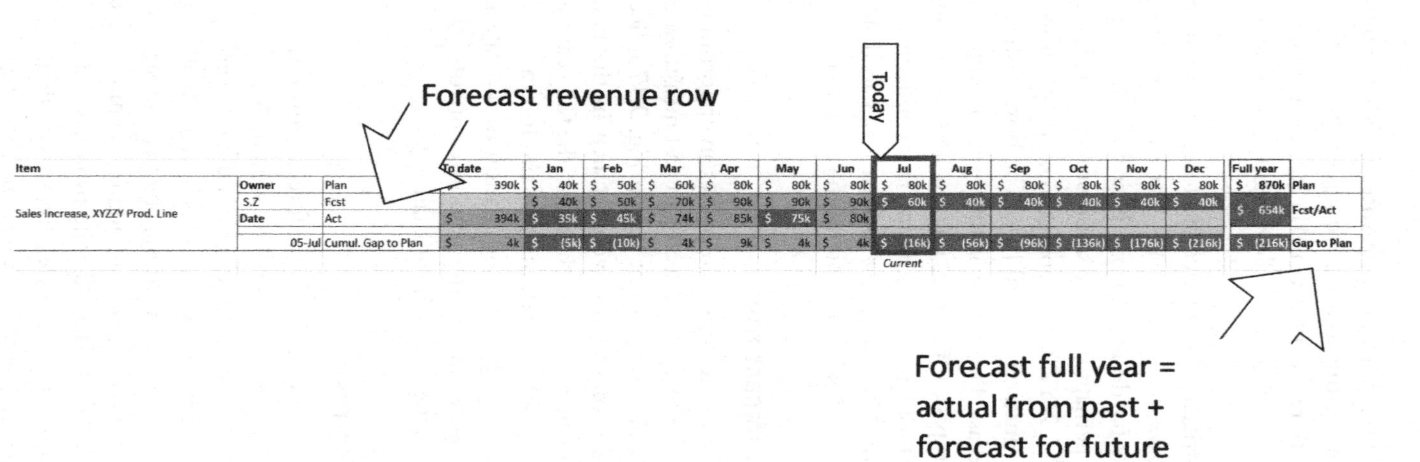

Item			To date	Jan	Feb	Mar	Apr	May	Jun	Jul	Aug	Sep	Oct	Nov	Dec	Full year	
Sales Increase, XYZZY Prod. Line	Owner	Plan	390k	$ 40k	$ 50k	$ 60k	$ 80k	$ 80k	$ 80k	$ 80k	$ 80k	$ 80k	$ 80k	$ 80k	$ 80k	$ 870k	Plan
	S.Z	Fcst		$ 40k	$ 50k	$ 70k	$ 90k	$ 90k	$ 90k	$ 60k	$ 40k	$ 40k	$ 40k	$ 40k	$ 40k	$ 654k	Fcst/Act
	Date	Act	$ 394k	$ 35k	$ 45k	$ 74k	$ 85k	$ 75k	$ 80k								
	05-Jul	Cumul. Gap to Plan	$ 4k	$ (5k)	$ (10k)	$ 4k	$ 9k	$ 4k	$ 4k	$ (16k)	$ (56k)	$ (96k)	$ (136k)	$ (176k)	$ (216k)	$ (216k)	Gap to Plan

Fig. 9.17 Forecasts in Bowlers expose gaps rapidly.

organization of an impending $200k shortfall that must be made up elsewhere. However, it also turns the Bowler red (dark gray in the print version) for the remainder of the year. No matter what the sales team does, they will lose. This encourages **More-is-Better Thinking**: just grind away knowing you're going to lose.

The replan row added in Fig. 9.18 strikes a balance of creating a contest we can win in weekly management, but still indicating that we didn't deliver the revenue from XYZZY that we needed to be successful. The replan is red (dark gray in the print version) because it shows that we are below plan. But forecast cells after August are now painted yellow (light gray in the print version), something between the red (dark gray in the print version) of outright losing (missing the forecast) and green (gray in the print version) of outright winning (meeting plan). It shows we have a gap that has been acknowledged and negotiated between knowledge staff and leadership.

9.5.5 Summary of Bowlers

The Bowler cuts the waste of **More-is-Better Thinking** by maintaining a definition of success with measured deliverables, clear ownership, and unambiguous targets. As with Action Plans, tracking forecast brings faster realization of impending issues. Similarly, the replan row allows recognition of negotiated performance gaps and creates a clear measurement of success going forward. It is an artifact well-structured for periods of value measurement in the later stages of a project; it is closely aligned to value and thus becomes more relevant as the work delivers value to the organization (Fig. 9.19).

9.6 The Success Map

Mr. Ohno was passionate about TPS,[c] He said you must clean up everything so you can see problems. He would complain if he could not look and see and tell if there is a problem.

Fujio Cho, President, Toyota Motor Company [9]

The *Success Map* is a single display of all three artifacts of this chapter and so is relevant for the life of the project. Each artifact includes a "Stop-Fix" alarm (see Section 14.3 for detail), an indication we will correct problems quickly. The Stop-Fix alarm indicates a subset of issues, those for which our structure ensures we have a means to address. In each of the artifacts, forecast being

[c] TPS is the Toyota Production System, Toyota's brand of lean thinking.

Item			To date	Jan	Feb	Mar	Apr	May	Jun	Jul	Aug	Sep	Oct	Nov	Dec	Full Year	
Sales Increase, XYZZY Prod. Line	Owner	Plan	$ 390k	$ 40k	$ 50k	$ 60k	$ 80k	$ 80k	$ 80k	$ 80k	$ 80k	$ 80k	$ 80k	$ 80k	$ 80k	$ 870k	Plan
		Replan								$ 64k	$ 40k	$ 40k	$ 40k	$ 40k	$ 40k	$ 654k	Replan
	S.Z	Fcst		$ 40k	$ 50k	$ 70k	$ 90k	$ 90k	$ 90k	$ 60k	$ 40k	$ 40k	$ 40k	$ 40k	$ 40k	$ 654k	Fcst/Act
	Date	Act	$ 394k	$ 35k	$ 45k	$ 74k	$ 85k	$ 75k	$ 80k								
	05-Jul	Cumul. Gap to Plan	$ 4k	$ (5k)	$ (10k)	$ 4k	$ 9k	$ 4k	$ 4k	$ (60k)	$ (100k)	$ (140k)	$ (180k)	$ (220k)	$ (260k)	$ (216k)	Gap to Plan
Sales Increase, Cluck-and-Neigh Prod. Line	Owner	Plan	$ 800k	$ 40k	$ 60k	$ 100k	$ 120k	$ 80k	$ 80k	$ 80k	$ 80k	$ 80k	$ 80k	$ 80k	$ 80k	$ 960k	Plan
		Replan								$ 64k	$ 40k	$ 40k	$ 40k	$ 40k	$ 40k	$ 744k	Replan
	S.Z	Fcst		$ 80k	$ 50k	$ 80k	$ 90k	$ 90k	$ 90k	$ 60k	$ 40k	$ 40k	$ 40k	$ 40k	$ 40k	$ 724k	Fcst/Act
	Date	Act	$ 464k	$ 45k	$ 45k	$ 74k	$ 120k	$ 100k	$ 80k								
	05-Jul	Cumul. Gap to Plan	$ (336k)	$ 5k	$ (10k)	$ (36k)	$ (36k)	$ (16k)	$ (16k)	$ (80k)	$ (120k)	$ (160k)	$ (200k)	$ (240k)	$ (280k)	$ (236k)	Gap to Plan

Fig. 9.18 Adding a replan row allows expectations to be adjusted so that there is always a contest we can win.

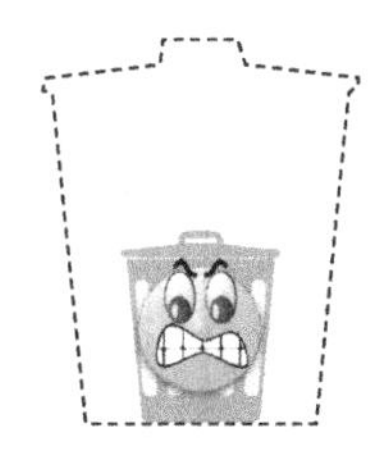

Fig. 9.19 Bowlers cut waste by tracking actual and forecast of aggregated data.

behind plan (or, if it exists, replan) is a Stop-Fix alarm. The team always has three ways to remedy the alarm:

- Correct the immediate issue, addressing the issue in isolation. If that doesn't work…
- *Replan* to address the problem while meeting the overall goal. If that doesn't work…
- Renegotiate stakeholder expectations and reflect reduced expectations in replanned goals.

With this approach, a project can be laid out from beginning to end with success mapped in a way that fits the growing maturity of the project, from change to traction to value, as shown in Fig. 9.20.

9.6.1 The ground view

The ground view displays the creation of work product from the perspective of the team doing the work. Fig. 9.21 shows an example, a high-level single-page canvas with an overview on the left and a *Success Map* on the right.

	Type	Time horizon	Visualization
	Artifacts of Change	Development	Action Plan
	Artifacts of Traction	Validation	Test Track
	Artifacts of Value	Sustainment	Bowlers

Fig. 9.20 The three artifacts of the Success Map.

Initiative	Upgrade Video Conferencing Capability	Status	4-Success Map tracking	Last update	1-May	Update due	

Guide (Sponsor)

Sponsor	Justin
Explain need	We need to be able to hold virtual meetings so we can fully include people remotely in meetings as standard
Need-by date	6 weeks
Owner name	Grainne
Team members	Grainne, Suneel

Focus

Problem or opportunity	We cannot communicate video information in most of our meetings with remote people
Metric	% of meetings that use video information
Current perf.	~10%
Target perf.	85%
Solve method	Informal problem solving in team

Solve

Root causes		Video information cannot be added outside of main site due to security issues
Solution		Choose application that meets IT security requirements that can be installed on all R&D mobile/desktop systems and those of closely related functions
Countermeasures	1	Select SW that is practical to install on a large number of machines
	2	IT to allow connection from inside and outside of main site
	3	

Action Plan

Deliverable	Next Action	Date	Owner	Plan	Replan	Fcst	Actual
Select application	Define criteria for selection	15-Mar	Grainne	16-May		16-May	
Prototype meeting	Install on 8 machines (remote and	15-May	Suneel	30-May		30-May	
Develop training material			Grainne	6-Jun		6-Jun	
Hold 2 training sessions, record 1			Grainne	13-Jun		13-Jun	
Ensure install of new SW on >85%			Grainne	4-Jul		4-Jul	

Test Track

Case	Dates	Lap 1	Lap 2	Lap 3	Lap 4	Lap 5
Design review, project XYZZY (prototype meeting)	Plan	1-May	15-May	22-May		
	Replan					
	Forecast	1-May	15-May	22-May		
	Actual					
Patent application meeting	Plan	11-May	25-May	1-Jun		
	Replan					
	Forecast	11-May	25-May	1-Jun		
	Actual					
Monthly project review	Plan	19-May	2-Jun	9-Jun		
	Replan					
	Forecast	19-May	2-Jun	9-Jun		
	Actual					
	Plan					
	Replan					
	Forecast					
	Actual					

Bowlers

Goal 1	Meetings held with video											
Date	1-May	1-Jun	1-Jul	1-Aug	1-Sep	1-Oct	1-Nov	1-Dec	1-Jan	1-Feb	1-Mar	1-Apr
Plan	2	5	10									
Replan												
Fcst	2	5	10									
Act												

Goal 2												
Date	1-May	1-Jun	1-Jul	1-Aug	1-Sep	1-Oct	1-Nov	1-Dec	1-Jan	1-Feb	1-Mar	1-Apr
Plan												
Replan												
Forecast												
Actual												

Fig. 9.21 The ground view summarizes the work on the left and on the right has a Success Map, a combination of three artifacts.

This enables stakeholders and those outside the core team to understand the context and determine if the project is on track in just a few seconds. And because the primary artifacts shift from Action Plan to Test Track to Bowlers as the project matures, the *Success Map* gives a view that is relevant from beginning to end. These three artifacts start every initiative defining an unambiguous path to success for the entire initiative, independent of whether the initiative is a new product, a new service, or a new process.

9.7 The helicopter view

The *helicopter view* is a canvas that shows many projects. First we must understand why we would need such a view. A team of knowledge staff is usually expected to be working on many types and instances of work product at one time. I recall once when we started to build a helicopter view for an R&D organization of about 100 people. We took 2 or 3 months to document every project and initiative. Originally, we thought there were 20 or so; we ended up identifying about 75. Having a large number of disparate initiatives and projects is the norm in knowledge work. Fig. 9.22 shows the concept that four separate initiatives/projects of widely differing types run using the same ground view format, understanding the real number, can be much larger than four.

The helicopter view summarizes many ground views to reveal how an organization is performing. The helicopter view pulls data from the many ground views. Done well, it pulls live data automatically, so these views are always accurate and up to date. Today, helicopter views are available in narrowly defined work. For example, if all your team's work is in Microsoft Project (a Gantt chart-based project management system), there are options to provide a rapid, error-proofed helicopter view of every project in the portfolio. Similarly, a software team using Agile project management such as Jira (https://www.atlassian.com) or VersionOne (https://www.collab.net/products/versionone) can use packages from the supplier to provide both ground and helicopter views. But for those groups that execute disparate types of work there is not today an accepted way to aggregate work the way disparate manufacturing work can.[d]

[d] For example, the QDIP (Quality-Delivery-Inventory-Productivity) is a single way to measure "success" in manufacturing cells of all types: assembly, machining, fabrication, packaging painting, and visually any other manufacturing process. Moreover, QDIP measurements can be aggregated for an entire manufacturing organization.

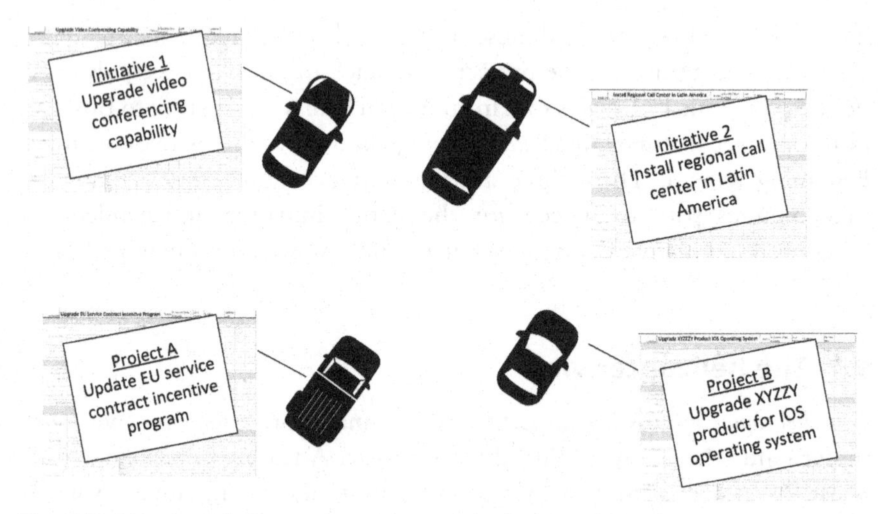

Fig. 9.22 Most knowledge work occurs as multiple ground views.

But what about the common case where a knowledge organization has multiple classes of work? The R&D organizations I have worked in simultaneously had project management systems for large projects, Action Plans for small ones, Agile Scrum projects for software, annual review management systems, patent management systems, and many other internal management systems customized for the company. Unfortunately, there are few if any tools for the general case. If general helicopter views are created at all, they are usually formed by manually processing the most important ground views—for example, cutting and pasting emailed Excel files. In those cases, manually stitching data together can result in views that are so out of date that they are unreliable for important decisions. As Keyte says, "Spreadsheets are notoriously difficult to centralize and maintain" [10].

An example helicopter view is shown in Fig. 9.23. I've found the helicopter view that is connected directly to ground views to be a marvelous source of data, able to help provide oversight. Excellent helicopter views directly pull data fully owned by initiative/project managers in seconds.

Like a traffic helicopter, the helicopter view monitors many drivers, but without directly controlling any. In this model, the initiative leader owns the data for his or her initiative—the leader is responsible for accurate and timely updates. As shown in Fig. 9.24, data ownership remains with the ground view—with the people who lead initiatives and projects. Only the owner can write data, but read access to that data is granted to all stakeholders for both the ground and helicopter views.

However elusive the comprehensive, rapid, error-proofed helicopter view may be, when it is available, it cuts the waste of **More-is-Better Thinking**

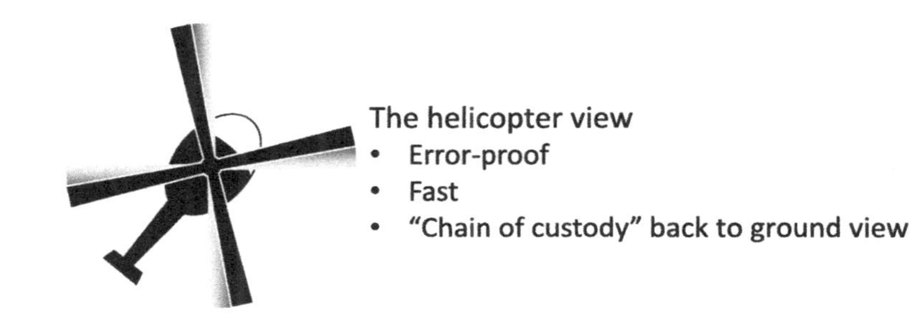

				Jan-2019	Feb-2019	Mar-2019	Apr-2019	May-2019	Jun-2019	Jul-2019	Aug-2019	Sep-2019	Oct-2019	Nov-2019	Dec-2019	Total	
			Plan	3,659	659	31,424	25,139	25,139	31,424	25,139	26,338	32,923	26,338	26,338	32,099	$286,617	
			Fcst	3,659	659	31,424	25,139	25,139	31,424	25,139	26,338	32,923	26,338	26,338	32,099	$286,617	
			Act	-	-	-	-	-	-	-	-	-	-	-	-	$0	
	Done		Plan	3,000	-	-	-	-	-	-	-	-	-	-	-	$3,000	Done
	100%		Fcst	3,000	-	-	-	-	-	-	-	-	-	-	-	$3,000	Done
			Act	-	-	-	-	-	-	-	-	-	-	-	-	$0	Done
	Active		Plan	-	-	34,000	27,200	27,200	34,000	27,200	28,532	35,666	28,532	28,532	35,666	$306,529	Active
	90%		Fcst	-	-	34,000	27,200	27,200	34,000	27,200	28,532	35,666	28,532	28,532	35,666	$306,529	Active
			Act	-	-	-	-	-	-	-	-	-	-	-	-	$0	Active
	Waiting		Plan	1,098	1,098	1,373	1,098	1,098	1,373	1,098	1,098	1,373	1,098	1,098	-	$12,902	Waiting
	60%		Fcst	1,098	1,098	1,373	1,098	1,098	1,373	1,098	1,098	1,373	1,098	1,098	-	$12,902	Waiting
			Act	-	-	-	-	-	-	-	-	-	-	-	-	$0	Waiting
	Stage	**Impl. Date**		**Jan-2019**	**Feb-2019**	**Mar-2019**	**Apr-2019**	**May-2019**	**Jun-2019**	**Jul-2019**	**Aug-2019**	**Sep-2019**	**Oct-2019**	**Nov-2019**	**Dec-2019**		**Backlog date**
Project XYZZY	Waiting	12-2018	Plan	1,098	1,098	1,373	1,098	1,098	1,373	1,098	1,098	1,373	$1,098	$1,098			Waiting
			Fcst	1,098	1,098	1,373	1,098	1,098	1,373	1,098	1,098	1,373	$1,098	$1,098			Waiting
			Act														Waiting
Project Cardie Beagle	Active	07-2019	Plan								666	833	$666	$666	$833		Active
			Fcst								666	833	$666	$666	$833		Active
			Act														Active
Project Heavy Gravy	Done		Plan	3,000													Done
			Fcst	3,000													Done
			Act														Done
NCC1701b	Active	02-2019	Plan			10,150	8,120	8,120	10,150	8,120	8,120	10,150	$8,120	$8,120	$10,150		Active
			Fcst			10,150	8,120	8,120	10,150	8,120	8,120	10,150	$8,120	$8,120	$10,150		Active
			Act														Active
Wacka Mole Project	Active	02-2019	Plan			13,700	10,960	10,960	13,700	10,960	10,960	13,700	$10,960	$10,960	$13,700		Active
			Fcst			13,700	10,960	10,960	13,700	10,960	10,960	13,700	$10,960	$10,960	$13,700		Active
			Act														Active
Project Beam Burger	Active	07-2019	Plan								666	833	$666	$666	$833		Active
			Fcst								666	833	$666	$666	$833		Active
			Act														Active
Jimmy Chew Project	Active	02-2019	Plan			10,150	8,120	8,120	10,150	8,120	8,120	10,150	$8,120	$8,120	$10,150		Active
			Fcst			10,150	8,120	8,120	10,150	8,120	8,120	10,150	$8,120	$8,120	$10,150		Active
			Act														Active

Fig. 9.23 The helicopter view combines multiple instances into one visualization.

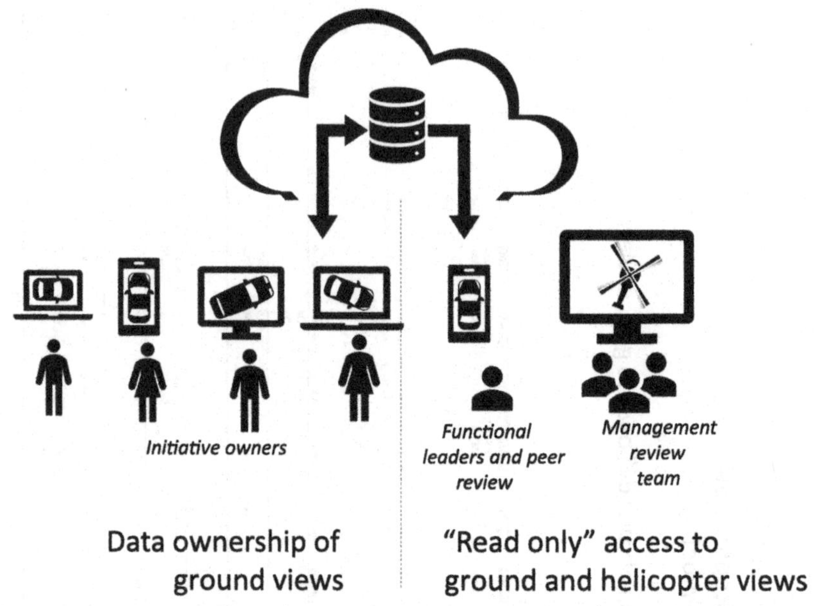

Fig. 9.24 Teams own ground views; leaders typically have access to ground and helicopter views.

by supporting more frequent and reliable leadership reviews. This drives faster remediation of issues, which then leads to higher accountability in the reported information. If you know your leadership team will review reported information frequently, you're probably going to make a greater effort to keep the data up to date. The availability of highly reliable data leads to the team having more data, and thus fewer issues escape the team's notice in the first place. As Fig. 9.25 shows, the improvements of reliable, fast helicopter views create a spiral of waste-cutting from leadership engaging more often to project/initiative managers actively driving down gaps to teams being aware of gaps and addressing more often without having to be directed

A universal measure for knowledge work to cut More-is-Better Thinking?

An intriguing area of knowledge work is creating a universal metric for work of all kinds. In manufacturing, there are well-known universal measures that can be applied to operations of every type: productivity, inventory cost, and defect rate are a few. These measures can be combined at every level, for example, a single manufacturing cell, a factory, all factories in a region, and globally for the whole organization.

Could knowledge work be scored the same way? I think so! The three artifacts of this chapter with Stop-Fix alarms can be applied to an enormous

Continued

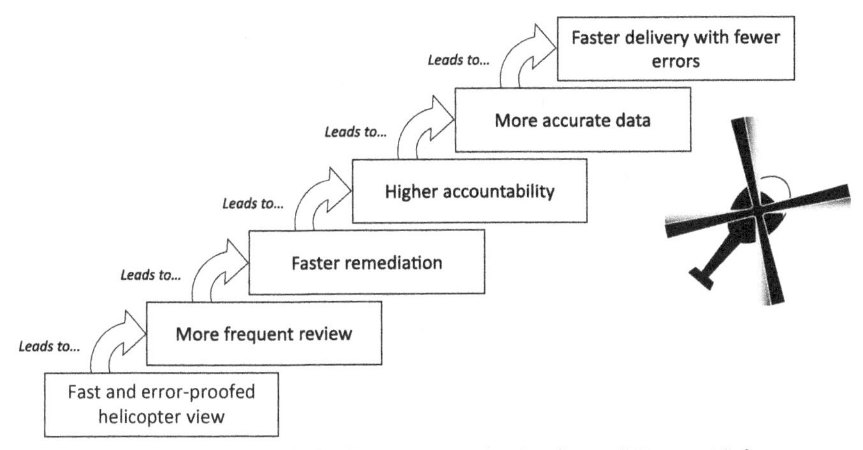

Fig. 9.25 Fast and error-proofed helicopter views lead to faster delivery with fewer errors.

breadth of knowledge work: small team projects, legal work, finance, product marketing, and so on. The requirement is that the plan be defined; from that point, it's easy to imagine a universal view that combines every type of work from a 10-hour, one-person document creation to a 30,000-hour product development. Imagine a department leader who could pull up all the work in his or her department with up-to-the-minute data because it pulls from what people work from. Unfortunately, most of today's workflow management software creates isolated work with incompatible measures, formats, and storage.

At two points in my career, we developed custom solutions that did just that. In both cases, we mixed many project types into a single format the way the Success Map can. The results were convincing. I was able to manage perhaps five times as much work because of the extreme clarity of Stop-Fix alarms funneled to a single helicopter view for the department. I spent more of my time helping improve work and less time fumbling through data. Unfortunately, I've yet to see a commercially available package that supports this and that is simple enough to install and sustain without demanding excessive IT resources.

9.8 Conclusion

In this chapter, we have worked to reduce **More-is-Better Thinking**, the wasteful mindset of grinding away at a task with no expectation of success. **More-is-Better Thinking** mixes hard work with lazy thinking, and often results in knowledge staff who are exhausted from constant busyness, all the while producing disappointing results. As shown in Fig. 9.26, the techniques

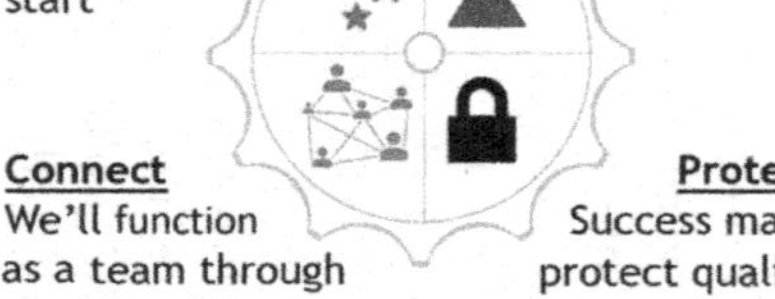
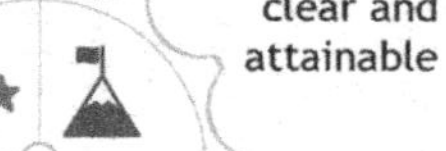
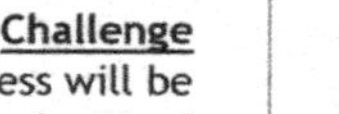

Fig. 9.26 Reducing More-is-Better Thinking simplifies, engages, and encourages experimentation.

in this chapter simplify workflow. The Test Track standardizes the measurement of traction, the Success Map measures performance to goals over the life of a set of work, and the helicopter view shows an entire portfolio at a glance. These techniques engage by creating shared vision of being able to finish tough work and doing so as a team. It also creates a measurable and clear challenge. Finally, the consistent use of a Success Map can raise performance to meet organization standards. And the Success Map forms an experiment, asking the team to lay out a set of goals over a long period of time and then constantly measuring performance to those goals to create a rich learning environment.

References

[1] https://quotes.deming.org/authors/W._Edwards_Deming/quote/10084.
[2] http://www.businessdictionary.com/definition/action-plan.html.
[3] W.E. Deming, Out of the Crisis, Massachusetts Institute of Technology, Center for Advanced Engineering Study, Cambridge, MA, 1986, 358.
[4] S.B. Wilson, M.S. Dobson, Goal Setting: How to Create an Action Plan and Achieve Your Goals, (2008) 5.https://www.questia.com/library/118621796/goal-setting-how-to-create-an-action-plan-and-achieve.
[5] C. Keyte, Developing Meaningful Key Performance Indicators, Intrafocus, 2014 Kindle Edition. Location 39.
[6] T. Ohno, Taiichi Ohnos Workplace Management, in: Special 100th Birthday Edition, McGraw-Hill Education, Kindle edition, 2012, p. 23.
[7] C. Keyte, Developing Meaningful Key Performance Indicators, Intrafocus, 2014 Kindle Edition. Location 329.
[8] C. Keyte, Developing Meaningful Key Performance Indicators, Intrafocus, 2014 Kindle Edition. Location 445.
[9] J. Liker, The Toyota Way: 14 Management Principles From the World's Greatest Manufacturer, McGraw-Hill 149 (2003).
[10] C. Keyte, Developing Meaningful Key Performance Indicators, Intrafocus, 2014 Kindle Edition. Location 552.

CHAPTER 10

Reduce Waste #4: Inertia to Change

Two basic rules of life are: 1) Change is inevitable. 2) Everybody resists change.

W. Edwards Deming [1]

10.1 Introduction

In this chapter, we will address the fourth waste, **Inertia to Change**, which is the waste that occurs when people behave in ways that present barriers to improvement, whether by intention or not. We'll begin by looking at the change model, a popular way to explain how people accommodate change. Then we will look at how leaders right-size change for those in their organization.

10.2 Change model

It's easier to act your way to a new way of thinking than to think your way to a new way of acting.

John Shook [2]

The change model demonstrates how adults deal with change. We have a keen interest in this topic because improvement brings change, often as much change as many people can deal with. We begin with two ways of thinking about driving change. In Fig. 10.1, which is based on John Shook's insights, the old way of thinking is compared to the new (lean thinking) way. In the old way, we sought first to change culture which, over time, would change values and attitudes; only then did behavior change. In Shook's new model, people must first try a new way of doing something—they must

Improve
https://doi.org/10.1016/B978-0-12-809519-5.00010-7

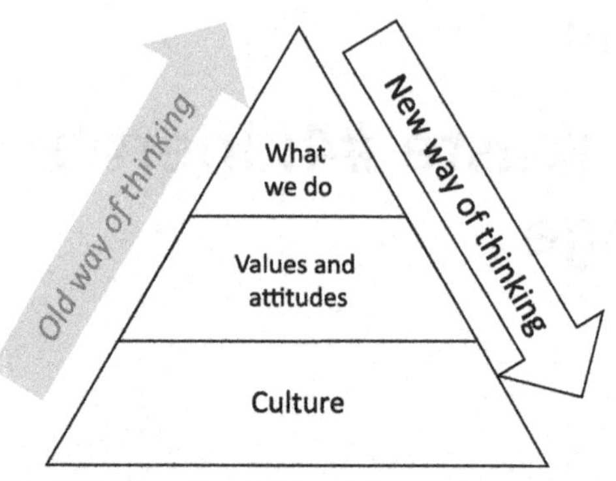

Fig. 10.1 Lean thinking: Behavior changes culture.

experiment and see the results. Then the values and attitudes will change. Finally, the culture of the organization will change [2, 3].

> *You cannot create experience. You must undergo it.*
>
> ***Albert Camus, 20th-century French philosopher***

A couple of examples may help. Let's say you used to work for an organization that held daily stand-up meetings—10- or 15-minute meetings first thing in the morning to get everyone on a project team aligned for the day. If you've seen it work, you understand how powerful that 10- or 15-minute meeting can be. It can save hours of waste by cutting the **Information Friction** that occurs when people don't talk through handoffs, changes, and unexpected events. Now, you work for an organization that has never held stand-up meetings. You try to change the culture by explaining the power of what you've seen. To your surprise, the people you talk with see it as yet another meeting in a day full of meetings. They might think people resolve things well with phone calls and emails. You can explain as many times as you like with the strong, logical arguments to no effect. Shook's "new way of thinking" from Fig. 10.1 recommends changing behavior first. In this case, pilot the technique with one team for a few months. The people on that team will then understand, allowing you to deploy it elsewhere. If it's well-applied, within 6 or 9 months other teams will want to use it. A year later, the culture of the organization will be to hold high-cadence stand-up meetings.[a]

[a] The *regular team stand-up* is the topic of Section 19.3.

Another example is "standard work," a topic that often invites yawns from experts and knowledge staff. How could an innovative person embrace standard work? Until you experience the way standard work simplifies tedious tasks, you cannot believe that it can cut so much waste and frustration. Again, someone can explain and argue, and they can point to examples or experience, but, outside of change agents, people are unlikely to make the mental leap. So, start by improving a relatively simple but frustrating process in your organization; approvals of any type—purchase, capital, or hiring—are usually a good place. As people see approvals get faster and less frustrating, they will understand the power of standardizing work. Slowly, the culture will change so that people will start asking of frustrating processes: why don't we create new standard work for this?

So, to bring about change, we must guide people to do something different and we shouldn't wait for them to understand all the implications before they start. Quite the opposite. It is precisely by doing something different that we learn. This is a form of experimental thinking: there is a hypothesis that, by taking a specific action, we will improve; when people join in and it works, they understand deeply what happened because they experienced it.

I hear and I forget. I see and I remember. I do and I understand.

Attributed to Confucius

Fig. 10.2 shows the typical feelings a person can have when experiencing change, referred to sometimes as the "change model." It is based loosely on

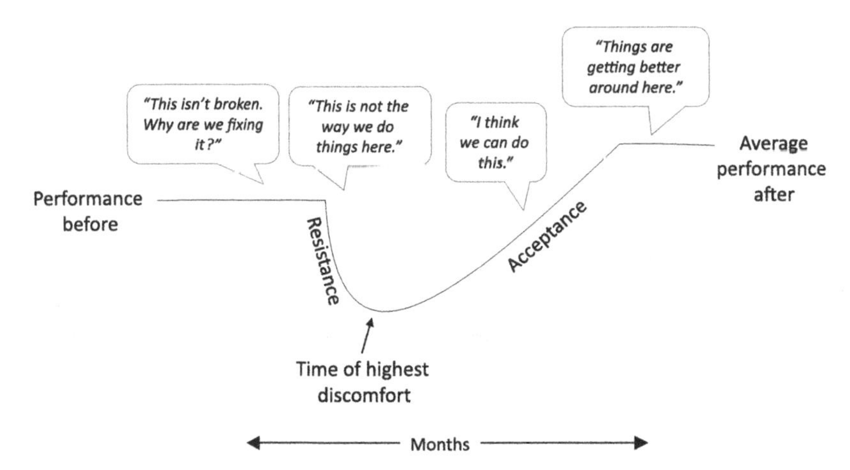

Fig. 10.2 The change model.

the work of Elisabeth Kübler-Ross, who identified five stages of grief: denial, anger, bargaining, depression, and acceptance; this model has been often applied to people dealing with change in a work environment. At the start, employees probably sense no need for the change. You may ask a team member to learn a new technique or lead their first improvement cycle. The more different what they are being asked to do is from what they have done in the past, the more likely they will be to resist the change. They might wonder whether it makes sense to change something that seems to work fine. Of course, the problem can be that something so familiar to them, something that has always been like this, might be working poorly in ways they don't comprehend. But it's not possible to comprehend something until you've spent time learning it–something you are unlikely to do until you're convinced you need it. This is what Shook meant in the quote that opened this section: it's hard to think your way into a new way of doing.

People who are satisfied with the way things are can never achieve improvement or progress.

Shigeo Shingo [4]

The second stage of the change model is resistance. At first, I needed only to consider doing something new. Now, I have to work hard on something I don't think we need. For a time, the experience is frustrating and uncomfortable. But if the technique is coached well, good results inevitably come. Seeing the sorts of results that are new to the person experiencing the change pulls down the barriers to change. The person starts to see that this really does work and, moreover, that the old way wasn't as good as it seemed. At the end of a successful change cycle, most people will see the benefits of the change. They emerge with a better understanding of the way it was and the way it can be. They are energized and their performance rises above what it was originally, because they are capable of deeper understanding and more effective action. Those who have led change know that resistance to change is normal; acceptance must be earned.

Resistance to change is normal. Acceptance must be earned.

10.3 Growth through challenge

> *When what they must do exceeds their capabilities, the result is anxiety. When what they must do falls short of their capabilities, the result is boredom...But when the match is just right, the results can be glorious. This is the essence of flow.*
>
> ***Daniel Pink (referring to Csikszentmihalyi's work) [5]***

The most talented people in your organization won't be satisfied if they are not growing. That's probably how they became the most talented people in your organization: a drive to know more, to do more, to be more. They rely on their leaders more in this area than most others, as measured by Google inside their organization (Section 6.4.3.2) [6]. As shown in Fig. 10.3, people grow when they are given the right amount of challenge: too easy and they get bored; too challenging and they are overwhelmed.

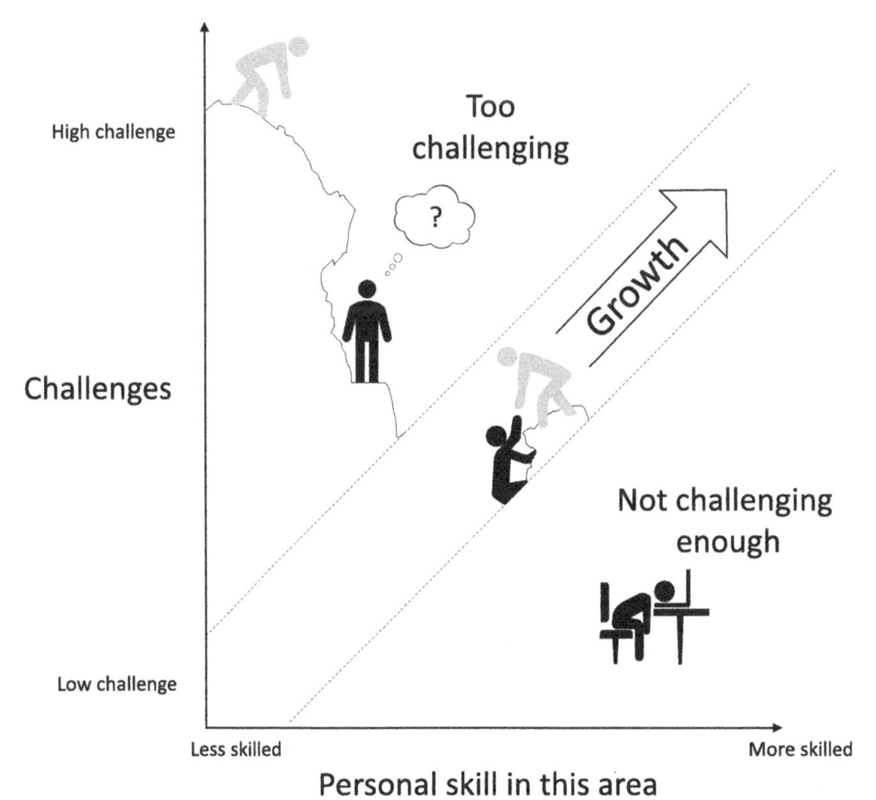

Fig. 10.3 Match challenge against individual capability to nurture growth in an area.

> *We are all growing and learning, and we all need teachers and coaches to help guide us. We say at Toyota that every leader is a teacher developing the next generation of leaders. This is their most important job.*
>
> ***Akio Toyoda, President, Toyota Motor Company [7]***

Mihaly Csikszentmihalyi describes *flow*, something closely related to growth, as what makes creative people satisfied with their work. He defines it as an almost magical state where creative people are so engaged in their craft that they feel a state of ecstasy, losing sense of time. To achieve flow, Csikszentmihalyi says the task must be difficult, but not so difficult that success is unlikely. If a task is just out of reach, we are pushed out of our comfort zone, which is just what's needed to acquire new skills rapidly.

Superior Managers for Superior Results

Several years ago, Gallup embarked on an ambitious analysis collecting information from almost 2 million employees working in more than 300,000 business units. It concluded that productivity hinges on the quality of managers, with a correlation so large that Gallup called it "the single most profound, distinct and clarifying finding" in its history of 80 years. "The study showed that managers didn't just influence the results their teams achieved; they explained a full 70% of the variance. In other words, if it's a superior team you're after, hiring the right manager is nearly three-fourths of the battle." They concluded that no other factor came close, not compensation and not the perception of the organization's leadership [8]. Jim Clifton, Gallup's chief executive said, "That blew me out of my chair."

Flow creates happiness for knowledge staff. Csikszentmihalyi quotes a former CEO of Lockheed Martin, Norman Augustin, as fitting a paradigm of a two-part model for CEOs: "I've always wanted to be successful. My definition of being successful is contributing something to the world...and... being happy while doing it...You have to enjoy what you are doing. You won't be very good if you don't" [9]. He also quotes Masaru Ibuka, co-founder of Sony, whose mission for Sony was "To establish a place of work where engineers can feel the joy of technological innovation, be aware of their mission to society, and work to their heart's content" [9].

A manager might ask, "How can I know the right next task for every person that reports to me?" You can't get it right every time. But, as Fig. 10.4 shows, the leader's role is to know his or her people, match tasks

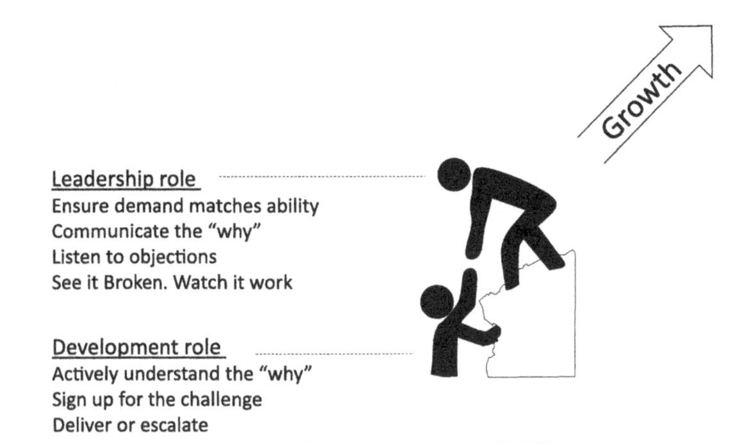

Fig. 10.4 Leaders have a unique role in "just right" task difficulty.

to their ability, and communicate the "why". Then he or she will stay connected to "See it Broken" when things are not going well and "Watch it Work" when people are learning new skills. The developing person has responsibilities as well: taking an experimental mindset toward a supervisor's advice and escalating when things aren't working, rather than just grinding through. Managers should invite critical thinking like Ohno, who, decades after he established himself as one of the world's most exceptional lean thinkers, still advised people not to follow his lead exactly: "You should think for yourself and come up with better ideas than mine" [10].

You cannot "Watch it Work" or "See it Broken" without going to look yourself. Ward and Sobek tell a story about the head of Toyota's Advanced Vehicle Development department, which was made up of 150 engineers; the department head didn't see a difference between helping people solve problems and helping people grow. He spent 95% of his time solving technical problems because, as he put it:

> *...every technical problem is also a personnel problem, because the problems don't get to me if my people know how to solve them. So all the time I'm solving technical problems, I'm also teaching someone [11].*

There is no greater satisfaction in leading than watching people improve. But it takes time—often years—to master new ways to work and think. As shown in Fig. 10.5, there is a natural progression from learning something all the way to changing it for the better.

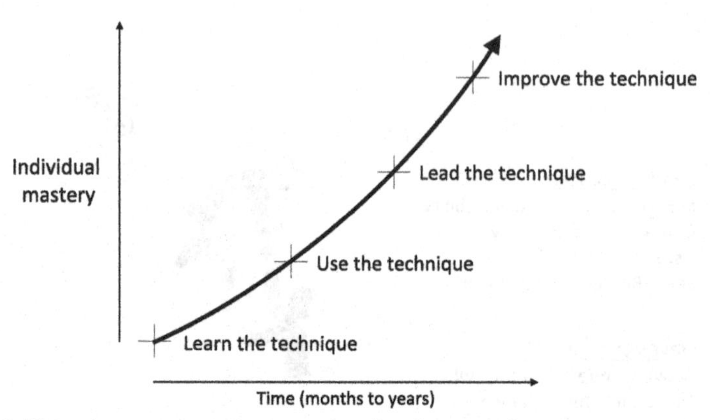

Fig. 10.5 Mastering new techniques and tools takes months or years.

Finally, find the positive. For some managers, positive feedback feels like the opposite of negative feedback so that they should be mixed in roughly equal measure. That simplistic view ignores a basic truth: it's an order of magnitude easier to respond to positive feedback: do more of what you're already doing. Responding to negative feedback is complex because it requires two steps simultaneously: stop doing something you naturally do and replace it with something unfamiliar. I learned this lesson when I (for reasons unclear to me now) decided I should take dancing lessons. Two people tried to help me, one almost always finding the positive and the second always the negative. You might think those were just two perspectives that produced similar results. Not at all. The first would point out what I was doing well and tell me to do that more—it was so easy because I needed only to repeat what I had done a moment before. On the other hand, one day after 30 minutes or so of the negative teacher, I remember stopping—literally standing still. She had told me so many things not to do, I didn't know what to do.

10.4 Change agents, wait-and-see people, and unwilling partners

The change model of Fig. 10.2 is a common one, but, as you might expect, some people are naturally welcoming to change, excited to take on unfamiliar ways of doing things. Those people are often called "change agents" (see Fig. 10.6). They welcome change; they add energy to improvement

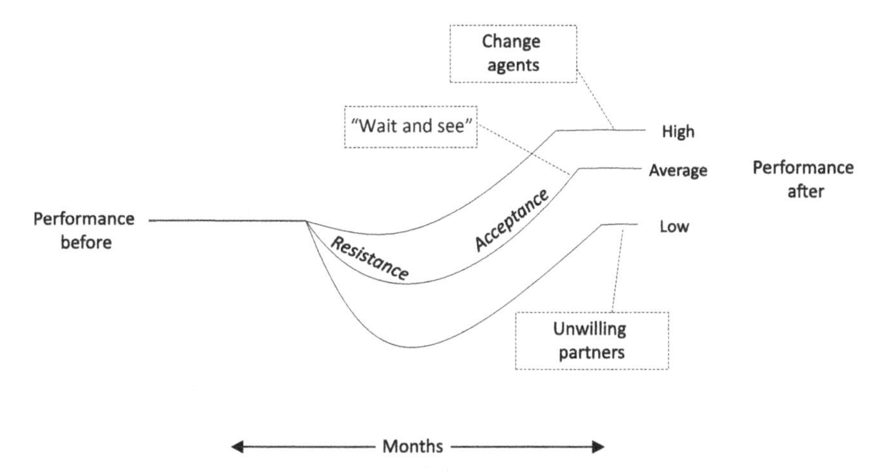

Fig. 10.6 Change agents adopt change more effectively than others.

events; they are excited about opportunities to improve. They keep driving for success even in the moments when things aren't going as well as expected. At the other end of the spectrum are "unwilling partners," people who avoid change. They may have a temperament that is uncomfortable with change. They may have low self-confidence and so be unwilling to display the transparency and vulnerability that group-driven change demands. Worse, they may have a level of arrogance or want to retain control and find team-driven change threatening. These people may never welcome change.

Change agents seem to create energy when they help bring change to an organization. They are excited about the possibilities. They are transparent with no fear of change. They are the people to whom you naturally go to discuss ideas with. At the other end of the spectrum are the unwilling partners: people who manage to deflate any excitement about how we can get better. They always seem to have a view that a new idea won't help, that we tried that before, and that no improvement from the past was ever as good as people think. So, when you bring change, start with change agents. It's not that change agents are perfect—maybe they are not as diligent as those less welcoming to change or perhaps they don't protect the brand as a Guardian (Section 7.2.4) would. Their reduced aversion to change makes them perhaps more susceptible to problems that are caused by excessive change. So, nothing here recommends populating a team entirely with change agents. I'm not sure such a thing is even possible, but if it were, I'd avoid it. Having a balance of change agents and wait-and-see people is good; sometimes the unwilling partners make up for their aversion to change by bringing high diligence to protect the brand during change.

Marty Cagan wrote: "Some people in your organization love change, some want to see someone else use it successfully first, some need more time to digest changes, and a few that will only change if they are forced to" [12]. David Hamer of the UK's National Health Service talks about three types of people. There are those who will accept change, which make up about 20% of your team. He adds that you know who those people are at the start. The bottom 20% of people oppose change: "You may win them over but not likely—they must be managed, not led" [13]. The other 60% are in the middle; they will embrace change after they see results.

If you are a change leader, you can view your mission at a high level as leveraging the 20% change agents to rapidly win over the 60% wait-and-see people before the 20% unwilling partners create doubt and distrust. Your new initiative may win your change agents over immediately, but the unwilling partners will likely push an alternative narrative that "what you're doing won't work here" or that "we tried that before and it didn't work." This is one reason why early improvement initiatives must be fast and take on minimal risk—if you start off with one or two failures, your unwilling partners won't miss an opportunity to bring those up. In the worst case, you'll lose confidence from the team as a whole; bringing change from that point on will be quite difficult.

10.5 Conclusion

This chapter has presented ways to reduce the **Inertia to Change** that **simplify**, **engage**, and **experiment**, as shown in Fig. 10.7. This includes understanding how different people have different reactions to change and differing needs for challenge. The Engagement Wheel in the center of Fig. 10.7 shows how to increase engagement by welcoming diverse viewpoints. And it's engaging to know your supervisor and your organization actively work to balance challenges against your skills to provide an environment where you can grow. Finally, observing your growth up the curve of learn-use-lead-improve will aid you and your supervisor to see what's working and what isn't, so that you can grow to your full potential.

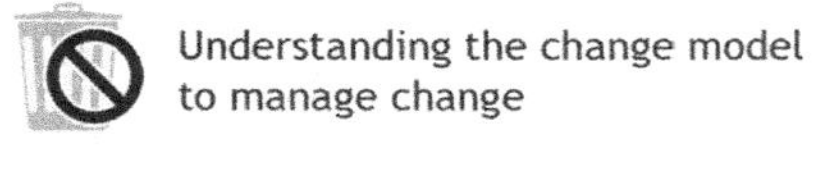
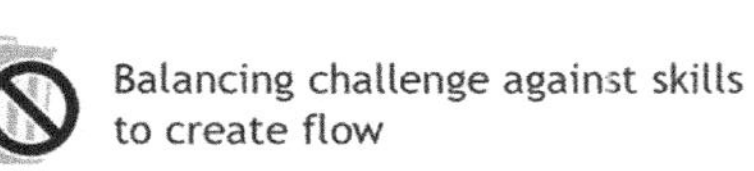

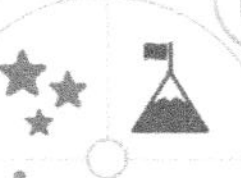

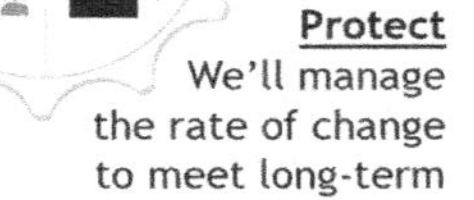

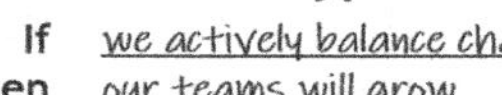

Fig. 10.7 Techniques of reducing Inertia to Change combine to simplify, engage, and encourage experimentation.

References

[1] Widely attributed to W. Edwards Deming.
[2] J. Shook, How to change a culture: lessons from NUMMI, MIT Sloan Management Review, Winter 15 (2) (2010).
[3] J. Shook, How NUMMI Changed Its Culture, John Shook's eLetters, Lean Enterprise Institute Knowledge Center, September 30, 2009. https://www.lean.org/shook/DisplayObject.cfm?o=1166.
[4] S. Shingo, The Sayings of Shigeo Shingo: Key Strategies for Plant Improvement, English translation, Productivity Press, 1987, P17.
[5] Pink, Daniel H. Drive. The Surprising Truth About What Motivates Us. Penguin Publishing Group. 2011, p. 117.
[6] D.A. Gavin, How Google Sold Its Engineers on Management, HBR, 2013.
[7] J.K. Liker, G.L. Convis, The Toyota Way to Lean Leadership: Achieving and Sustaining Excellence Through Leadership Development, (McGraw-Hill, 2012).
[8] S. Walker, The Economy's Last Best Hope: Superstar Middle Managers, Wall Street J. (2019). Updated Mar 24, 2019, Available at: https://www.wsj.com/articles/the-economys-last-best-hope-superstar-middle-managers-11553313606.
[9] TED, Flow, the Secret to Happiness, https://www.ted.com/talks/mihaly_csikszentmihalyi_on_flow?language=en, 2004.
[10] T. Ohno, Taiichi Ohnos Workplace Management, Special 100th Birthday Edition, (2012) pp. 178–180. McGraw-Hill Education, Kindle edition.
[11] A.C. Ware, D.K. Sobek II, Lean Product and Process Development, second ed., Lean Enterprise Institute, 2014, 32.
[12] M. Cagan, Inspired. How to Create Tech Products Customers Love, second ed., John Wiley & Sons, 2018, 292.
[13] David Hamer, https://www.youtube.com/watch?v=eCBlar81O8o.

CHAPTER 11

Reduce Waste #5: No-Win Contests

A key Toyota tenet is "Respect for People," the conviction that all employees have the right to be successful every time they do their job.

John Shook [1]

11.1 Introduction

In this chapter, we will address **No-Win Contests**, a waste common in knowledge work. Our primary aim will be to eliminate oversubscription. We'll also discuss multitasking (a nearly ubiquitous companion to oversubscription), negotiation with knowledge staff vs prescription, and structured escalation. The good news is that **No-Win Contests** are usually a direct result of leadership, which means that when leaders are engaged, it's one of the easiest wastes to cut. All of the **8 Wastes** are, of course, in some way related to leadership, but **No-Win Contests** are probably within their most direct control.

No-Win Contests require people to do something they cannot do well within the allotted time. Left without the option of success, they will do something. Give good people two jobs to do in the time they can do one well, and they will almost always fail: most will either do one well and leave the other undone, or do both, but poorly. There are occasions where people will somehow win: they will work Herculean hours or perhaps they will just get lucky. But there is no path to sustained success.

This is what it's about. It's about winning. Now I'm not talking just military. If you're in a business, it's about winning what you do. If you're an educator, it's about educating kids. There's some standard of winning in whatever you do.

General Stanley McChrystal [2]

No-Win Contests heap waste upon waste for knowledge work, as shown in Fig. 11.1.

Improve
https://doi.org/10.1016/B978-0-12-809519-5.00011-9

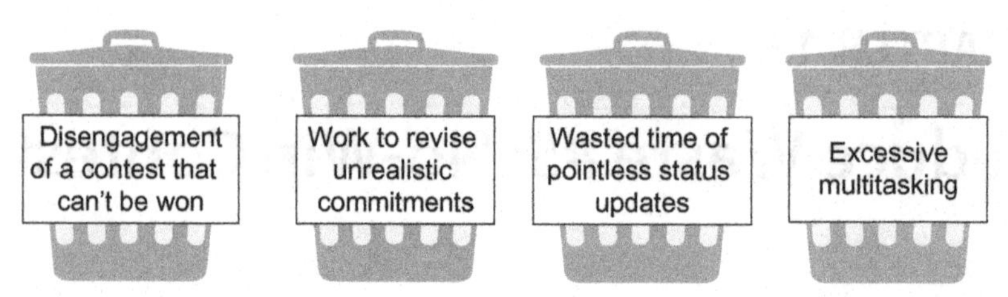

Fig. 11.1 The waste of No-Win Contests is varied and large, but often among the easiest to reduce.

- Disengagement

  Engagement is muted when people are given only the choice of which way to lose. They won't marshal the focus they would if there was a path to success.
- Revising unrealistic workplans

  No-Win Contests frequently present as unrealistic workplans—schedules the team doesn't believe or results they have not bought into. This will be discovered by a larger audience when dates pass without expected results. Once the team has failed to perform to plan, they will likely have to provide a revised workplan. Of course, the forces that caused the initially unrealistic planning are still at work—if anything, they are intensified—so that the revised plan will likely also be unrealistic; the next time a date is missed, another iteration of this wasteful cycle will be required.
- Pointless status updates

  No-Win Contests often manifest with large numbers of open but inactive tasks because new tasks continue to come into people's workloads before old tasks are finished. Every open task has a hidden cost—for example, updates to stakeholders and customers who mistakenly believe work is ongoing. If we have 20 tasks that are open, that's 20 times someone will want to know how things are going. Worse, inactive tasks, the ones for which updates are most wasteful because nothing has happened, often take longer to update because you must respond to the disappointment when customers and stakeholders learn their work is stalled. This erodes trust, and this means that people want more explanation, more evidence, and more remediation plans.

11.2 Excessive multitasking

Opening tasks, initiatives, and projects when other work is unfinished increases multitasking—more questions on more topics, more tasks started

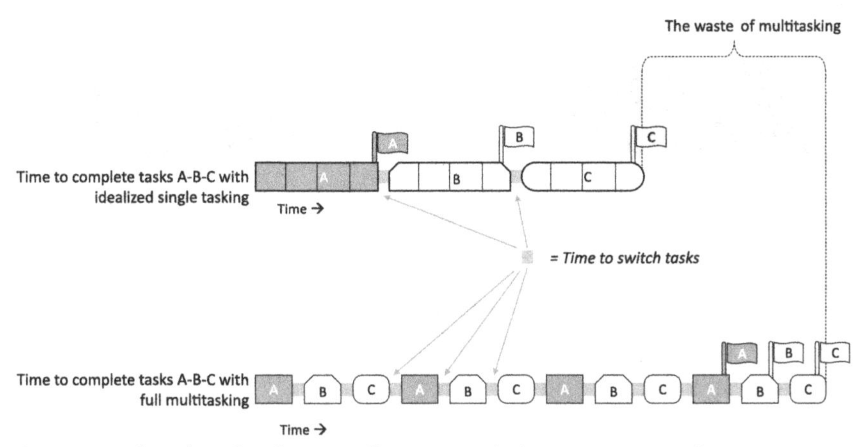

Fig. 11.2 Multitasking has large inefficiencies including context switching.

and not finished. A common misconception is that multiple tasks can be done serially as fast as they can be done in parallel. In reality, multitasking brings many inefficiencies.

Two of the inefficiencies of multitasking are time for context switches (Fig. 11.2) and the distraction of open tasks. The context switch is familiar; for example, after a phone call interrupts complex work, it takes time to build the context of the work back up in your mind; more complex work takes more time. A person with twice as many open tasks will have more interruptions and his or her efficiency will decline as more and more time goes into context switching.

Another inefficiency is the Zeigarnik effect [3], the phenomenon in people where simply having one task open reduces efficiency in processing another. This effect is shown in Fig. 11.2 by each of the four slices of multitasked work taking just a little longer to process compared to the single-task time. The result is the total time required can be much longer with multitasking.

11.3 Oversubscription: The No-Win Contest collides with More-is-Better Thinking

The **No-Win Contest** is a waste caused primarily by an organization's leaders. **More-is-Better Thinking** comes from knowledge staff willing to grind away with little hope of succeeding. The two feed upon each other, spiraling waste up over time. Staff working without a plan for success almost always disappoint by being late or having to redo work to correct errors; work backs up in long queues. This causes more open tasks, more multitasking, and more time to review and remediate disappointing results. This slows down work further. Managers see the staff as incapable of meeting their commitments; staff agree

Fig. 11.3 When leaders create No-Win Contests and staff concede with More-is-Better Thinking, it creates oversubscription.

to whatever they are told because they sense it doesn't matter what they say (Fig. 11.3). It creates a traffic jam: work creeps along, never on schedule; managers keep putting more cars on the highway, hoping that will get things moving. Of course, the traffic jam just grows, as does the frustration of all involved.

None of this is necessary. One of the first steps of improving knowledge work is *Ruthless Rationalization*: never accept work that, when properly done, exceeds the team's capacity. This delivers results. I've watched again and again how teams go from 30% on schedule to over 80%, while improving quality and enjoying work more. Much of this improvement comes from simply cutting the waste of oversubscription—both the concrete wastes of (1) excessive multitasking, (2) revising unrealistic schedules, and (3) pointless updates, as well as the more abstract wastes such as loss of engagement from everyone feeling they are losing.

What do highways and knowledge work have in common?

Knowledge work and highways share at least one thing: both can start more than they can finish. Entrance ramps can quickly take on more traffic than the roadway can handle. Similarly, knowledge organizations can take on more work than they can keep up with. Of course, both have the option of increasing capacity: for the highway it's adding more lanes, and for the knowledge organization it's adding more people. That brute-force option works; unfortunately, it's expensive and takes a long time to implement.

Start with a simpler option: when flow stalls, limit the traffic coming onto the highway. In California, metered entrance ramps are common. It works even though it's a simple first-come-first-served solution. We can make the approach even more powerful in knowledge work because we can add prioritization so only the most important work gets attended to.

Ruthless Rationalization: never accept work that, when done well, exceeds team capacity.

11.3.1 Theory of Constraints (ToC) and oversubscription

Applying the Theory of Constraints (ToC) from Section 6.4.5, we manage the bottleneck to eliminate oversubscription. Fig. 11.4 shows the bottleneck varying depending on the type of work; the precise definition of the bottleneck can vary, but all are some sort of throughput measurement. So, if you're working to end oversubscription, start by identifying the relevant throughput. If you lead a project team, measure how often projects are completed on time. If the work is on recurring tasks such as diagnosing patients, the bottleneck might be best measured by the number of patients diagnosed per day. If it's revenue, add up how revenue is accumulating vs the plan. If it's a series of initiatives to reduce cost, perhaps the best measure is the total forecasted cost savings the team delivers (vs the realized cost savings, which may take years to measure). These are all leading indicators. Lagging indicators are also important: did the project deliver new customers or do the cost-reduction projects create quality problems? But you cannot manage oversubscription according to metrics that deliver in years. Oversubscription is best treated by managing leading indicators to create smooth flow and validating over time using lagging indicators.

11.3.2 A structure that recognizes oversubscription

One measure of whether an organization systematically oversubscribes is how it reacts to "traffic jams." Does it meter incoming traffic or do the new "cars" get on the highway at will? Look first at the resourcing process: managing how resources are assigned to new work is the most meaningful

Common type of knowledge work	Description	Typical bottleneck
Project/initiative	Accomplish clear goals over a defined period of time with a specific team	Schedule
Recurring work	Accomplish well-defined tasks at a regular cadence	Throughput rate
Revenue build	Initiatives designed to meet revenue needs	Revenue delivered to plan
Cost reduction	Reducing material or labor costs in operations	Total cost savings delivered to plan

- Theory of Constraints
  - Define goals clearly
  - Identify the "bottleneck" constraint
  - Minimize the effect of the bottleneck: prioritize work, improve efficiency, and smooth demand
  - Subordinate all other decisions to the bottleneck
  - Open the bottleneck—add resources in every form

Fig. 11.4 The bottleneck constraint must be managed to eliminate oversubscription.

way to limit work and avoid oversubscription. We'll call this *resource rationalization*: how we decide what we will do.

It's easy to confuse prioritization with rationalization. Prioritization or "ranking" is where the organization decides what is most important. Ranking alone is ineffective. If an organization diligently ranks all its work, but still insists all work must be done, nothing real changes. Ranking is relevant to the degree it results in rationalization, which is to say, how low-priority tasks are denied resources. Resource rationalization is where the prioritization rubber meets the road. We will talk more about the difference between rationalization and prioritization in Chapter 21 when we take up Kanban task management.

The first question to ask of our organization is: what mechanism stops resourcing new work when active work is unacceptably late? In other words, is there any sort of "one car per green" mechanism (see side bar "What do highways and knowledge work have in common?" above)? For those organizations regularly experiencing the effects of oversubscription, failing to ask this question is tantamount to willfully oversubscribing. If you want to break this cycle, start by measuring on-time delivery of knowledge work product; then, ruthlessly rationalize: limit incoming work until the team achieves high on-time delivery. Oversubscription won't be able to hide for long.

If your organization is suffering from oversubscription, you don't need to achieve 100% on-time delivery; in fact, you probably don't want to because that usually creates an overly conservative resourcing process. Knowledge work always has unknowns, and we want people to take wise risks, so seek an on-time rate of perhaps 80% or 85%. Of course, we won't be satisfied by simply throttling back the workload. The whole purpose of this book is to discuss many techniques and tools to increase productivity in knowledge workflows. But start by getting the work under control to gain quickly the benefits of winnable contests.

11.4 Escalation: Guarantor of a winnable contest

A winnable contest creates a balance of engaging factors. It creates inspiration because the staff believe they can be successful: they can satisfy customers and clients, and they can deliver value for their organization. It creates connection because people will pull together when given a reasonable chance of success. The challenge of a winnable contest only engages when it can be won within the rules: according to regulations and with high quality. Offering a contest that can only be won with the sacrifice of work-product quality is no winnable contest, at least not in the eyes of the sort of people you want in your organization. My experience was that every challenge should be paired with an injunction to play by the rules. If you only emphasize challenge, it can be read as a subtle wink at lapses in diligence.

A robust escalation process is the final guarantor of a winnable contest. Knowledge work planning is subject to ambiguity: unknown technical and commercial issues, complex estimations, and changing customer needs illustrate why commitments cannot be made with mathematical precision. Every commitment that knowledge staff make brings risk. Knowing there's a leader who can be depended upon to be fair and to help if things go wrong reduces those risks.

Containing issues requires a guaranteed path of escalation. If a person *Identifies* a new issue, they have the duty to place those issues in one of three categories: *Delegate, Escalate, or Action*. This allows every person in a team to create a mechanism here called the IDEA of containment (Fig. 11.5):

Identify: I see a new issue. This issue is something that:

Delegate: someone in a team I lead needs to resolve, or

Escalate: a sponsor or peer needs to resolve.

Action: I need to resolve myself or,

Notice in Fig. 11.5 that there is no path where the person identifying the issue is free of responsibility. That may seem like common sense, but sometimes in knowledge work people can identify issues and then do nothing or something useless like complaining, murmuring that "we do nothing right" or bragging that "I knew this would never work." The path of clear

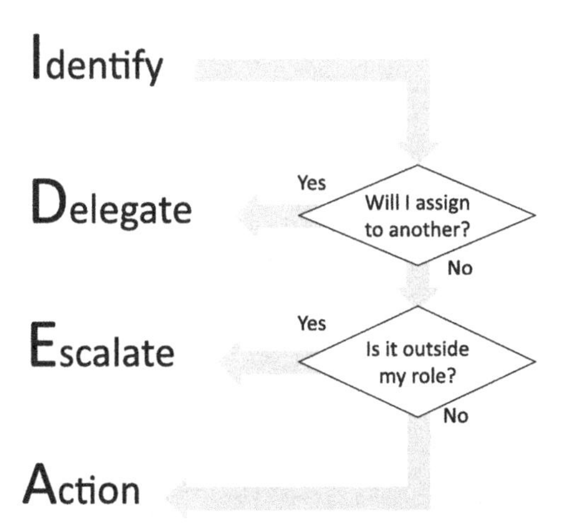

Fig. 11.5 The IDEA of containment: Identify and then Delegate, Escalate, or Action.

escalation means no person in the organization has an excuse to ignore a problem.

11.5 Conclusion

In this chapter, we have addressed one of the most common wastes in knowledge work: **No-Win Contests**. We also addressed that this waste, more than the others in the **8 Wastes**, is within the direct influence of an organization's leadership. The first approach is to eliminate oversubscription by throttling back new work when the group is overloaded. Another is to limit multitasking by minimizing the number of open tasks. As shown in Fig. 11.6, these techniques **simplify** work by identifying bottlenecks so that we know we are focused on improving those things that will make a difference. Creating clear ownership and rules for owners to escalate when they cannot create a path to success builds a winnable contest. Eliminating oversubscription **engages** with a path to success, collaboration to create optimal workloads, the ability to win each contest, and allowing time for high-quality work. It also creates **experiments**, for example, measuring completion to schedule as an indication of oversubscription level.

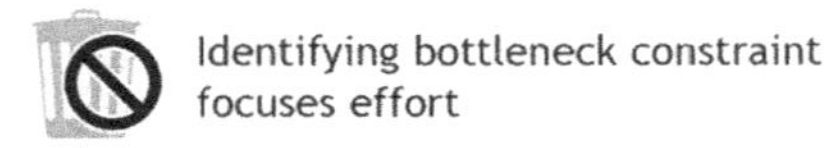

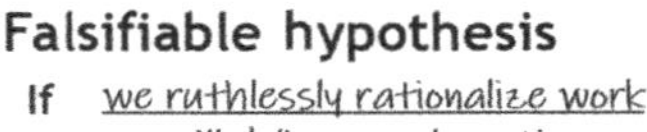

Fig. 11.6 Cutting the waste of No-Win Contests **simplifies**, **engages**, and **experiments**.

References

[1] John Shook, How NUMMI changed its Culture, John Shook's eLetters, Lean Enterprise Institute Knowledge Center, https://www.lean.org/shook/DisplayObject.cfm?o=1166 (September 30, 2009).
[2] Stanley McChrystal 14:22 in https://www.youtube.com/watch?v=p7DzQWjXKFI.
[3] W.J. Friedman, The Zeigarnik Effect and Completing Everything, Retrieved from: https://www.mentalhelp.net/articles/the-zeigarnik-effect-and-completing-everything, 2010.

CHAPTER 12

Reduce Waste #6: Inferior Problem Solving

The first step toward solving a problem is to clarify what is not understood, i.e., to make clear distinction between what we know and understand and what we do not.

Shigeo Shingo [1]

12.1 Introduction

In this chapter, we will investigate identifying and reducing the waste of **Inferior Problem Solving**. This waste presents when knowledge staff solve the wrong problem or treat symptoms rather than solving a problem at its roots. It also occurs when staff put in place ineffective countermeasures, something that remains hidden when they fail to track results. Smart, hard-working, and well-meaning people jump to conclusions when they don't understand the problem or assume a countermeasure will work because it worked on something similar in the past. It's partly human nature to see a pattern that doesn't exist, and partly a lack of diligence: it's easier to move on to solving when we ought to spend more time understanding the evidence.

This waste is unique in that it is core to expertise. **Information Friction** and **Discord** slow knowledge staff down doing what they do. But problem solving is what they do. When knowledge staff don't solve problems well, who else can? Problem solving and implementing countermeasures are among the most obvious contributions of knowledge work to an organization, so when knowledge staff fail to deliver, the disappointment can be severe. And, unlike other wastes, it's hard to hide **Inferior Problem Solving**. The customer or stakeholder who complained about the problem at the start will likely keep complaining until it is resolved.

Improve
https://doi.org/10.1016/B978-0-12-809519-5.00012-0

12.2 Formal problem solving

Formal problem solving includes processes that organize people to solve problems in a methodical manner. It starts with needs and flows to root causes [2]. Fig. 12.1 shows six steps of formal problem solving. It borrows heavily from the Toyota A3 problem-solve method described by John Shook in "Managing to Learn" [3]. There's nothing magic about the steps; Shook includes 5 "A3" Problem Solves tucked in the end flap, each of which has variation in these steps. The key is to tell the whole story methodically on one page, while adapting the visualization to the problem at hand.

Each person on the problem-solve team must see the real problems with their own eyes. In lean manufacturing, this is often called going to *Gemba*, a Japanese word that roughly translates to "the actual place" or "the place where the action happens." Notice, in Fig. 12.1, that the left-hand side of the problem-solve process emphasizes that you must observe the effects of the problem or "See it Broken." The right-hand side advises to observe the countermeasures in place or "Watch it Work" with your own eyes. Too often, lean thinking focuses on the "See it Broken" aspect of going to *Gemba*, to the point where the "Watch it Work" aspect can be almost ignored. This can cause a lack of sustainment—we install countermeasures and are subconsciously convinced they will work; then we neglect to return to *Gemba* to ensure the problem is truly resolved. It's not that we consciously decide our countermeasures are guaranteed to work. That's obviously foolish. The mistake is that we fail to decide consciously there are too many unknowns to trust our solutions. To help with this, we will avoid the lean manufacturing term "go to *Gemba*" in favor of the more specific "See it Broken" and "Watch it Work."

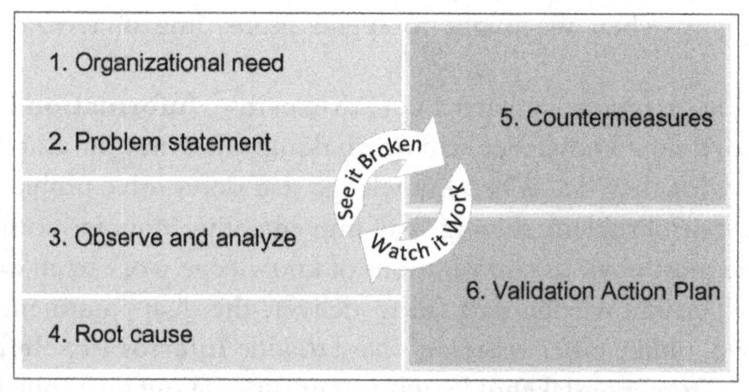

Fig. 12.1 A Problem-Solve Canvas: a single page with about six sections can communicate the whole problem from why it matters to what we will do.

I remember many years ago being repelled when I first heard the term *formal problem solving*. I imagined a stodgy, boring process that would steal my creativity. In fact, it's just the opposite. Formal problem solving seeks to eliminate needless creativity spent in the tedious parts of problem solving, specifically so that more creativity can be applied to the interesting parts. It's based on understanding the most common reasons smart, hardworking, and well-meaning people fail to solve problems well; it seeks to create workflows that reduce those failures. We begin with the highest respect for the experts who solve problems. The starting assumption is a compliment to the team: they will do this well if their expertise and creativity are focused on solving the right problem.

This chapter presents basic problem-solving methods, those that assume that the team in aggregate knows the answer to the problem, though likely no individual has the whole answer. We won't take up problems so complex that the team lacks the needed expertise; those cases are less common and often specific to the domain of knowledge work, and so are outside the scope of this discussion. Our methods tease out the answers we already have; they discover the "wisdom of the team." These methods make no promise to be right, but they always provide an opportunity to learn. They start with the solutions that seem most likely to the team of experts, and in later steps, use validation to create an experiment. If the solutions fail validation, the team repeats the process with newfound knowledge. This, of course, demands that the necessary expertise is represented in the problem-solving team. If you don't have the right experts in the room, the results will be disappointing.

The next six sections of this chapter will present each of the six steps in Fig. 12.1, and in doing so, will step-by-step build a Problem-Solve Canvas. This is a variation of the canvas of Section 8.5 tuned to the needs of a problem-solving process.

12.3 Step 1: The organizational need

- *Never accept the status quo,*
- *Find problems where you think none exist,*
- *Work is more than people in motion,*
- *Perceiving and thinking are not the same thing*

Four concepts of problem identification from Shigeo Shingo [4]

Organizational need states the symptoms of the problem. There is no attempt to provide a solution or even define measurable goals. The need states the pain points the problem is causing. An example of organizational need is shown in Fig. 12.2.

Defining the need can be the most difficult step to do well in the problem solve because it must accurately identify the bottleneck of hidden waste. Recall from Section 6.4.5 Goldratt's principle of the bottleneck: that there are many good things to do, but only a few will matter. Selecting the one or two problems that need to be solved from among the dozens that present themselves requires business and domain acumen.

"See it Broken" begins with personally observing the effects of the problem, so you can understand without the filtering of other people. This understanding leads to insights that enable you to provide valuable guidance. If you don't see for yourself, you will likely provide irrelevant direction and disappoint those whom you lead.

12.4 Step 2: The problem statement

> *The first step of any problem-solving process, development of a new products, or evaluation of an associate's performance is grasping the actual situation, which requires "going to Gemba."*
>
> ***Jeffrey Liker [5]***

The problem statement translates the pain people feel to a measured description of what needs to be solved. Those who understand the need communicate the symptoms with no attempt to solve the problem. Those who can solve the problem must listen to the symptoms and resist the natural urge to start solving because when you start solving, you stop listening. Fig. 12.3 is an example of a problem statement derived from the need in Fig. 12.2. Notice that it quantifies symptoms and states measurable goals; also, the team is defined.

1. Organizational need
For the last two years, employee access to data through shared directories has not worked well. Data transfers during business hours are so slow they are unusable by remote employees. This has made standard work, esp. in Sales, hard to maintain. Sales faces special challenges because of travel and because many salespeople work from home. As a result, most of their use is remote from our server.

Fig. 12.2 Start with leaders defining needs, not solutions.

2. Problem statement			
Symptoms	Up to 30-second delay to access files for people in remote offices during peak hours 5–10-second delays are common at all times		
Goals	Reduce average time to access files	From up to 30 s to 1 s	Measures
	Maximum cost per employee per month	Max. $20/person for subscriptions	
	Max. cost install/purchase incl. IT at $100/hour	$60k	
	Time to complete work	Installed by 31 December	
Team	Lead: Manager Web Services. Team: 2–3 IT experts; 4 remote users (2 from sales, 2 from R&D); 1–2 on-site users.		

Fig. 12.3 The problem statement: A measurable problem with quantified goals.

The first pitfall of bad decision making… is to subconsciously make the decision first and then cherry-pick the data that supports it [6].

Ray Dalio

Organizations spend enormous amounts of time and energy debating, exploring, and trying solutions—yet, how often is it clearly asked and answered, "Just what problem are we trying to solve?" Simply clarifying what we mean when we say "problem" can be powerful.

John Shook [7]

One of the most common weaknesses in problem statements is unquantified problems. On the left side of Fig. 12.4 are three vague statements; on the right are the same three issues stated in a measurable way. Good problem statements are quantified to the extent possible.

As shown in the bottom right of Fig. 12.4, good problem statements have three attributes:

- Concise, specific, and clear

  The statements are concise, specific, and clear. They are free of emotion.

- Quantified goals and current state

  The goals are quantified, as is the current state. This allows a straightforward calculation of the gap the problem solve should fill. People usually

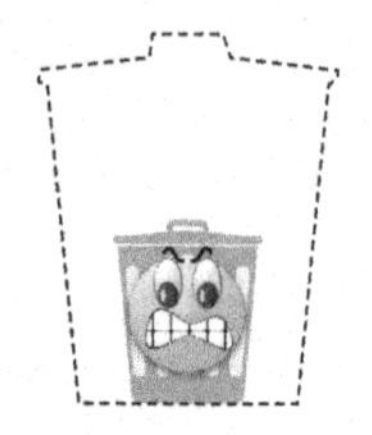

Fig. 12.4 Vague problem statements create waste.

resist quantification because of the risks brought by unknowns. Expect to have to ask more than once and to help people new to this approach. When someone cannot get started quantifying, my advice to experts usually starts: "At this point nothing is quantified. You know more than nothing. Just write what you know."

- Built iteratively with consensus

  The problem statement is often built iteratively. It often requires time for those feeling the pain points to digest the quantification; it takes time for those solving the problem to gain understanding of how the problem is affecting people. And it takes times to collect data to measure the current state. Normally, the problem statement should be beyond its third or fourth iteration before solving starts in earnest.

12.5 Step 3: Observation and analysis

> *If our identification of root cause is even slightly incorrect, then our countermeasure will be completely out of focus.*
>
> ***Taiichi Ohno [8]***

After the problem statement comes observation and analysis; this step ends with the root causes specified by a problem-solving team in consensus. Begin with careful observation: See it Broken with your own eyes. Then move to analysis: understanding observations well enough to state the root causes, the cause that applying countermeasures to will prevent the problem from happening again. There are many techniques used to analyze; we will discuss a few in this section including the 5 Whys, the Fishbone diagram, and Trystorming.

12.5.1 Observe: See it Broken with your own eyes

Think and speak based on verified, proven information and data.

- *Go and confirm the facts for yourself*
- *You are responsible for the information you are reporting to others*

Tadashi Yamashina, President, Toyota Technical Center [9]

"See it Broken" is the experts seeing with their own eyes the effects of the problem on people at the place where the problem is happening. In lean manufacturing, that place is ordinarily on the factory floor—if a machine is producing defects, throw off the blinders of the conference room and have the conversation next to the machine.

Without seeing it broken yourself, your insights will be subject to the *Illusion of Explanatory Depth* [10], the phenomenon where people believe they understand something much better than they do, something revealed to themselves when they try to explain that thing. According to Steven Sloman [11], "We think the source of the illusion is that people fail to distinguish what they know from what others know." Only by going to "See it Broken" can you ensure you are comprehending what's actually wrong, without the filtering of others.

In knowledge work, defining the place where you can "See it Broken" is complex. If a customer is complaining, the place is probably by the customer's side, while they're experiencing the failure. If it's an engineer who is suffering with an antiquated approval system, take 20 minutes and watch them use the system, probably in their office. If a project team is falling behind, join the daily project meetings for a week. Problems that seem impenetrable from the conference room or the end of an email chain can become obvious when you see the pain of the problem with your own eyes. So, before you start analyzing, use your own eyes to "See it Broken."

> *In my Toyota interview, when I asked what distinguishes the Toyota Way...the most common [answer] was "go to Gemba" whether I was in manufacturing, product development, sales, distribution or public affairs. You cannot be sure you really understand any part of any business problem unless you go and see for yourself.*
>
> ***Jeffrey Liker [12]***

12.5.2 5 Whys

> *As a matter of fact, we could say that the Toyota Production System came about as a result of the sum of, and as the application of, the behavior by Toyota people to scientifically approach matters by asking "Why?" five times.*
>
> ***Taiichi Ohno [13]***

The 5-Whys method is perhaps the simplest method of formal analysis in lean thinking: it drives from symptoms to root cause by asking "Why?" again and again. It takes asking "Why?" roughly five times [14, 15]. The simplest format is a face-to-face meeting writing on a whiteboard or a virtual meeting typing text onto a shared screen. The facilitator does the writing. The less the facilitator is tied to the problem, the easier it will be for him or her to be unbiased. If you have a sensitive problem, say where the team is experiencing obvious disappointment from the organization, get someone from outside the team to facilitate.

> *The most important skill to practice is questioning. Asking "Why" and "Why not" can help turbocharge the other discovery skills [16].*
>
> ***The Innovator's DNA***

The facilitator starts by asking the first why question, which is some version of "Why are we experiencing this symptom?" Guiding the team toward systemic issues and away from blaming individuals, the facilitator keeps asking why until the team reaches one or more root causes, which, once remedied, will eliminate the problem.

Here's an example of a 5-Whys discussion:

1. Why do we have a revenue gap of $200k in December?
   - Our sales in California and Oregon are down $180k.

2. Why do we have falling sales in California and Oregon?
   - We lost our main sales rep who went to a competitor.
3. Why did we lose our main sales rep to a competitor?
   - In the exit interview, he said our products were inferior and it made selling harder.
4. Why did he feel our products were inferior?
   - He said the last straw was when his main customer returned 3 months of inventory due to a quality defect and then stopped ordering our product.
5. Why did they return 3 months of inventory?
   - We printed the packaging without meeting California standards for environmental marking.
6. Why did we print in a way that didn't meet California standards?
   - There was a mistake in the product marking instructions software.
7. (a) Why did we make a mistake in software development?
   - The programmer was new and the process for ensuring conformance to California environmental regulations was out of date.

   (b) Why did a mistake in our software get through without detection?
   - We don't have testing for printing in all regions. One of the regions we don't test is California.

The flow of the 5 Whys can take twists and turns as we search for a reason that blames no one, but instead addresses an issue from the perspective of there being something wrong in the organization. It's a lazy way of thinking to start with the assumption that "the problem is someone's fault": someone isn't smart enough or doesn't care enough. The facilitator helps the team navigate around counterproductive answers: those that blame people, complain about past wrongs, or that argue with the inevitable. I once heard John Shook joke that this is when we stop looking for the "5 Whys" and start looking for the "1 Who."

Avoid causes that result in "try harder" countermeasures such as reminding people to think about what they are doing, repeat training sessions, or read documents more carefully. You'll often find branches that choose between someone not trying hard enough and the task being unnecessarily complicated. For example, if someone made an error reading a document, start with the causes around the document being too complex vs. the person reading not concentrating enough. After exhausting causes around unnecessary complexity, consider causes of reliable mistake detection. In this model, some rate of mistakes in complex work is inevitable, so search for root causes from the lack of rapid detection of an error.

> *A relentless barrage of "why's" is the best way to prepare your mind to pierce the clouded veil of thinking caused by the status quo. Use it often.*
>
> ***Shigeo Shingo [17]***

Very often the whys will have multiple answers, as this example had two in the 7th Why. This can happen in any step, causing the 5 Whys to branch to many answers. Don't be surprised when one 5-Why yields three or four relevant root causes. If the path is getting diluted by too many branches, help the team focus on the main paths. Once you get all the way to the end of one path, then come back and finish the next most important. My experience is it's best to write down all the branches people suggest. If you don't, some team members may feel the facilitator is steering the outcome. However, if you follow the main path to its end, people will usually see which branches are less relevant.

12.5.3 Fishbone

The Fishbone [18, 19] or Ishikawa diagram gives a view of all the potential sources of a problem at a glance. The power of the Fishbone is that the team can see the breadth of causes under investigation. A common weakness in problem solving is unconsciously favoring a cause before it's been validated. The Fishbone diagram is a quick view of the sources the team is investigating. It's meant to be iterative, constantly updated as the team learns. It also makes it easier for all team members to contribute where they see gaps (Fig. 12.5).

12.5.4 Ideate-Aggregate analysis

Section 7.3.3.3 presented a method to aggregate ideas from a team. Chapter 7 focused on reducing **Discord** and so the method was used to bring alignment. However, it's also an excellent tool for problem solving. When you have a room of experts, it can take a lot of time to collect and explain ideas. Some people are more extroverted and often they will take over the discussion, unintentionally pushing out the ideas of their more introverted teammates. Others will argue or squelch creativity with summary judgment: "We tried that. It doesn't work." The aggregation technique from Section 7.3.4.2 creates an efficient way to meld together many solutions and so include all voices.

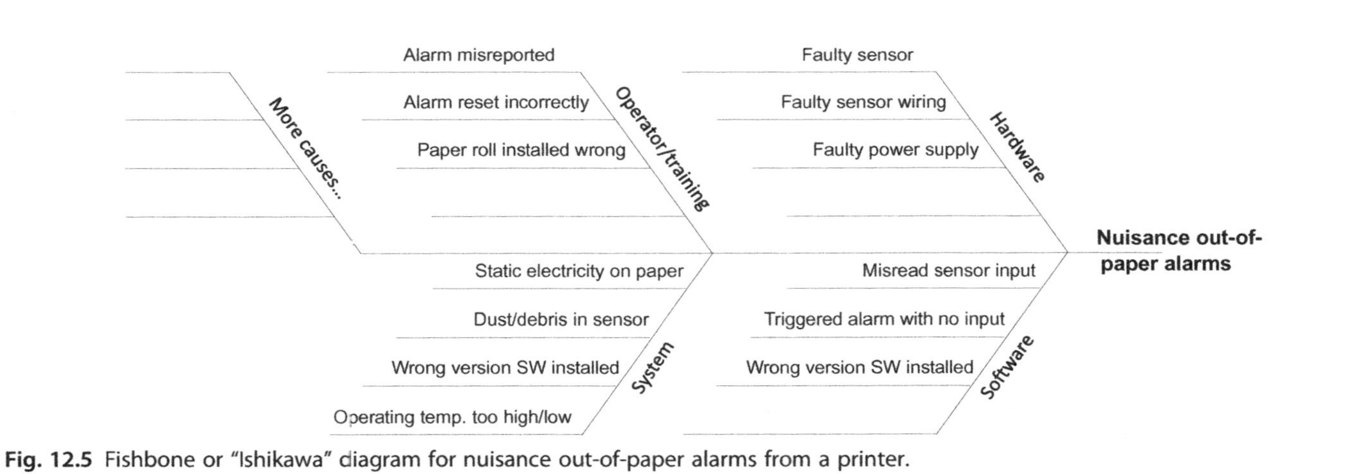

Fig. 12.5 Fishbone or "Ishikawa" diagram for nuisance out-of-paper alarms from a printer.

I've used group aggregation often with a large group to distill the wisdom of the team quickly. The first hypothesis is that the solution is probably represented in the team. One person will see through the impenetrable noise to find the pattern no one else sees. Another has a rapid way to collect the data we need. Someone else knows how to modify code to try the idea. Yet another has a hunch on a competing root cause they want to dive into. If the first hypothesis is that the solution is represented somewhere in the team, the first goal is to distill the wisdom of the team. This is the power of formal problem solving: to release the creativity and expertise that are already there.

The power of formal problem solving: to release the creativity and expertise that are already there.

12.5.5 Trystorming

Instead of sitting around conference tables and trading arguments whenever doubts arise, how about actually giving new ideas a try and seeing what happens?

Shigeo Shingo [20]

Trystorming is a Toyota technique that creates early prototypes rapidly. Cut out nonfunctional mechanical models of plywood and 2 × 4s. Hack in a software quick fix that addresses a critical feature. Throw together a landing site on a web page you can present to a few key customers. The old "brainstorming" method had people around a conference table tossing out ideas. Trystorming adds that you should prove your idea with a crude prototype. Trystorming works. It works in software. It works in hardware. It works in services.

Faced with having to prove an idea at the end of a week, it's amazing how people apply their creativity. Tom Gilb tells a story where a defense industry software set that served many thousands of people took hours to respond to relatively simple queries when people expected answers in minutes. The code set was massive and solutions the team initially identified would take months even to prototype. The team first believed there was no practical Trystorm. Except there was. In the space of 1 week, they took the critical use case—when one of a handful of senior officers needed a response to support an active operation—and moved that user to the front of the queue. So the great majority of users still waited hours, but the handful of critical users got their answer in minutes. It relieved an 11-year pain point in a few days [21]. I watched many times where a Trystorm seemed impossible at first and the team figured out how to try out ideas in 2 or 3 days.

Trystorming is "like Brainstorming on steroids"...what Trystorming does is it gives people a safe playground...for them to test their ideas out in real time besides just brainstorming.

Bob Petruska [22]

Trystorming combines brainstorming with rapid prototyping to watch ideas work or fail quickly. A bad idea that might have consumed months of work is abandoned in days. A good idea that proves itself through Trystorming builds consensus at an astonishing rate; technical people, commercial people, and managers stand back and Watch it Work. It can propel a solution to a high priority in a day or two.

Trystorming has a few steps [23]:

1. Define a problem and multiple approaches to solve it.
2. Distill out the core advantage of the approaches. Create the simplest prototype possible that embodies those advantages.
3. Try out combinations of ideas and measure results within a day or two.
4. Repeat one or two times.

12.5.6 Kaizen events

The Kaizen event is traditionally a 3–4-day meeting to solve a problem. It borrows from the Japanese term *Kaizen*, which roughly means *incremental improvement*. Sometimes, the event is referred to as simply a "Kaizen," an abbreviation that is avoided here because it causes confusion between a single event and the broad mindset of continuous improvement.

12.5.6.1 Kaizen events and manufacturing

The Kaizen event has roots in lean manufacturing. A cross-functional problem-solving team arrives on Monday at the manufacturing area that needs to be improved—at *Gemba*. It may be an assembly cell, an incoming inspection area, or a machine shop. The team focuses for a few days on creating one cycle of "Kaizen," one incremental improvement, generally using a well-defined "tool," selected depending on the problem they are attacking.

Table 12.1 Matching common problems on the factory floor to well-known Kaizen events.

Problem	Tool
Reduce waste in a process	Value Stream Mapping (VSM) [24, 25]
Organize a manufacturing area	5S [26]
Shorten time to change over manufacturing line	Single Minute Exchange of Dies (SMED) [27]

These tools are essentially formal problem-solving workflows customized for a narrow problem. Most have a long heritage, starting at Toyota. Table 12.1 shows a few of the dozens of tools commonly available in lean manufacturing.

12.5.6.2 Kaizen events and knowledge work

The Kaizen event is used in knowledge work [28], but not to the extent it is in manufacturing. The primary reason is probably because of the extended time it takes to realize results in knowledge work. On the factory floor, measurable outcomes are often realized in a few days; for example, if it takes 15 minutes to produce a widget and the target is 5 minutes, it wouldn't be unusual to see the time reduced to 10 or 12 minutes during a 1-week Kaizen event. In addition, there probably will be a clear path to get to, or at least near, the target. When dealing with problems in knowledge work like solving technical issues or expanding distribution into a new region of the world, it takes months to see meaningful results. For example, if we meet to improve how we manage projects, it would be months before any meaningful data could be gathered on how much our new techniques improved projects.

A second reason is that factory teams are much more likely to be colocated than knowledge staff. Thus, getting the team together for a week requires little travel. Knowledge staff are normally distributed, so the cost of getting together is higher. As I write this, I have just returned from a Kaizen event in knowledge work: five people had to travel from the United States to Europe for the event.

12.5.6.3 The 5S event as an example

Another reason Kaizens are used more in manufacturing is that there is so much variation in knowledge work that creating a fixed event that can be used in other places is challenging. For example, consider the factory-based "5S" event [26, 29], an event named for the five steps that transform an unorganized area:

(1) Sort—evaluate each item in an area for utility. Get rid of what you don't have to keep.

(2) Straighten—place items so they are easy to find when needed, for example, hanging tools on a peg board inside an outline of the tool so it's obvious when it's out of place.

(3) Shine—clean the area.

(4) Standardize—create a set of rules to maintain orderliness such as scheduling the last 15 minutes of each day to put all tools away and sweep the area.

(5) Sustain—put in place measures to keep the area organized. For example, create a few simple measures like "all tools are put away" and "area is swept clean." Have someone from another department measure the area once a week. Post results immediately.

The 5S is surprisingly effective on the factory floor. It works more or less the same way in a small production cell as it does in a large shipping department: follow the standard 5S steps with a knowledgeable facilitator and you're almost certain to get good results. However, I have found no equivalent for knowledge work. I've seen many attempts to try to "5S" file storage on a personal device or "5S" someone's office area, but I've yet to see anything like this be sustained. I'm sure there are ways to use 5S in knowledge work as with other manufacturing Kaizen events, but they aren't obvious.

12.5.6.4 Places Kaizen events work especially well in knowledge work

Even if knowledge work improvements are founded on Kaizen events less than factory improvement, there still are places where Kaizen events fit well, including the following:

- Process change
- Project scheduling
- Problem solving

What these areas share is that each requires uncountable interactions within a team over a short time. Having 15 people together allows those interactions to occur quickly. For process change, you need to have represented in the event those who run the process and those who depend on it being run well. For project scheduling, you need all the people who have substantial commitments to the project. For the problem solve, in addition to the domain experts needed to create solutions, you need people who have experienced pain from the problem. Each of these tasks can be greatly accelerated by getting everyone in a room for a few days, to plot a prototype process, to schedule the project on a wall, and to Trystorm solutions.

Facilitating a Kaizen event is a skill in itself. Balancing the need to listen to all voices against the need to keep moving to the goal is difficult, as is knowing when to dig further into a topic as a group vs. parking that topic to take up another day. Facilitators must also keep the Kaizen event "flat" so that organizational rank doesn't disengage junior members of the team. If you want to have a Kaizen event that you're new to, bring in an experienced facilitator. Kaizen events are expensive and time-consuming; poorly led they bring no value, or worse, they teach people that "this doesn't work here."

> *We view errors as opportunities for learning. Rather than blaming individuals, the organization takes corrective actions and distributes knowledge about each experience broadly. Learning is a continuous company-wide process.*
>
> ***The Toyota Way document, 2001 [30]***

12.5.7 Putting it together

Methods such as the 5 Whys, the Fishbone, the Ideate-Aggregate, and Trystorming allow a team to create a broad analysis that can and should be summarized in a small space. An example analysis for our canvas is shown in Fig. 12.6.

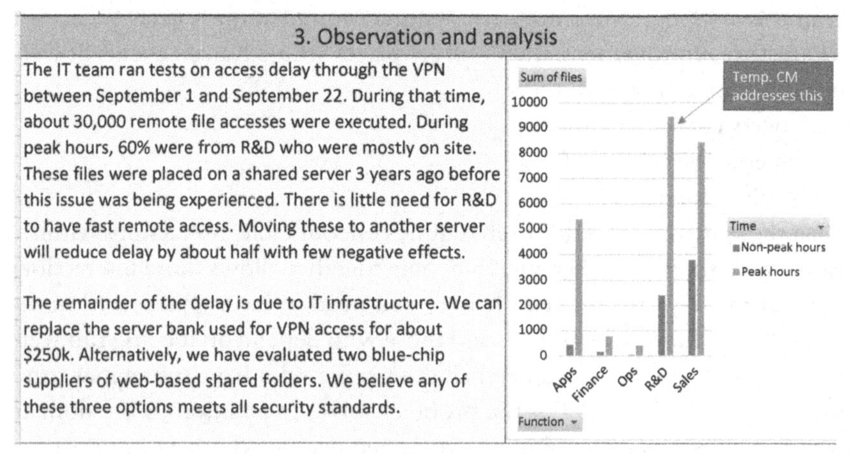

Fig. 12.6 Observing and analyzing the problem.

12.6 Step 4: Root cause

Having observed and analyzed the problem, it is now time to state the root cause(s) as currently understood. The *root cause* is "a cause that, once removed from the problem fault sequence, prevents the final undesirable event from recurring" [31]. A couple of root causes from our example in Fig. 12.6 are shown in Fig. 12.7.

12.7 Step 5: Countermeasures

Now that we have declared the current understanding of root causes, we now look to address them. Despite the name "formal problem solving," we use the term *countermeasure* rather than *solution* because no problem is ever truly solved. A problem can be addressed well enough that we feel little pain from it. But at some point, it or something like it will likely reappear. So, solving is an unending journey of recognizing symptoms, root causing, and implementing countermeasures.

Typically, several countermeasures will be needed to address the problem well enough to meet the goals of Step 2. Some of those will be "temporary," steps that reduce the effects of the problem, and others will be "permanent," steps that treat the root cause. An example is in Fig. 12.8; the first countermeasure is temporary and the second is permanent.

4. Root cause		
	Symptom	Root cause
A	30-second delays during peak hours	High R&D use of shared drive. R&D data can be moved to another server with little effect since they are on site
B	5–10-second delay during all times	Our server bank is underpowered for the amount of file loading across the VPN

Fig. 12.7 Root cause: actionable sources of the problem.

5. Countermeasures		
	Countermeasure	How does this meet the goals?
1	(Temp) Move R&D data to secondary server (will increase delays for R&D, but this is acceptable b/c team is almost always on-site)	Will reduce overall access delay to 10–15 seconds. Can be done in 1–2 weeks
2	Add bank of high speed servers and upgrade VPN	Reduces access time to 3–4 seconds. No subscription. About $250k & 8 months to install

Fig. 12.8 Countermeasures, effective responses to the root causes.

6. Success map							
Expected result	Next Action	Next Date	Owner	Plan	Replan	Fcst	Act
Decision on CMs (proposals: 1 & 2)	Schedule decision meeting	10-Oct	Raoul	17-Oct		17-Oct	
Implement CM1			Brad	24-Oct		24-Oct	

Test Track	Dates	Rcv team signed up	Data Set Selected	Data Set Moved	Access Time <=5 sec	--
SW Team Adoption	Plan	10-Oct	10-Nov	15-Dec	15-Jan	
	Forecast					
	Actual					
Application Team Adoption	Plan	4-Nov	5-Dec	9-Jan	9-Feb	
	Forecast					
	Actual					

Fig. 12.9 An abbreviated Success Map to implement the first countermeasure: Action Plan and Test Track.

12.8 Step 6: Success Map

The final step is implementation. We will use two elements of the Success Map from Section 9.6. The Success Map in Fig. 12.9 has only the Action Plan to manage the change and a Test Track to ensure implementation in two teams. A Bowler will be appropriate in some cases, but here is probably unnecessary.

12.9 Canvas View

> *Toyota is very strict about having managers and associates go to great lengths to put key information on one side on an A3-sized [or 11 × 17] piece of paper.*
>
> ***Jeffrey Liker [32]***

Each of the above six steps can be shown as a Canvas View (refer to Section 8.5). This view is based on traditional problem-solving processes such as that used by Toyota in the "A3 Problem Solve" (Fig. 12.10).

12.10 Closing remarks on problem solving

There are a few other points to make about formal problem solving.

12.10.1 Experimentation = Iterations

A common mistake in formal problem solving is rushing to get to the conclusion. In Shook's book, "Managing to Learn," the story starts with

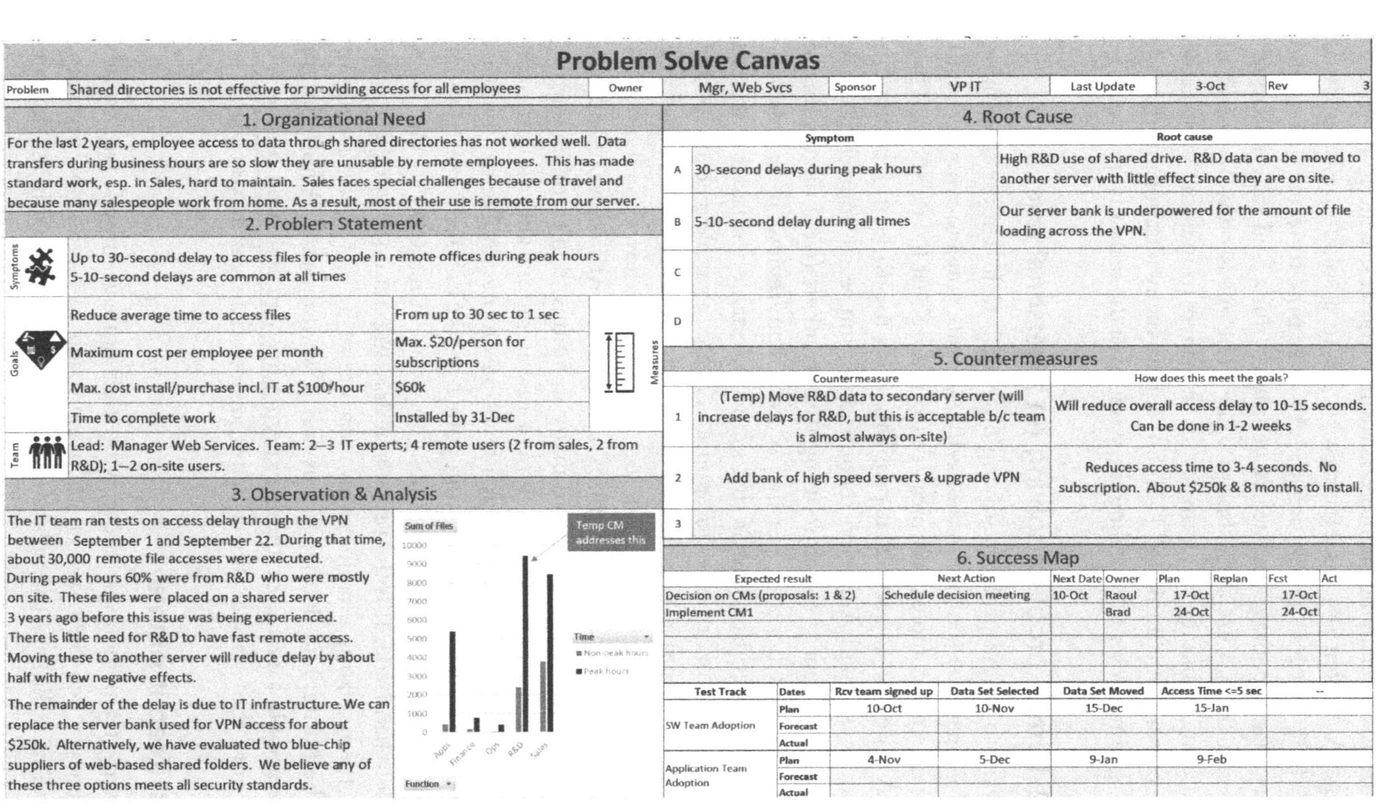

Problem Solve Canvas

Problem	Shared directories is not effective for providing access for all employees	Owner	Mgr, Web Svcs	Sponsor	VP IT	Last Update	3-Oct	Rev	3

1. Organizational Need

For the last 2 years, employee access to data through shared directories has not worked well. Data transfers during business hours are so slow they are unusable by remote employees. This has made standard work, esp. in Sales, hard to maintain. Sales faces special challenges because of travel and because many salespeople work from home. As a result, most of their use is remote from our server.

2. Problem Statement

Symptoms: Up to 30-second delay to access files for people in remote offices during peak hours
5-10-second delays are common at all times

Goals	Measures
Reduce average time to access files	From up to 30 sec to 1 sec
Maximum cost per employee per month	Max. $20/person for subscriptions
Max. cost install/purchase incl. IT at $100/hour	$60k
Time to complete work	Installed by 31-Dec

Team: Lead: Manager Web Services. Team: 2–3 IT experts; 4 remote users (2 from sales, 2 from R&D); 1–2 on-site users.

3. Observation & Analysis

The IT team ran tests on access delay through the VPN between September 1 and September 22. During that time, about 30,000 remote file accesses were executed.
During peak hours 60% were from R&D who were mostly on site. These files were placed on a shared server 3 years ago before this issue was being experienced.
There is little need for R&D to have fast remote access. Moving these to another server will reduce delay by about half with few negative effects.

The remainder of the delay is due to IT infrastructure. We can replace the server bank used for VPN access for about $250k. Alternatively, we have evaluated two blue-chip suppliers of web-based shared folders. We believe any of these three options meets all security standards.

4. Root Cause

	Symptom	Root cause
A	30-second delays during peak hours	High R&D use of shared drive. R&D data can be moved to another server with little effect since they are on site.
B	5-10-second delay during all times	Our server bank is underpowered for the amount of file loading across the VPN.
C		
D		

5. Countermeasures

	Countermeasure	How does this meet the goals?
1	(Temp) Move R&D data to secondary server (will increase delays for R&D, but this is acceptable b/c team is almost always on-site)	Will reduce overall access delay to 10-15 seconds. Can be done in 1-2 weeks
2	Add bank of high speed servers & upgrade VPN	Reduces access time to 3-4 seconds. No subscription. About $250k & 8 months to install.
3		

6. Success Map

Expected result	Next Action	Next Date	Owner	Plan	Replan	Fcst	Act
Decision on CMs (proposals: 1 & 2)	Schedule decision meeting	10-Oct	Raoul	17-Oct		17-Oct	
Implement CM1			Brad	24-Oct		24-Oct	

Test Track	Dates	Rcv team signed up	Data Set Selected	Data Set Moved	Access Time <=5 sec	--
SW Team Adoption	Plan	10-Oct	10-Nov	15-Dec	15-Jan	
	Forecast					
	Actual					
Application Team Adoption	Plan	4-Nov	5-Dec	9-Jan	9-Feb	
	Forecast					
	Actual					

Fig. 12.10 The Problem-Solve Canvas combines the elements of the formal problem solve into a single view.

Porter, the manager being mentored, rushing the A3, which "reveals, as early-stage A3s often do, his eagerness to get to a solution as quickly as possible" [33]. But his mentor uses the A3 over time to teach "habits and mind-sets that encourage and teach people to think and take initiative" [33]. It's not the A3 itself that brings value. The A3 is only an artifact of the underlying thinking. It is a means for managers to mentor "root-cause analysis and scientific thinking" [33]. So, expect many iterations of the A3, each one getting closer to revealing the wisdom of the team that produces it.

12.10.2 Formal problem solving on a personal level

If you want to try A3 problem solving on a small scale, choose something personal—a gap you see in your behavior, an issue you and your boss cannot align on, or a complex technical/domain issue that has persisted. Then use the A3 method to solve that problem. This use-to-learn technique is similar to how David Anderson recommends you teach yourself Kanban Project Management [34] before using it with others (we'll discuss this in Section 21.5.4). Choose a person to help you review it—a coach—perhaps your supervisor, but if that's uncomfortable, consider a trusted colleague, a mentor, or even a spouse. Create a simple version of the A3 perhaps with just a pencil, and start with defining the need as you perceive it, the problem as you see it including how you'll measure improvement, and your observations and analysis. At first write only on the left-hand side. Good A3s have many iterations; don't bother starting the right-hand side until you've done a few iterations and the left-hand side is solid. Meet with your coach a few times to iterate. Remember this is an experiment and you are writing your hypothesis. Don't worry about getting the right answer the first time; you won't. It may take weeks to get just the left-hand side. It is the process of drawing out one hypothesis after another, then creating unambiguous measures to be reviewed a month or two in the future. Did the hypothesis fail? If so, no matter. Adjust and try again. Through a series of these experiments you will find answers—perhaps not "the" answer, but a better answer than you have today.

12.10.3 Keep it simple

The solving team, typically those with high domain knowledge, creates transparency when they lay out their approach to solving the problem. Of course, it is difficult to explain new insights to those with less domain knowledge: it's hard to make something complicated easy to understand. It's hard, but valuable, something like when a doctor explains a malady to a nonmedically trained patient. The doctor may need to read 30 pages from a medical textbook, but the patient needs only enough information to be confident the diagnosis is accurate and make treatment decisions. As I frequently say to domain experts, "You're the doctor; think of your coworkers with less domain knowledge as the patients."

12.10.4 Solve-Select process for the biggest problems

This problem-solve process is designed for cases where the leadership hands off the problem to a team that resolves it. The leadership may be involved only at the start (to ensure the need is well-stated) and at the end (to evaluate the completed Success Map). This works well for most problems. However, in the most important problems faced by an organization, leaders will often need to be involved throughout the process, especially in helping to select the countermeasures. This ensures that the needs of the business will be considered carefully before countermeasures are implemented. In this case, the solving team will deliver not **the** countermeasures, but rather a menu of **multiple** countermeasures from which a leadership team will choose. This variation of problem solving will be called the "Problem Solve-Select" process and is presented in Chapter 18.

12.11 Conclusion

Formal problem solving provides many ways to cut the waste of **Inferior Problem Solving** as shown in Fig. 12.11. The quantified problem statement, specific root causes, and countermeasures provide clarity. The Problem-Solve Canvas provides a standard way to create Single Point of Truth (SPoT). It also increases engagement by creating a shared vision to be outstanding at solving problems, to create an environment where people collaborate at every stage, providing clear and attainable goals, and building a culture of solving problems well. Formal problem solving creates an experiment via the Success Map. If solutions fall short, this will be identified to provide opportunities to learn and improve.

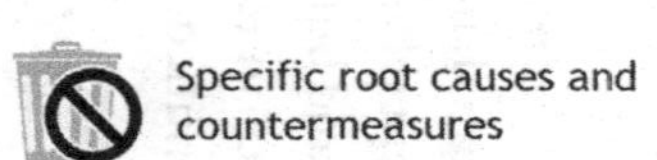
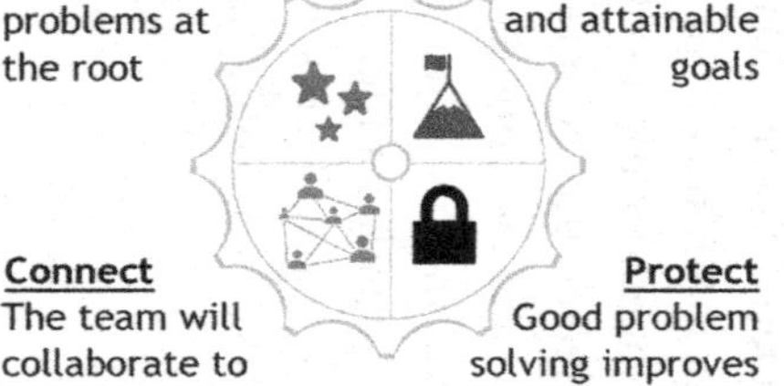

Fig. 12.11 Simplify, engage, and experiment with formal problem solving.

References

[1] S. Shingo, The Sayings of Shigeo Shingo: Key Strategies for Plant Improvement, English translation, Productivity Press 1987, p. 29.

[2] J. Shook, Managing to Learn. Using the A3 Process to Solve Problems, Gain Agreement, Mentor, and Lead, The Lean Enterprise Institute, 2008, p. 44.

[3] J. Shook, Managing to Learn. Using the A3 Process to Solve Problems, Gain Agreement, Mentor, and Lead, The Lean Enterprise Institute, 2008.

[4] S. Shingo, The Sayings of Shigeo Shingo: Key Strategies for Plant Improvement, English translation, Productivity Press, 1987, p. 17.

[5] J. Liker, The Toyota Way, 14 Management Principles From the World's Greatest Manufacturer, McGraw-Hill, 2004, p. 224.

[6] Ray Dalio, Principles, Simon & Schuster, 2017, p. 237.

[7] J. Shook, Managing to Learn. Using the A3 Process to Solve Problems, Gain Agreement, Mentor, and Lead, The Lean Enterprise Institute, 2008, p. 32.

[8] T. Ohno, Toyota Production System, Diamond Press, Tokyo, 1980. first published 1978; John Shook translation.

[9] 10 Management Principles of Tadashi Yamashina, President, Toyota Technical Center, The Toyota Way, p. 225.

[10] L. Rozenblit, F. Keil, The misunderstood limits of folk science: an illusion of explanatory depth, Cogn. Sci. 26 (5) (2002) 521–562.

[11] Freakonomics Podcast "How to Change your Mind", Episode 379. http://freakonomics.com/podcast/change-your-mind/ (about 60% through).

[12] J. Liker, The Toyota Way, 14 Management Principles From the World's Greatest Manufacturer, McGraw-Hill, 2004, p. 223.

[13] T. Ohno, Taiichi Ohnos Workplace Management, Special 100th Birthday Edition, 2012, McGraw-Hill Education, Kindle edition, p. 176.

[14] P.R. Williams, Visual Project Management, Think for a Change Publishing, 2015, p. 32.

[15] T.G. Zidel, A Lean Guide to Transforming Healthcare. How to Implement Lean Principles in Hospitals, Medical Offices, Clinics, and Other Healthcare Organizations, Quality Press, 2006, p. 91.

[16] J.H. Dyer, H. Gregersen, C.M. Christensen, The Innovator's DNA, HBR, 2009. December 2009 Issue.

[17] S. Shingo, Kaizen and the Art of Creative Thinking – The Scientific Thinking Mechanism, Enna Products Corporation and PCS Inc., Vancouver, WE, 2007. Available from: https://books.google.com/books?id=JVOWDwAAQBAJ&pg=PA110&lpg=PA110&dq=shingo+%22a+relentless+barrage%22&source=bl&ots=jugnDBXYF1&sig=ACfU3U2ha_ga-qbyq1USPSON-99DsKMcJA&hl=en&sa=X&ved=2ahUKEwi-o_jNrKvlAhUxUt8KHZJ-CDgQ6AEwEHoECAkQAQ#v=onepage&q=shingo%20%22a%20relentless%20barrage%22&f=false.

[18] https://asq.org/quality-resources/fishbone.

[19] T.G. Zidel, A Lean Guide to Transforming Healthcare. How to Implement Lean Principles in Hospitals, Medical Offices, Clinics, and Other Healthcare Organizations, Quality Press, 2006, p. 90.

[20] S. Shingo, The Sayings of Shigeo Shingo: Key Strategies for Plant Improvement, English translation, Productivity Press, 1987, p. 158.

[21] Quantify the un-quantifiable: Tom Gilb at TED x Trondheim. November 3, 2013. https://www.youtube.com/watch?v=kOfK6rSLVTA.

[22] Bob Petruska author of Gemba walks for Service Excellence, "Ramblings of a Lean Practitioner", Business901 Podcast, Transcript available at https://business901.com/wp-content/uploads/2014/10/Ramblings-of-Lean-Practitioner.pdf.

[23] http://leansixsigmadefinition.com/glossary/trystorming/.

[24] J. Liker, The Toyota Way, 14 Management Principles From the World's Greatest Manufacturer, McGraw-Hill, 2004, p. 275.
[25] K. Martin, Value Stream Mapping: How to Visualize Work & Align Leadership for Organizational Transformation, YouTube, 2013. https://www.youtube.com/watch?v=5YJYMLaV9Uw.
[26] J. Liker, The Toyota Way, 14 Management Principles From the World's Greatest Manufacturer, McGraw-Hill, 2004, 150 (defines 5S).
[27] J. Liker, Toyota Way Field book. A Practical Guide for Implementing Toyota's 4Ps, McGraw-Hill, 2006, pp. 71–72.
[28] T.G. Zidel, A Lean Guide to Transforming Healthcare. How to Implement Lean Principles in Hospitals, Medical Offices, Clinics, and Other Healthcare Organizations, Quality Press, 2006, pp. p115–p146.
[29] Six Sigma Daily, https://www.sixsigmadaily.com/5s-sort-set-shine-standardize-sustain/.
[30] J. Liker, The Toyota Way document, in: The Toyota Way, Toyota Motor Corporation, 2001, p. 250.
[31] P.F. Wilson, L.D. Dell, G.F. Anderson, Root Cause Analysis: A Tool for Total Quality Management, ASQ Quality Process, Milwaukee, WI, 1993, pp. 8–17 (From Visual Project Management, p. 31).
[32] J. Liker, The Toyota Way, 14 Management Principles From the World's Greatest Manufacturer, McGraw-Hill, 2004, p. 157.
[33] J. Shook, Toyota's Secret: the A3 report, MIT Sloan Management Review 50 (2) (2009). Summer.
[34] D. Anderson, Deep Kanban, worth the investment? in: London Lean Kanban Day, 2013. https://www.youtube.com/watch?v=JgMOhitbD7M.

CHAPTER 13

Reduce Waste #7: Solution Blindness

We shall not cease from exploration,
and the end of all our exploring
will be to arrive
where we started
and know the place for the first time

TS Elliot

13.1 Introduction

In this chapter, we will discuss Waste #7, **Solution Blindness**, the tendency to proceed with solutions when new information becomes available that recommends otherwise. Where **Inferior Problem Solving** created poor solutions, **Solution Blindness**, by contrast, continues investing in initially solid solutions that don't hold up over time. Perhaps the information was originally too difficult to dig up or perhaps things changed. Whatever the cause, **Solution Blindness** results when teams dismiss information that detracts from the favored solution or fail to take reasonable steps to collect such information.

It's easy to get emotionally invested in a solution—it's almost required to get buy-in from an organization. And once you are invested, it's then easy to dismiss information that challenges the solution and to overvalue anything that supports it. **Solution Blindness** helps explain how products are developed that fail to deliver on their promises, a few of the more spectacular examples being Windows Phone, Google Glass, and the 2002 Segway [1], whose head predicted it would be "to the car what the car was to the horse and buggy" [2]. **Solution Blindness** is common in large part because there are powerful incentives to dismiss negative information, chief among them being the *sunk-cost fallacy*: we've spent too much to stop now [3].

Improve
https://doi.org/10.1016/B978-0-12-809519-5.00013-2

Of course, you've never spent too much to stop an idea that isn't going to work.

Solution Blindness comes not from people being foolish or careless, but rather from having difficulty challenging basic assumptions. Eliyahu Goldratt put it well when he explained why we don't need more intelligence to be good scientists: "…we have enough. We simply need to look at reality and think logically and precisely about what we see" [4]. This chapter presents ways to avoid **Solution Blindness** by responding logically to information that comes to light after work starts on a solution.

Consider this simple example. We "know" that customers want our products and services to be easier to use. So, let's start an initiative to make them easier to use—something that takes half the time out of using our website. How could that solution be wrong? Actually, it can be wrong in many ways, a few of which are shown in Fig. 13.1. First, some employees of our customers may *ask* for an easier website, but the customer as a whole may not *value* ease of use enough to pay more for it. If someone uses a website a few times a month, it might be frustrating, but still not rise to a high priority. Second, we might be fixing the wrong thing—we might make the website easier to get product data, but what if the customer's primary complaint is the difficulty in entering and tracking service requests? Third, it might be diverting attention from a more important problem such as a competitor who will shortly release a new product or service that will be dramatically better than ours while we are tinkering with our website. These are just three cases of how a truism such as "customers want ease of use" can seem unassailable at first, but turn out to be wrong enough to create a lot of waste.

We know customers want our products/services to be easier to use, so we should make our website easier to use. What could be wrong with that?

(1) They might prefer better ease of use, but they might not make a buying decision based on that. In other words, it might not be worth paying for

(2) The effort to improve usability might fix a few "nice to haves," but not address critical ease-of-use complaints

(3) There might be a competitor on the horizon with a new offering so disruptive that website usability improvements might not be noticed

Fig. 13.1 What we think we "know" almost always includes unfounded assumptions that can create waste.

13.2 The falsifiable hypothesis and Solution Blindness

Become a learning organization through relentless reflection and continuous improvement.

Toyota Way Principle #14 [5]

The waste of **Solution Blindness** increases when people become emotionally invested in a solution because it makes it more difficult to recognize inconsistencies in early results. This is the tendency for us to create beliefs: that solutions are good and will make things better, that this is important, and that we must do it. Beliefs engage our desire to win.

Molding a belief into a falsifiable hypothesis (from Section 6.4.3) immediately challenges assumptions. Saying that something will make things "better" is a belief. Quantifying the amount of that "better" and the time when we'll be able to measure it converts belief into hypothesis. The key difference: a hypothesis has enough information to be disproven. In other words, it's *falsifiable*. Stating a falsifiable hypothesis is risky precisely because it can be disproven. But it is that risk that creates the opportunity to learn—an opportunity that is missed when people are satisfied with beliefs, opinions, and guesses.

Why refuse to specify [success]? Because while you are failing to define success..., you are also refusing to define failure, to yourself, so that if and when you fail you won't notice, and it won't hurt.

Jordan Peterson [6]

13.3 A simple example of Solution Blindness

Solution Blindness can creep in even when we only slightly misunderstand an issue, because it can cause critical problems to be hidden at the time a solution is crafted. Consider the example of Fig. 13.2: we have a new service we want to offer that we will call XYZZY. Developing XYZZY presents a common scenario: there is a well-meaning champion who is certain the world needs it. He's found two customers who are bought in: Alpha and Samgis. Both need the service in the early summer, so our champion has found $375k in revenue that can be won with a $100k project. For all anyone could tell at the start, XYZZY was a well-crafted solution.

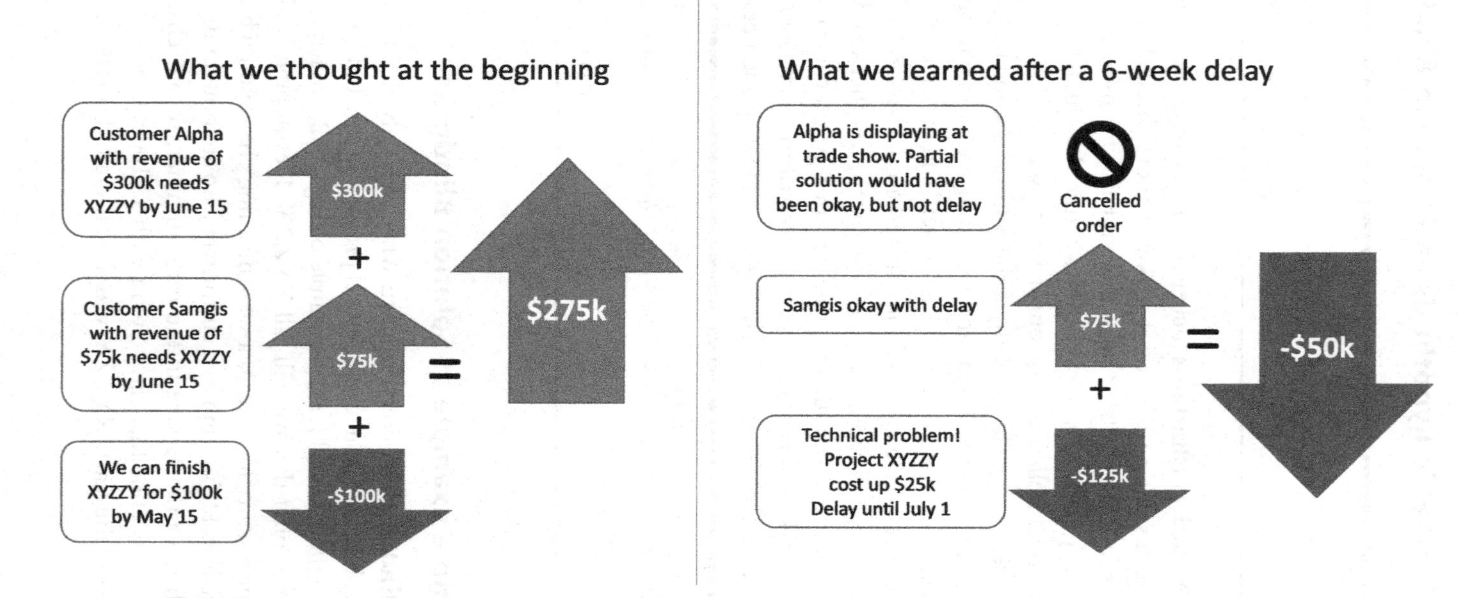

Fig. 13.2 A slightly misunderstood problem leads to misguided measures.

A few weeks pass, and let's say we know now the project will be about 6 weeks late. Here's an opportunity to collect information by asking Alpha and Samgis the effect of the delay from their perspective. Had Samgis been asked, we would have found that 6 weeks wasn't a serious problem for them. The story is different with Alpha, who would have informed us they needed XYZZY for a trade show and timing was tight; in fact, if the project is even a little late, they will go in a different direction. Moreover, Alpha would have told us they only needed a partial solution for the show. So, if Alpha had understood the delay, we could have restructured the project to meet their need. But we didn't ask Alpha and the development team continues working on XYZZY, blind to the fact that Alpha will back out due to what seemed an ordinary delay. When they do, they will take 75% of the revenue so that the project will cost $50k more than it gains (Fig. 13.2, right-hand side). All that work for XYZZY just to lose.

This is a simple example of **Solution Blindness**. We had a strong solution, but something changed over time; here, the project went from "on time" to "6 weeks late." That's information that wasn't available at the start of the project. The information we should have sought was what effect that delay would have on our two key customers. Sometimes **Solution Blindness** comes when we ignore new information or, as in the case with Alpha, when we don't make a reasonable effort to gather information coming from the change.

There's something emotional about a favored solution, especially after work has begun. No matter how diligent we are before we create a plan and no matter how solid a solution seems based on known data, things can change. There just isn't time to collect every relevant speck of information before starting work. Where **Inferior Problem Solving** focused on the waste from poor solutions, **Solution Blindness** focuses on the waste that occurs when diligent problem solving produces clever solutions that eventually fail because of things we didn't know—perhaps could not have known—at the start. Our primary ally in this struggle will be the falsifiable hypothesis.

13.4 Value proposition

The value proposition (prop) is a form of a falsifiable hypothesis that is oriented to new products and services. They typically have three or four elements like those shown in Fig. 13.3. The value prop is written to address some common assumptions that are often proven wrong. It asks the team to quantify the value the new service or product will offer to a customer.

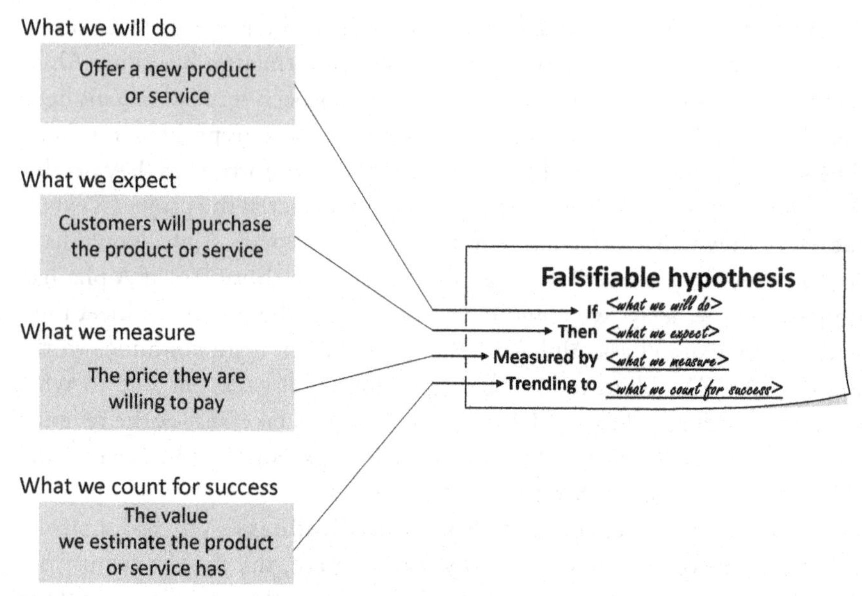

Fig. 13.3 A value proposition shown as a falsifiable hypothesis.

It requires clear delineation between what's proposed and what competitors offer or are likely to offer soon. It requires that the vehicle to deliver this solution be defined and that there is transparency for the proof. It's a falsifiable hypothesis; there is enough information to disprove these four key assumptions. It fights **Solution Blindness** by continuously testing the key assumptions throughout the development of new products and services.

The value proposition is a systematic way for new product and service developers to "See it Broken" and "Watch it Work." At the beginning, take the value prop to the customers and see for yourself that they want it; in other words, from the customer point of view, something is broken. Later, when the new product or service is first implemented, return to the customers to see the value prop working, again, from the customer's point of view (Fig. 13.4).

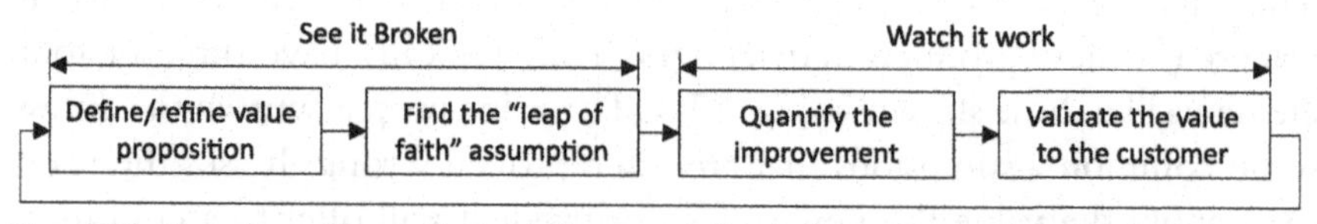

Fig. 13.4 Experimenting with the value prop: See it Broken; Watch it Work.

The people doing the work should "See it Broken," not just the customer-facing employees like those in Sales and Applications. It's easy to sit at a desk 1000 miles from the nearest customer and work on the problems you imagine a customer having. But it's not possible to sharply understand the problem if you don't see it yourself. When customer views are passed from person to person, they have the accuracy of telephone tag. So, go see something broken yourself, and after you believe it's fixed, go Watch it Work yourself. When you watch someone use your creation, you will always be surprised.

When you watch someone use your creation, you will always be surprised.

Get closer than ever to your customers. So close that you tell them what they need well before they realize it themselves.

Steve Jobs [7]

Arranging for noncustomer-facing staff to visit customers can be difficult. It's trouble for the person to travel and it's trouble for customer-facing colleagues to set such visits up. Finding the right customer for a visit of this sort isn't always easy—many customers don't have the time or inclination to evaluate your new products or services in their early forms. But others will be more than willing; they may perceive benefit from engaging early or they may just enjoy collaborating. These are the people you want knowledge staff to visit. When staff who are solving the problem can observe the pain the customer experiences, it works. Again and again, I've seen noncustomer-facing knowledge staff return from customer visits with exciting new insights born of a newfound empathy for customers dealing with long-standing problems or newly discovered needs. One of my favorite stories is when one of our staff finally went to see a customer after months of being reminded to do so. He was working on a difficult feature for months with little feedback from a potential user. He showed it to her and almost instantly she didn't like what she saw. At the visit, she got frustrated with our implementation, stood up and walked our developer to a competitive product and said, "You guys do it a *stupid* way." Pointing at the competitor's solution, she said, "This is what I need!" It took months to code our feature and all it did was frustrate her; but being there showed us that what she really wanted was so simple, he implemented it in a few weeks.

We must learn what customers really want, not what they say they want or what we think they should want.

Eric Ries [8]

13.5 The Lean Startup and the Minimal Viable Product (MVP)

In the Lean Startup, Erik Ries made famous the Minimal Viable Product or MVP. The MVP is the fastest way to get feedback from a customer on a product or service used in a real-world situation: a low-feature version of a planned website, a new temperature controller that works in the one mode this customer needs, or a remote video-streaming device that works only for the one file type a key customer needs. An MVP is less than the first product to be commercially launched. Ries defines it as follows: "The minimum viable product is that version of a new product which allows a team to collect the maximum amount of validated learning about customers with the least effort" [9,10].

The MVP is an experiment, created to learn what customers need. Experimentation, what Ries often calls "validated learning," places some portion of the value proposition in front of the customer and asks for a decision: will you purchase this today? According to Ries, "Validated learning is the process of demonstrating empirically that a team has discovered valuable truths about a startup's present and future business prospects. It is more concrete, more accurate, and faster than market forecasting or classical business planning" [11]. Tom Agan said, "Lean innovation is not a better innovation process; rather it's a more efficient learning process" [12].

Here's a simple example: from time to time, I'm annoyed that my battery electric shaver doesn't run when it's plugged in charging. I'm sure the designers have some good reason for this design, but if someone asked me, "Would you like a razor that ran when charging?" I'm sure I'd say yes. Now, let's say someone asked me to pay $100 for a razor with this feature vs $80 without. I'd probably say, "This only comes up a few times a year… I can live without that." Asking someone what they want is quite different from offering them a choice but asking them to pay for it.

13.5.1 MVPs: The space between mock-up and prerelease

When Ries used the word *viable* in MVP, he meant that it would work in the environment for which it is intended [13]. MVPs are not half-baked ideas with buggy code and unreliable hardware—they are minimal in the sense that they solve problems in a limited subspace. However, in that subspace, they solve those problems well. Also, an MVP is less than the first product to be commercially launched. Traditional prototypes, which are prerelease versions of the commercially viable product, take often about 80% of the project to create. That's far too much time to be useful for early learning. But neither is an MVP a wire frame drawing or a nonfunctional mock-up. Those representations have their place, but they should not be confused with an MVP because they don't solve real problems.

The MVP is created to solve a real problem, something a mock-up cannot do. At the same time, it need only solve problems for a subset of customers that will provide rapid learning. The central assumption of the MVP is that there is space between mock-ups and prerelease versions that is rich with learning opportunities. This space, often ignored when people are developing products and services, provides a great deal of learning (much more than mock-ups) and does so with a relatively small amount of effort (much less than prerelease units). See Fig. 13.5.

The curve of learning vs effort in Fig. 13.5 is an "up elbow" curve, one where a lot of learning comes fast. This is the place an MVP brings a great deal of value. But some things in life are "down elbow" where the initial

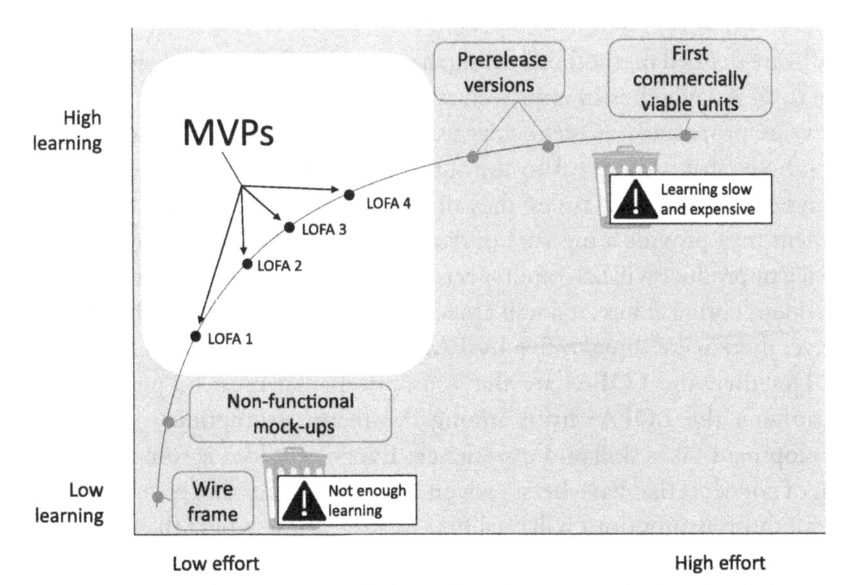

Fig. 13.5 The central assumption of MVPs: the richest space for learning is between nonfunctional mock-ups and traditional (alpha/beta) prerelease versions.

learning is slow and then grows later. A simple example is a college degree where you spend years developing a foundation and then, in the last year or two of your bachelor's degree, learn a great many useful things in a short time. "Down elbow" curves are a poor fit for MVPs. So, be discriminating: consider the whole of your work to identify those aspects that are "up elbow"—where you can piece together something quickly and try it out—and focus MVPs in these areas.

In their book, Lean Product and Process Development, Ward and Sobek provide an outstanding illustration of watching customers use evolving products. The company Menlo Innovations created "High Tech Anthropologists" (HTAs) who travel to customer sites for days to watch end users in their native environments [14]. They seek to understand deeply the full context in which their products are used. HTAs work to identify pain points for their users—shortcomings of their products that indicate a likely place they can add value. They watch users and they interview them. They return several times as the team evolves from understanding how the product is used to identifying problems that users are enduring. Menlo Innovations created a wide range of standard work to guide teams through all phases of learning (https://menloinnovations.com/services/high-tech-anthropology).

13.5.2 Leap-of-Faith Assumption (LOFA)

MVPs are defined methodically using the concept of a Leap-of-Faith Assumption (LOFA), which is an unproven assumption that is a necessary support for the value proposition (Osterwalder used the name "Business Killers" as "the hypotheses that are critical to the survival of your idea" [15]). LOFAs are intended for the critical issues; they don't cover every way that a new development may provide a measure of disappointment. If a LOFA isn't true, the service or product will fail, often spectacularly. If the assumption being disproven doesn't bring failure, it doesn't meet the definition of a LOFA. In my experience, there were three to five LOFAs in most projects.

Together, the LOFAs are the assumptions that must be proven early. Identifying the LOFAs from among the many assumptions in any new development takes skill and experience. Every new idea is some combination of concepts that have been proven and assumptions that have not; only a few of those assumptions will qualify as LOFAs, those where the penalties of error will be too severe to be managed later. Those teams that identify their LOFAs and diligently prove them out with MVPs will be more likely to find problems rapidly.

We can apply the concept of LOFAs anywhere in knowledge work; it's not limited to product development. Let's say, for example, that we want to open a new applications office in Korea to support our many Asian customers. What might some LOFAs be? Let's say we pulled the Asia sales team and the US applications team together for 2 days to evaluate this. At that event, we came to consensus that the two primary LOFAs were that we can:

1. Train and sustain a team in Korea with the needed technical ability; and
2. Install the IT infrastructure to transfer the large amounts of data needed to support customers there.

So, we then set about methodically to quantify both. For LOFA #1, how large would such a team be? Could we hire a portion of that team and bring them to the US? Could one of our US applications team go to Korea for 3 or 5 months? For LOFA #2, how much data do we move today? Can IT measure the equivalent in today's US-based structure to prove that? Can we hire a third party in Korea to set up a server and measure data transfer time? This isn't a radical approach—most likely anyone thinking about building out their applications group to a foreign country would think through such issues. The difference is not the kind of thinking, but rather the degree to which we (1) quantify hypothesis, (2) drive to build consensus among the whole team, and (3) diligently test the hypotheses early—before we lease space in Korea for 5 years, sign hiring contracts with a full team, and commit to the Korean customers.

13.5.3 MVPs across the life of a project

MVP is a mindset of experimentation that fights **Solution Blindness** through the life of the project. Cisco called this approach "rapid iterative prototyping" [16], where increments of value "are viewed as probes—as learning experiences for subsequent steps." Here we iterate from experiment to experiment, each time able to deliver more value to the customer, each time experimenting to validate more precisely the value proposition. The closer the launch date, the more the value proposition is validated [17]. By launch, the high-risk assumptions should be nearly removed, making a successful launch likely. Compare this to the traditional project where the team works for 12 or 18 months to release the product, perhaps sending out an alpha or a few beta units 3 months before launch. By the time they hear back from the customer, launch is so near that no significant changes can be made without disappointing everyone involved—late, partial, and even cancelled launches are all possible. All that work for a project that loses.

13.5.4 A Test Track for MVPs

The Test Track (Fig. 9.14) is an ideal way to track LOFAs. Each column or "lap" is one LOFA starting on the left with the first LOFA to validate. The LOFAs are validated in stages. Usually a few can be easily validated, the next can be harder, and the next harder still. Think of each level of LOFA as a kind of insurance. With the first LOFAs, the cost is low and value high. The cost of each stage increases and, at some point, validating more LOFAs is too expensive to be worth doing; at this point, it's time to release beta units or commercial launch. Notice how in Fig. 13.5, LOFAs 1–4 become progressively more difficult to validate, but taken together they span the gap between the mock-up and the traditional alpha and beta prototypes.

As shown in Fig. 13.6, each row in the Test Track can be one customer where those MVPs are deployed. This deals with one of the complexities of MVPs: tracking how multiple customers are planning to use MVPs and how those plans are maturing. Using the Test Track allows those outside the core team to understand this at a glance.

13.6 Conclusion

This chapter has presented a view of **Solution Blindness** as a waste independent of **Inferior Problem Solving**. The need addressed here is dealing with the common situation where well-crafted solutions, or at least solutions that seemed well-crafted at the time, don't hold up. Even solutions that are created and implemented by diligent, dedicated, and skilled teams don't always work. The response is to maintain an experimental mindset, to be willing to accept new information with an open mind, even with it contradicting our dearest assumptions. As shown in Fig. 13.7, this approach

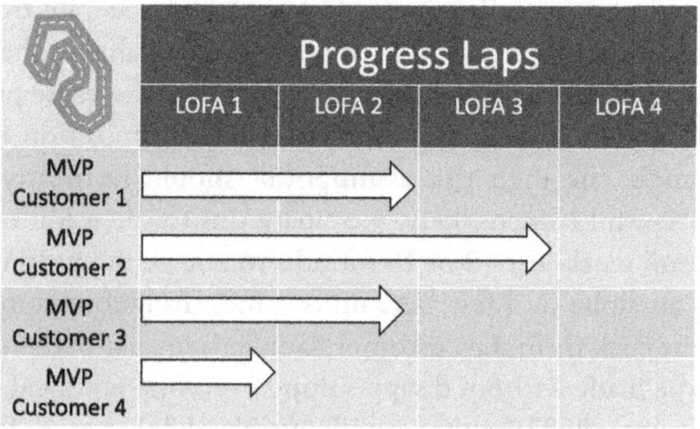

Fig. 13.6 The Test Track is ideal for tracking MPVs.

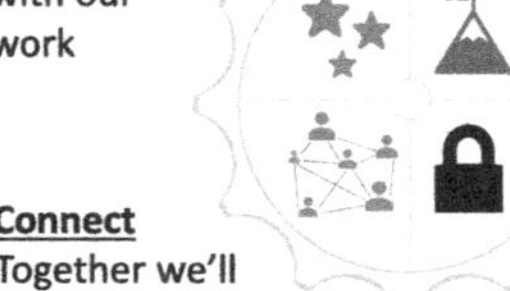

Fig. 13.7 Simplify, engage, and encourage experimentation with techniques that fight **Solution Blindness**.

simplifies, **engages**, and **experiments**. We can simplify with a structured falsifiable hypothesis. By creating Minimal Viable Products (MVPs), we can accelerate the learning process. And we can use the Test Track as a standard way to report on MVP success. These methods are engaging: they inspire a team with the promise of delivering work that delights customers, by working as a team to understand customer needs deeply and to create a culture that works diligently to understand problems and our solutions to them. Experimentation is created by the combination of "See it Broken" and "Watch it Work."

References

[1] J. Golson, Well, That Didn't Work: The Segway Is a Technological Marvel. Too Bad It Doesn't Make Any Sense, Wired Magazine, January 16, 2015. https://www.wired.com/2015/01/well-didnt-work-segway-technological-marvel-bad-doesnt-make-sense/.
[2] The 10 Biggest Tech Failures of the Last Decade, Time Magazine, May 14, 2009. http://content.time.com/time/specials/packages/article/0,28804,1898610_1898625_1898641,00.html.
[3] R.L. Leahy Ph.D., Letting Go of Sunk Costs. How to Escape the Past. September 24, 2014. https://www.psychologytoday.com/us/blog/anxiety-files/201409/letting-go-sunk-costs.
[4] E. Goldratt, The Goal: A Process of Ongoing Improvement, North River Press, 2004 Introduction to First Edition.
[5] J. Liker, The Toyota Way: 14 Management Principles From the World's Greatest Manufacturer, McGraw-Hill, 2004, p. 40.
[6] J. Peterson, 12 Rules for Life. An Antidote to Chaos, Random House, 2018, 276.
[7] Steve Jobs as quoted in 101 of the Best Customer Experience Quotes, Blake Morgan, Forbes, April 3, 2019. Can be viewed at https://www.forbes.com/sites/blakemorgan/2019/04/03/101-of-the-best-customer-experience-quotes/#5805375645fd.
[8] Eric Ries as quoted in Entrepreneur Magazine, Want to Know What Your Customers Really Think? Try Working Side by Side With Them to Solve Problems, Dan Behrendt, February 12, 2018. https://www.entrepreneur.com/article/308367.
[9] E. Ries, The Lean Startup: How Today's Entrepreneurs Use Continuous Innovation to Create Radically Successful Businesses, Crown Business, 2011, p. 77.
[10] E. Ries, The Lean Startup: How Today's Entrepreneurs Use Continuous Innovation to Create Radically Successful Businesses, Crown Business, 2011, p. 77.
[11] E. Ries, The Lean Startup: How Today's Entrepreneurs Use Continuous Innovation to Create Radically Successful Businesses, Crown Business, 2011, 38.
[12] T. Agan, The secret to lean innovation is making learning a priority, HBR (January 23, 2014).
[13] A. Osterwalder, Y. Pigneur, G. Bernarda, A. Smith, Value Proposition Design, How to Create Products and Services Customers Want, John Wiley & Sons, 2014, p. 222, 228.
[14] A. Ward, D. Sobek, Lean Product and Process Development, second ed., Lean Enterprise Institute, Inc., 2014 Kindle Edition. Locations 5605ff.

[15] A. Osterwalder, Y. Pigneur, G. Bernarda, A. Smith, Value Proposition Design, How to Create Products and Services Customers Want, John Wiley & Sons, 2014, p. 202.
[16] R. Luecke, Managing Projects Large and Small. The Fundamental Skills to Deliver on Budget and on Time, Harvard Business School Publishing Corporation, 2004, 117.
[17] A. Osterwalder, Y. Pigneur, G. Bernarda, A. Smith, Value Proposition Design, How to Create Products and Services Customers Want, John Wiley & Sons, 2014, p. 178.

CHAPTER 14
Reduce Waste #8: Hidden Errors

"It can be a shock to the system to be actually expected to make problems visible," said Ms. Newton…who joined Toyota 15 years ago…"Other corporate environments tend to hide problems from the bosses" [1].

14.1 Introduction

In this chapter, we will treat **Hidden Errors**, the waste created when unidentified mistakes in work product are passed on. The emphasis here is on errors that escape detection. Sometimes people view the problem of **Hidden Errors** as a people problem: *Why did someone make a mistake? Is the person not dedicated? Not diligent? Not competent?* That's a poor place to start. We will begin with our core assumption from Section 4.4: most waste comes from good people trying to do the right thing. So, rather than blaming people for making errors we feel they should have avoided, we will focus on robustness: creating knowledge workflows that are tolerant of a reasonable error rate.

We will take two approaches to deal with **Hidden Errors**. The first is to create a culture that doesn't tolerate errors being passed on: a culture that, as soon as a major error is detected, will stop and fix it. We won't take the lazy way out, passing errors on and then complaining about why our organization doesn't do better. This first method will here be called the "Stop-Fix" alarm, which applies to knowledge work the Toyota Production Systems pillar of autonomation: "when a problem occurs, the equipment stops immediately, preventing defective products from being produced" [2]. The second approach is mistake-proofing: creating workflows that cannot generate certain errors, such as when an e-commerce website won't let you place an order without a validated credit card number. In neither case will we attempt to stop every error; that's impractical. Both approaches must be targeted to those errors that (1) are common and (2) result in

Improve
https://doi.org/10.1016/B978-0-12-809519-5.00014-4

unacceptable quality of work. Both Stop-Fix and error-proofing take significant effort, so they must be applied prudently.

Stop-Fix: define the defects we will not tolerate and then don't tolerate them.

14.2 Two mindsets

The core of reducing **Hidden Errors** is understanding why the error remained hidden versus why someone made an error. Knowledge work is too complex and varying to proceed error-free. We need systems that are tolerant of this basic fact. Let's consider a simple example: a technical salesperson missed a customer deadline to provide a quote, which resulted in the customer selecting another supplier. The first reaction might be to blame the salesperson, perhaps questioning his or her competence, dedication, or diligence. It's not that those reasons are never true; there are times when a problem is with the person. However, one premise of this book is that this is a poor starting assumption because few knowledge staff demonstrate these flaws to the degree that they become a primary cause of a failure. Yet these are often among the first places managers and peers may probe.

Knowledge work is too complex and varying to proceed error-free. We need systems that are tolerant of this basic fact.

Our mindset will be that some number of errors is certain to occur, so the first question to ask is: what could the organization do to quickly detect, correct, and prevent these errors from being passed on to the customer? In our example, is there a standard workflow that salespeople use to manage quote activity? Is there an orderly flow of quotes from Sales to Operations to Customer Service that doesn't require a heavy push from Sales to move the quote through every stage? We start with systemic questions such as: how many quotes are late in the organization? What can we do to reduce that number? The 5 Whys of Section 12.5.2 is a good tool to get through these types of questions in a few minutes. Other times, more complex problem-solving processes may be necessary. Only after we have exhausted the search for systemic causes do we start looking at the individual.

The following section focuses on the case where a system error has occurred. Here, attention will be directed toward identifying errors we will monitor closely, detecting those errors quickly, and correcting them before proceeding. The primary tool is the "Stop-Fix" alarm discussed in Section 14.3. Section 14.4 will discuss how to mistake-proof workflow, which is to say, how we can change workflow to prevent errors from being generated.

No process can apply either Stop-Fix alarms or error-proofing to every potential error. Both of these approaches are expensive in time and mindshare, so, in the spirit of Goldratt's bottleneck (Section 6.4.5), they must be applied to a narrow set of errors: errors that repeat often enough to pull the overall quality of work down. So, we will first apply them to the most obvious errors, and as the organization improves over time, more cases can be treated. Applying the Stop-Fix alarm and mistake-proofing is a never-ending journey that today begins with the one or two most important cases that we can identify.

14.3 Stop-Fix alarms

Lean thinking began before the turn of the 20th century with automation of a textile loom. The innovation was not in the weaving, which was already automated, but rather in the *autonomation* principle being applied to a then common defect in automated weaving: thread breaks. Threads broke so often that one worker could do no more than monitor a single loom, constantly watching for breaks—if a break was detected, the worker halted the machine and repaired the thread before much defective cloth was produced. It was Sakichi Toyoda, the founder of Toyoda Loom Works[a] and sometimes called the "King of Japanese Inventors" [3], who revolutionized the automatic loom with a long string of inventions including "a device that detected when a thread broke and, when it did, it would immediately stop the loom. You could then reset the loom and, most importantly, solve the problem..." [4].

[a] Sakichi Toyoda founded the Toyota Motor Company in the 1930s (see https://global.toyota/en/company/vision-and-philosophy/guiding-principles).

The Toyota Production System (TPS), which is based on the philosophy of the complete elimination of all waste in pursuit of the most efficient methods, has roots tracing back to Sakichi Toyoda's automatic loom [2].

Toyota website

This is called *autonomation* or, in Japanese, *Jidoka* (see Fig. 14.1). This not only improved quality (preventing machines from producing cloth without all threads), but, more importantly for Sakichi Toyoda, dramatically increased productivity; with autonomation, one worker could monitor 20 machines. Sakichi Toyoda's concept of autonomation is now recognized as one of the two pillars of the Toyota Production System, though its invention in the automatic loom predates the system by half a century; thus, autonomation is commonly viewed as the beginning of lean thinking.

Build a culture of stopping to fix problems, to get quality right the first time.

Jeffrey Liker, The Toyota Way

Toyota auto factories made famous the andon system. Here, any line worker can pull an overhead rope switch to signal the identification of a quality defect. Pulling the chain does two things: it turns on an overhead "andon" light that indicates where the problem is and calls a supervisor to that point to resolve the problem. If the line supervisor cannot resolve the problem with in a minute or so, the production line stops. You can watch this in a video from Toyota's MotoMachi factory [5].

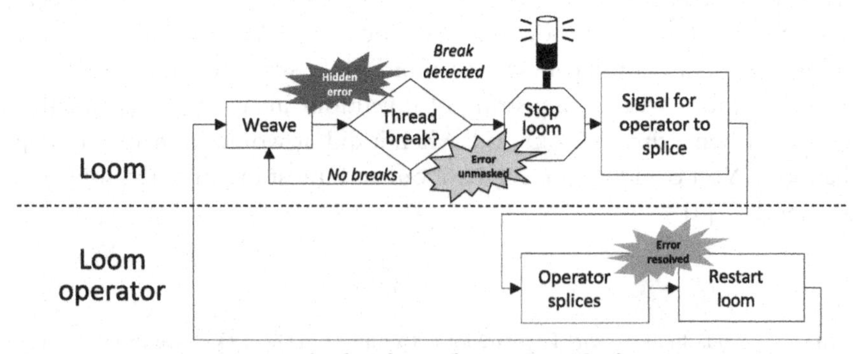

Fig. 14.1 Autonomation was the first form of Stop-Fix in the late 1800s.

Today, in lean factories you will often see dozens of machines, each with a "stack light" mounted above the machine as shown in Fig. 14.2. If a worker detects a quality defect originating from the machine, he or she can flip a switch that causes the stack light to go from green to red and stop the machine from producing more parts. That light gives an unambiguous signal as to whether that machine is capable of producing or not. The lights are mounted high so that, from almost any point in the factory, a person can survey the entire floor, viewing dozens of stack lights. Assuming almost all the stack lights are green, in a few seconds, line supervisors and technicians will see a red light and then swarm to the machine.

Stop-Fix in knowledge work is the equivalent of the stack light on the factory floor (see Fig. 14.3). It's one of the most powerful concepts in knowledge work. It begins with the Stop-Fix alarm, an unambiguous signal that work product has failed or is likely to fail to meet the relevant standard—an event likely to delay a project, a miss in forecasted revenue, or a looming compliance violation. Effective Stop-Fix demands that the alarm is evaluated at a high cadence and, when the alarm is "on," we stop and fix the error. Perhaps one of the best examples of autonomation outside the factory floor is the home printer. There are several defects the printer is able to automatically detect, for example, out of paper, out of ink, and paper jam. When any of these errors are detected, the printer stops printing, turns on a light, and beeps. Printing resumes only after a person has resolved the problem. In the 1980s, many home printers had no paper jam detection. I remember once setting my PC to print the final copy of my master's thesis and leaving the

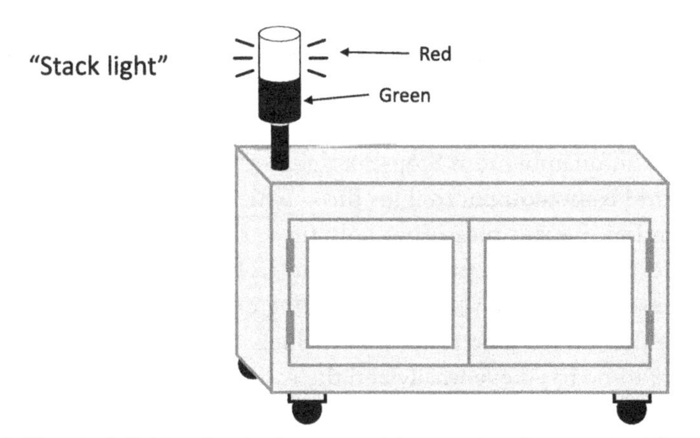

Fig. 14.2 The stack light or "andon": an unambiguous signal to stop and fix.

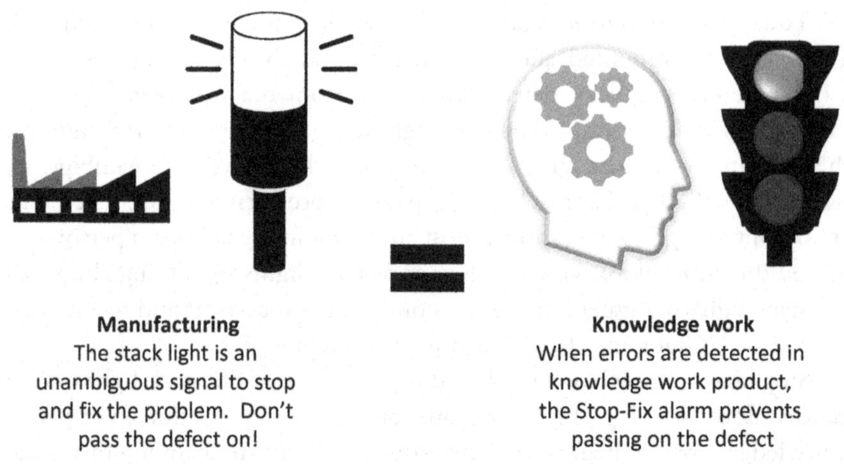

Fig. 14.3 The stack light is to manufacturing what the Stop-Fix alarm is to knowledge work.

house, then hours later returning to find a printer jam had produced a pile of shredded paper. I was delighted when I could buy a printer with autonomation for paper jams, though I wouldn't know it by that name for many years.

Stop-Fix may seem like common sense for knowledge work; however, it's anything but. Perhaps it's because experts are comfortable with ambiguity that they see little need to invest the effort to bring the clarity. Perhaps it's because their memory of past problems focuses on "corner cases," the handful of issues of high complexity that are beyond the reach of Stop-Fix, which is intended to treat more pedestrian problems. But in an organization of any size, a great many errors in knowledge work are pedestrian. Simple omissions and ordinary mistakes steal time and mindshare; they inject needless delay. But to many experts, that isn't how it seems:

- Asked for an unambiguous Stop-Fix signal, the natural reply seems to be, "My world is too complicated for this," which of course it isn't because, by definition, we are measuring only those things that are standard—in other words, the simple errors that occur in large number.
- Asked to stop and fix the error before going on, the natural reply seems to be, "We don't have time to fix this," which, of course, we do because it will have to be fixed eventually and the sooner it's fixed, the less effort it will absorb.

14.3.1 Stop-Fix improves quality and productivity

The mindset of Stop-Fix can be expressed in two parts:

(1) Define those errors we will not tolerate.

(2) Don't tolerate them.

One of the most common misconceptions about the Stop-Fix mindset is that it improves quality, but reduces productivity. Forget this foolish model of trading off quality for productivity, the thinking that says, "We don't have time to do it right." How can doing something wrong save time? When I hear this remark, I realize it means something quite different from what the speaker intended: "We will tolerate this error." Confront that flawed perspective, and you will often find people will blame someone—say, too much pressure from the boss or other people won't do their jobs properly. But, of course, the person you are talking to is tolerating the error, enabling this practice by knuckling under and passing on the defect. It's hard to be patient with people who complain that others don't care about quality, but then quietly pass along defects hidden in their work product.

> *We now understand [Stop-Fix alarms] to be the avoidance of making defects… back in those days people only saw that it improved productivity.*
>
> ***Taiichi Ohno [6]***

Each time an effective Stop-Fix is installed, waste is reduced. No time spent to rework the quotation because one specification was wrong. No time spent explaining to the customer and the boss why we screwed up. And the small increment of time can be invested so that, over the course of years, there will be dozens or hundreds of Stop-Fix alarms paying for themselves again and again, raising satisfaction and reducing tedious work.

Stop-Fix creates improvement only when it's applied to errors that occur frequently enough to offset the cost of installing the Stop-Fix detection. It seems to be human nature to start with the most spectacular failures from the recent past, probably because these errors were so painful they come to mind quickly. For example, let's say we had a patent application that ran afoul of the patent examiner because of an obscure rule in publication, something we never experienced before and are unlikely to experience again. We could expend large amounts of energy on preventing office actions related to this obscure publication rule for little benefit. Remember that all the energy spent here would detract from more pedestrian but more frequent errors like filling in applications completely. Avoid applying Stop-Fix to complex errors that occur rarely. Understanding that people may be biased toward the more spectacular errors from recent memory, it's best to guide selection of Stop-Fix errors using data: a history of errors that actually happened.

Improve quality, you automatically improve productivity.

W. Edwards Deming [7]

So, we will proceed understanding that autonomation is a powerful tool that is underused. The good news is that the method delivers quick results, and that people can acclimate rapidly after they see results. Fig. 14.4 shows the four steps of Stop-Fix.

The Stop-Fix alarm is the enabling constraint to the problem "How should we react?" This is often a difficult problem for knowledge staff, especially junior team members. The disincentives to act are often large, for example, being seen as rocking the boat. Without a Stop-Fix alarm, the question of "Should we react at all?" is not always clear. The Stop-Fix is a contract between leadership and staff that simplifies the call to action. With it, the problem of *whether* to react is immediately resolved; moreover, the problem of *how* to react is greatly simplified.

Virginia Mason uses Stop-Fix to save lives

A Patient Safety Alert system instituted by Virginia Mason in 2002 was inspired by the Toyota stop-the-line practice empowering any worker to halt the assembly online to prevent a defect...Since the program's inception more than 15,000 Patient Safety Alerts have been called, resulting in countless cases where harm was prevented—in some cases, fatal harm [8].

1. **Identify the errors we will not tolerate**
2. **Create a means of detecting those errors rapidly and with high certainty**
3. **Generate an unambiguous signal if an error is detected**
4. **Resolve the Stop-Fix error quickly**

Fig. 14.4 Stop-Fix: Identify the intolerable errors and never tolerate them.

14.3.2 The playbook

The "playbook" is a short list of actions the team should take when Stop-Fix errors are detected. It's short because the number of types of Stop-Fix errors is small and the most common responses to them will fit a pattern. The playbook may be an explicit document, but more likely it is a collection of responses based on experience. For example, the playbook for a project delay might involve the following:

1. **Reduce workload**

   Reevaluate if the function or feature causing the delay is critical; if not, consider delaying it to keep the other work on schedule. Review other areas for functions or features that may be noncritical and thus can be delayed until a later version to make up time.

2. **Add resources**

   Search for more resources to work on the area causing delay. Is there another person in the organization who can help? Can we bring in an outside resource?

3. **Increase focus**

   Can we focus more of the team's effort on this project until the delay is resolved? Can we deprioritize other work temporarily?

The playbook is always short, meant to cover the common cases—let's say 80% of the time. If the team cannot respond to the Stop-Fix alarm with the playbook, they may need to innovate new plays or escalate to a higher level. However, they must respond until the alarm is extinguished, which occurs only when the forecast date improves enough to demonstrate that the Stop-Fix alarm has been resolved. For example, if we fall behind but secure a new a resource *today* that will enable us to catch back up within 2 weeks, we turn off the alarm *today*—the day the forecast showed success sufficiently likely that no other response was required. We don't wait two weeks leaving the alarm on when no other action is needed to address the issue. There must be a means to turn off Stop-Fix quickly. Stop-Fix only works if you keep red rare through a culture that responds quickly. An alarm that cannot be turned off is no alarm.

There must be a means to turn off Stop-Fix quickly. An alarm that cannot be turned off is no alarm.

14.3.3 Yellow vs red

The ability to detect Stop-Fix alarms implies also the ability to detect alarms of a lesser urgency—alarms that reveal we are not meeting goals, but in a manner that does not require the workflow to be corrected immediately. In a printer, a paper jam is a Stop-Fix alarm, as is an empty ink cartridge; however, low-ink is only a warning because printing can continue in its presence. Something similar occurs in knowledge work, where we may be off the original plan in a way that was renegotiated with leadership; because initial expectations were missed, it's not a "green," but it's certainly not a Stop-Fix alarm. For example, let's say a product design element created more issues than expected, but the topic was reviewed by the company leadership and correction is planned for the beginning of next year. This condition cannot be represented as green because the plan was missed. But it's not red either because there is no immediate action required. This is the place for the yellow warning.

This configuration is common in sales. At many companies, the year begins with a plan and only by satisfying that plan can sales be "green." But if an unrecoverable event comes, it can result in taking the forecast down so that it's below the plan. If the team is meeting the updated forecast, that would be yellow, indicating the reduced forecast has been accepted by the leadership team. By contrast, if performance continues to worsen so that the team is missing the already-reduced forecast, that's red: a Stop-Fix condition (Fig. 14.5).

A Stop-Fix Alarm for a Rock Band

In the 1980s the rock band Van Halen had a contract rider that served as a Stop-Fix alarm: venues had to provide M&Ms for the band with all the brown M&Ms removed. The story goes that they were playing in venues where they were concerned that the stage couldn't hold the weight of their equipment. The contract included a rider with the oddest requirement: M&Ms (absolutely no brown ones) [9]. If the band arrived at the venue and there were no M&Ms or there were M&Ms but with brown ones, it revealed the venue either hadn't read the contract carefully or if they had, they hadn't followed the demands; the band took action because that caused doubt that the stage could safely hold their equipment.

Fig. 14.5 Three states: Green (no alarm), Yellow (running alarm), and Red (Stop-Fix).

14.3.4 A culture of Stop-Fix

The culture of Stop-Fix requires a high-cadence review, as shown in Fig. 14.6.

14.3.4.1 Complete the next increment of work

A culture of Stop-Fix begins by viewing contributions to work product in increments, understanding that errors are likely to be hiding in the work we just completed. It's a blameless review, understanding that knowledge work is too complex to proceed error-free. This encourages transparency: a willingness to allow others to see into work that is in progress.

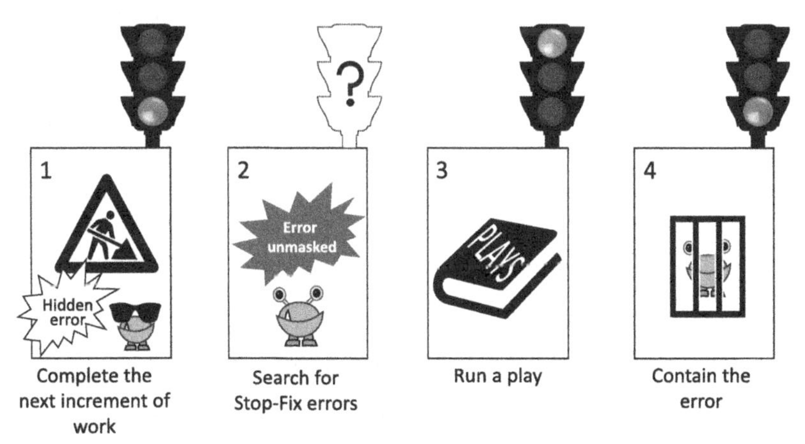

Fig. 14.6 The culture of Stop-Fix demands we check work product often and react quickly to error detection.

14.3.4.2 Search for Stop-Fix errors

The team meets, perhaps daily or weekly, to inspect the last increment of work for Stop-Fix errors. In a project, these errors are most often events that delay the project past the committed delivery date. In sales management, this may be shortfalls that are likely to cause the team to miss revenue targets. If a Stop-Fix alarm is identified, proceed to the next step; otherwise, no other action is required until the next cycle of the high-cadence reviews.

14.3.4.3 Run a play

At this point, an error has been uncovered that must be responded to. First, pick from plays in the playbook. If that doesn't work, the team should innovate a response for the given situation. If that doesn't work, the matter must be escalated. All this needs to happen quickly—within hours or perhaps a couple of days.

14.3.4.4 Contain the error

After a successful play, the error will be resolved, or at least contained. Work can return to normal.

14.3.5 Stop-Fix: Triage

The benefit of the Stop-Fix alarm is rapid response: creating a culture of nimbleness that catches errors before they escape. The first goal is triage: we need not address the longer-term issue at the moment that the problem is discovered. If a project is late because we didn't understand the customer requirements correctly, we first need to get the project back on track. That may require taking steps that are unrelated to the long-term issue of why we got so far in a project without understanding customer needs. For triage, we may hire an extra programmer to compress work so we can finish on time. Stop-Fix alarms often lead to temporary measures—taking an alternate route if the bridge is out on the main route (see Fig. 14.7). Of course, the bridge should be fixed at some point in the future, just as we should understand customer needs early—but that's not the focus of the Stop-Fix alarm.

14.3.6 The necessary elements of a Stop-Fix alarm

Stop-Fix alarms are effective only to the extent that they change behavior. So, what separates effective and ineffective Stop-Fix alarms? As shown in Fig. 14.8, there are at least five requirements. Stop-Fix alarms must be easy to use because they must be evaluated at a high cadence. If they are complex to calculate or interpret, they will not be used for long. They must be mistake-proofed: the first time a team takes time out to address an erroneous

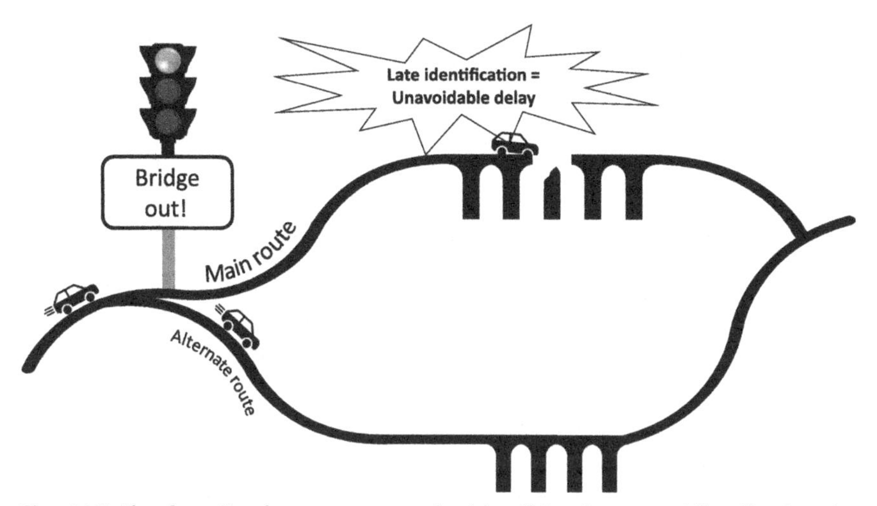

Fig. 14.7 The Stop-Fix alarm cuts waste by identifying issues rapidly, allowing the fastest response possible.

Stop-Fix alarm—say, data was keyed in incorrectly—may be the last time they take it seriously. It must be available to all of the people whom we expect to respond to it; otherwise it may not be considered legitimate. It must be up to date. Finally, it must be balanced and relevant, which is to say a Stop-Fix alarm must indicate a condition that is worthy of stopping the team to address. The measure should be value as defined in Section 5.4 or directly

Fig. 14.8 Five elements of a Stop-Fix alarm.

related to value; avoid abstract measures. For example, missing a delivery date promised to a customer is concrete and relevant. Using calculated metrics like #tasks/day may aim at the same goal, but may be too abstract to unite the team in a call to action.

14.4 Mistake-proofing complements Stop-Fix alarms

Mistake-proofing creates workflows that cannot be done incorrectly or, if they are, the errors they create are so obvious that they cannot be overlooked. The seat belt is a common mechanical example; it's very difficult to use it incorrectly. Another example is web forms that require an entry to proceed such as requiring the user phone number before the "OK" button enables. Like Stop-Fix alarms, mistake-proofing prevents "a defect from being propagated through the process" [10]. The two approaches are complementary: mistake-proofing prevents **Hidden Errors** where as Stop-Fix finds **Hidden Errors** and demands rapid resolution.

Mistake-proofing has a cost. There is an unending list of errors that can occur in knowledge work. Consider a simple project running for a few months with a few people from a few functions (e.g., Marketing, Operations, and Finance). The numbers of errors and omissions are uncountable: poor estimations, incomplete work, miscommunication, and ignoring customer changes are just a few. For almost all knowledge work, full mistake-proofing is unattainable. And the list of potential errors is ever growing. International regulations about data privacy and breaches in data security are just two examples of areas that are changing rapidly. Aside from that, the necessary tacit knowledge in knowledge work by definition implies there will always be large swaths of work that cannot be mistake-proofed.

So, we must look at mistake-proofing through the lens of return on investment. We start with history: where do we see the majority of errors? Now identify those errors that we can predict and will likely occur often enough to demand our attention now. Then we estimate the waste those errors cause and compare it to the cost of creating mistake-proofing. Focus on the good investments: areas with high waste reduction and low cost to implement.

Let's consider a simple example: mistake-proofing the entry of a credit card number into an online form as shown in Fig. 14.9. As a first step, we might ensure that 16 characters are entered. This is among the simplest of things to check. Then we might ensure only numerical digits can be entered. We can proceed from there to query the card issuer and determine that the

Error	Mistake-proofing
Too many or too few digits entered	Require 16 digits
Non-numeric digits entered	Ignore non-numeric characters
Invalid credit card number	Instant credit card validation with issuer
Invalid user has credit card number, but not physical card	Require expiration date and CCV
Invalid user has the physical card	Require PIN
Invalid user guesses until PIN entered correctly	Limit user to three tries for PIN and then lock account
...	...

Fig. 14.9 Mistake-proofing is a never-ending process, such as minimizing errors when verifying a credit card account.

number is legitimate: the account is open, it matches the name and shipping address on the order, and so on. This continues until the mistake-proofing is impractical.

Let's consider a common example from knowledge work: calculation of return on investment (ROI) as shown in Fig. 14.10. ROI calculations are common in knowledge work. An expert needs to help the leadership of the organization determine if an investment is wise; various forms of ROI provide a simple way to express monetary return. Let's start with a workflow that proposes the expert calculate some sort of ROI. Perhaps our history is that people pick all sorts of ROI calculations, many of which are irrelevant for the investment being considered. So, the first step in mistake-proofing might be to specify the type of ROI calculation, such as net present value (NPV). But people make errors keying in their own formulas, so our next step might be to provide a template so everyone uses one fully debugged formula. But then we find people are using wildly varying foreign exchange (Fx) rates, so our next step might be to provide a standard set of company exchange rates.

Error	Mistake-proofing
Using invalid type of calculation	Specify net present value (NPV)
Keying in the formula incorrectly	Provide an NPV template
Using out-of-date exchange rates	Put standard exchange rates in template
...	...

Fig. 14.10 An example of error-proofing is applied to a return-on-investment (ROI) calculation.

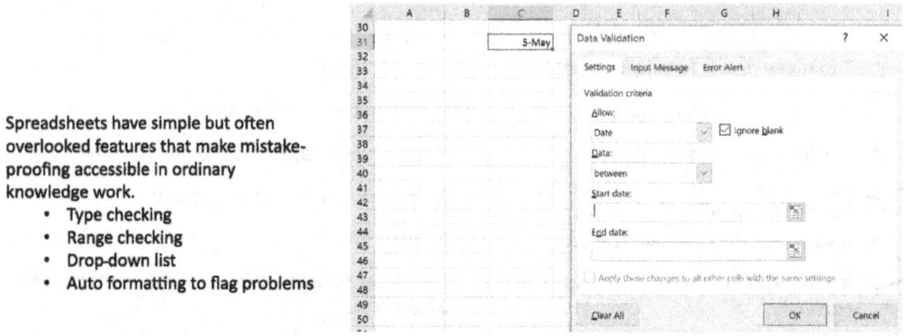

Spreadsheets have simple but often overlooked features that make mistake-proofing accessible in ordinary knowledge work.

- Type checking
- Range checking
- Drop-down list
- Auto formatting to flag problems

Fig. 14.11 Microsoft Excel makes many mistake-proofing tools accessible in ordinary knowledge work.

The journey never ends. Each time we find an error occurring frequently enough to deserve attention, we search for means to mistake-proof that case. The journey may not end, but we don't need to get to the end; the organization will get better and better at calculating ROI. So, it may never be perfect, but one day the remaining problems will be small enough that they don't need to be addressed.

Spreadsheets like Microsoft Excel are a common tool used to gather input from users. Spreadsheets have many virtues, especially that they are capable of displaying data in a large range of ways, and they can be built by people without deep software expertise. When we are collecting data from thousands of people, we will want to invest in a more elegant framework, such as a web page-connected database. But in knowledge work, needs are often changing rapidly and perhaps only a few dozen people are providing data. The extensive IT and database support needed for a web page is often impractical.

Another advantage of spreadsheets is that they have built-in mistake-proofing features (Fig. 14.11). Using data validation in Microsoft Excel®, we can limit the type (e.g., only dates), and for numerical entries we can limit the range (e.g., only dates between January 1, 2015 and January 1, 2025 or integers less than 1000). Optionally, we can ensure users select from a drop-down list with specified entries (e.g., "yes" or "no"). And spreadsheets can visualize Stop-Fix alarms through automatic formatting functions that change the display depending on cell value, for example, using 72-point red letters to display the word "ERROR."

Zidel presents a "Mistake Proofing Worksheet," standard work where mistakes can be analyzed quickly to find opportunities to prevent reoccurrence. It's essentially a mini-problem solve for simple issues: describe the

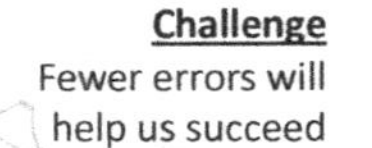
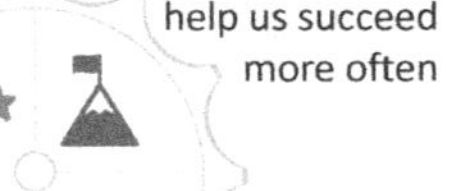

Fig. 14.12 Stop-Fix alarms and mistake-proofing simplify, engage, and encourage experimentation.

problem, provide some background and analysis, then list suspected root causes. Finally, the person can propose mistake-proofing for that mistake [11].

14.5 Conclusion

Hidden Errors cause waste that grows the longer the errors are hidden. This chapter has presented methods that address this waste. As shown in Fig. 14.12, Stop-Fix alarms simplify with clear calls to action. Mistake-proofing simplifies by stopping errors from being created, thus removing the need to revise erroneous knowledge work product later. And playbooks document standard responses when common errors are detected; the playbook is intended to cover perhaps three-quarters of the cases—a large benefit, especially to less experienced members. These techniques engage a team that can envision creating work product with fewer errors and correspondingly happier customers. It also provides a clear requirement for work that often suffers from ambiguous demands. Finally, if the entire organization works to a higher standard, its people can lift the quality of work along with the organization's reputation.

References

[1] M. Fackler, The Toyota Way is translated for a new generation of foreign manager, The New York Times (February 15, 2007), https://www.nytimes.com/2007/02/15/business/worldbusiness/15toyota.html.
[2] https://global.toyota/en/company/vision-and-philosophy/production-system/.
[3] P.R. Williams, Visual Project Management, Think for a Change Publishing, 2015, p. 32.
[4] J. Liker, The Toyota Way, 14 Management Principles From the World's Greatest Manufacturer, McGraw-Hill, 2004, p. 14.
[5] Toyota Andon concept MotoMachi factory, Japan, March 28, 2016, Youtube. https://www.youtube.com/watch?v=r_-Pw49ecEU.
[6] T. Ohno, Taiichi Ohnos Workplace Management: Special 100th Birthday Edition, McGraw-Hill Education, 2012, p. 62. Kindle Edition.
[7] https://blog.deming.org/w-edwards-deming-quotes/large-list-of-quotes-by-w-edwards-deming/.
[8] C. Kenney, Transforming Health Care. Virginia Mason Medical Center's Pursuit of the Pursuit of the Perfect Patient Experience, Virginia Medical Productivity Press, 2011.
[9] https://www.npr.org/sections/therecord/2012/02/14/146880432/the-truth-about-van-halen-and-those-brown-m-ms.
[10] R.J. Pryor, Lean Selling: Slash Your Sales Cycle and Drive Profitable, Predictable Revenue Growth by Giving Buyers What They Really Want, AuthorHouse, 2015. Kindle Edition.
[11] T.G. Zidel, A Lean Guide to Transforming Healthcare. How to Implement Lean Principles in Hospitals, Medical Offices, Clinics, and Other Healthcare Organizations, Quality Press, 2006, p. 97

CHAPTER 15

Standardize workflow

> *Where there is no Standard there can be no continuous improvement.*[a]
>
> ***Taiichi Ohno [1]***

15.1 Introduction

Standard workflow creates a single-best way to do the tedious parts of knowledge work, for example, the structure around solving problems or the details around approvals. It can seem dull, and certainly standardizing has its dull moments, but not nearly as many dull moments as work without standards: renegotiating tedious steps again and again followed by the useless innovation to reinvent something tedious; and then, recovering from the inevitable mistakes of the reinvention. What makes standardizing inviting is how we can do something tedious in a tenth of the time it used to take and so use the saved time on something that fascinates us: innovation, mastering a craft, nurturing a budding leader, or delighting a customer.

15.2 Building a foundation

> *There is something called standard work, but standards should be changing constantly. Instead, if you think of the standard as the best you can do, it's all over. The standard is only a baseline for doing further kaizen.*
>
> ***Taiichi Ohno [2]***

Continuous improvement accepts an organization where it finds itself and then enables that organization to improve step-by-step, transforming into something that would hardly have been imagined at the outset. But, before

[a] Ohno used the word *Kaizen*, Japanese for continuous, incremental improvement.

Improve
https://doi.org/10.1016/B978-0-12-809519-5.00015-6

a knowledge organization can start its first cycle of improvement, there must be a foundation from which it can take that first step: the ability to execute repeated tasks in the same way. That begins with the simple step of identifying the best way we know today and using that best way each time we do a task. This is standard work.

One of the most prevalent misconceptions about standard work is that it steals creativity, turning people into automatons, mindlessly doing work they used to enjoy. In fact, standard work addresses the least creative parts of our jobs, making them easier, and that allows more time for those parts of the job that require creativity. For example: I'd much rather be working on an exciting new product or visiting a customer than working on the budget, trying to determine how many PCs we can afford to replace this quarter. I was spending too much time on just such topics when we created a "visual budget" that standardized how managers requested budget, how Finance compared forecast to actual, and what data was required to gain approval for requisitions. By creating this standard work, we reduced the tedium of budget management by >75% and we nearly eliminated the number of times we exceeded our budget (a knowledge work "defect" that can take a lot of time to explain to the boss!). Both aspects allowed us to increase the time we spent on more interesting things.

There are several other misconceptions about the role of standard work. In traditional "top-down" thinking, process may have been pushed down from the management and then used to blame people when things didn't go well. In lean thinking, standards make requirements clearer and provide more opportunity for driving change. In traditional thinking, process creates paperwork—useless instructions and make-work reports. In lean thinking, standard work is a lightweight mechanism to build a foundation that empowers everyone to improve upon it. Standard work also flattens the organization: "No one is above the standards we created!" When done well, standardizing builds relationships among the team because all pull together, all doing things the same way so that over time they can (1) find what works and do more of it and (2) find what doesn't and get rid of it.

Standard work creates clarity that allows the people doing the work to improve the process. When the organization is diligent in documenting the standard way to do something and that standard way doesn't work, it's quickly seen that that standard is the problem. When that way is not defined, it's easier to blame the person doing the work, or the management, or bad luck. In this way, standard work, whether on the factory floor or in knowledge work, allows people doing the work to see gaps, make changes, and observe improvement. Standard work is just the guidelines created and maintained by the people doing work for the people doing the work.

When someone gives an opinion that it would work better by doing it another way, you should immediately try it. That way, the decisions that were made become the rules made by you, for yourself.

Taiichi Ohno [3]

In deploying continuous improvement, you may encounter some of the misconceptions of traditional thinking shown in Table 15.1. Unfortunately, there is some truth in that old thinking: process has often been applied to reduce creativity, to rob people of opportunity to make decisions, to create make-work, and to frustrate personal growth. You will doubtless find a few change agents in your organization who will support using standardization to bring improvement from day one. But most people must be won over, which will happen when they see lean thinking work, fixing problems that have held them back. And you probably will find some portion of the

Table 15.1 Traditional thinking vs lean thinking about process.

	Traditional thinking about process	Lean thinking about standard workflow
Creativity	**Process** makes work boring	**Standard workflow** makes simple decisions easy so that I can focus more time on tasks that demand creativity and expertise
Decisions	**Process** takes away my ability to make decisions	**Standard workflow** makes my role in decision-making clear
Workload	**Process** adds to my workload with useless paperwork	**Standard workflow** reduces workload by cutting the waste of reinventing the way we do tedious work
Relationships	**Process** isolates people, commanding them to "Just do what you are told"	**Standard workflow** enables teams to improve how they do things
Career growth	**Process** stunts career growth by teaching a check-the-box mentality and wasting time on paperwork	**Standard workflow** provides opportunities to lead real change as groups improve themselves

organization that, no matter how much benefit lean thinking creates, will hold to the traditional thinking in Table 15.1.

15.2.1 Where does the standard start?

> *Standardized work at Toyota is a framework for kaizen improvements. We start by adopting some kind—any kind—of work standards for a job. Then we tackle one improvement after another, trial and error.*
>
> ***Taiichi Ohno [4]***

It is a premise of lean thinking that an existing workflow, however weak, can be improved in increments to become outstanding over time. But without an existing process, there is no starting point. This is not as demanding a requirement as it might seem at first—you certainly don't need an ISO-9000 documented process at the outset. Bear in mind that everyone has a process for doing repeated tasks, though that process is often only partially documented and may vary considerably between individuals. The starting requirement is only that there's an initial process as it is today, even if that process is undocumented or only partially documented.

Even when there is a documented workflow, organizations usually discover that what people actually do differs from what is written down. This is a common vestige of traditional management: a supervisor or designated expert may have documented a process that he or she felt the team should execute. That documented process may have poorly represented what the team did at the time or it may have been initially accurate, but was not updated as the process changed. In either event, process documentation may not represent the current state. This is why we begin with a broad team specifying what people *actually* do rather than what they *should* do.

15.2.2 Merging different ways of doing things into one

Some organizations will, unintentionally, do the same things in different ways. This can happen when teams are merged together through acquisitions, or when multiple teams started with a common process and then drifted apart over time. Normally, the long-term goal is common process across the organization. However, if the current state has multiple processes, there are potential negative effects of quickly forcing all teams together. If a manager dictates all teams will use the process used by one of the teams, it's just another form of forcing solutions down from the top, something we avoid in lean thinking. If the management directs the teams to "settle it themselves," demanding they come to consensus on which process to move forward with, it may create unhealthy competition as teams battle, each promoting their own methods. Often the best alternative is to drive change methodically: narrow the scope to select the most urgent need around one of the process variants. In the initial improvement cycle, focus on one team, but include a few people from other teams. Over time, future improvement activities can expand and gradually create a common process across the organization. This can increase harmony among the teams, but has the disadvantage of taking longer to complete.

Stabilizing the foundation

Kenneth Noonkester is the global leader for product development at H&T Widgets. He has identified project management process as the target for his next improvement cycle. His US team manages their projects with Gantt charts from Microsoft Project; his Singapore team uses paper Kanban boards (see Chapter 21); his British team uses an internally developed spreadsheet. He talks with Tucker, H&T's lean thinking expert, about what process to use as a starting point.

Noonkester first suggests they include all the project managers in one event. He wants the three groups to start by working out which method they will use: Microsoft Project, Kanban boards, or the Brits' spreadsheet; then they can add improvements together. Tucker warns him off this path: "It's likely the event will wind up a battle as each group strives against the others to secure their method as the global standard. We want to create an environment where the team pulls together."

Noonkester responds, "Well then, the Gantt chart seems like the strongest method to me. Let's move Singapore and London to the US team's method."

Tucker advises against this path as well: "If we select the method by management fiat, we'll have a difficult time gaining adoption with Singapore and London in the months to come. Every time they have a late project, they'll be tempted to blame it on being forced into change from above."

Noonkester is frustrated. "If we can't start from where we are because some teams will disagree and we can't create a workable starting point because some teams will disengage, what do we do?"

Tucker guides Noonkester to assume this is a management problem. "For years the company didn't attend to creating a common tool for managing projects. As a consequence, each of the various sites invested a great deal of energy creating their tool set. It's only natural that people

Continued

Stabilizing the foundation—cont'd

who have passion for their craft will feel ownership of what they've created—let's build on that passion."

Tucker suggests a two-part plan that will generate improvement with minimal friction. "Let's focus the first improvement cycle on one of the three teams." Based on revenue needs and low performance to schedule, they decide the US team has the most urgent need. They create a problem statement based on the issues in the US and build a team of US R&D engineers and a couple of salespeople. They also invite a project management consultant to get an outsider's view. Finally, they invite two engineers from the London team and two from the Singapore team to join. They target process improvement for the US team in the next 6 weeks, followed by several months of piloting to ensure the improvements are real and sustained. The second part of the plan is to create a global process, which may be 6–8 months in the future.

15.3 The ground view

Most people believe that process is simply a defined workflow—like writing a recipe for a cookbook: just tell people what they should do and expect them to do it. Actually, cooking recipes are a poor analogy for process in an organization because recipes are written for a single person; a better analogy would be recipes in a restaurant chain. Consider the famously stringent quality control processes McDonald's demands of their franchisees. Sure, the recipe is part of standard work, but there is much, much more standard work [5].

Another analogy is a complex process familiar to almost every adult: managing traffic speed. In this analogy, the "organization" is the state Department of Motor Vehicles (DMV) plus all the drivers in the state. Driver **workflow** (roughly, the "recipe") is mostly to stay within posted speed limits. Of course, simply posting speed limits is inadequate to manage vehicle speeds. Consider the other things the state does to ensure drivers remain within the speed limit:

- Requiring driver's school and licensing. This is the **training and certification** to qualify a person to execute the workflow.
- Requiring a speedometer in vehicles. The speedometer is the **scorecard** used by the driver to help determine if he or she is within the speed limit.

- Equipping law enforcement with radar speed detectors. This forms a **reporting system** used to gather information on a vehicle's speed.
- Issuing and enforcing speeding citations. Speeding tickets are the **countermeasures** most often used to bring the process back to within specifications. Other countermeasures include increasing insurance rates for those caught speeding and statewide point systems where too many violations over a period of time result in license suspension and other penalties.

These components of process work together as shown in Fig. 15.1. Here we see that workflow and training are just the start. The dashboard speedometer is constantly monitored by the driver and becomes a "scorecard," informing the driver about how well he or she is managing this part of the process. A reporting system (roadside use of radar speed detectors) measures process compliance. When compliance is violated, there is a "review system" where a policeman stops a car to explain the details of the violation when a driver is caught speeding. This often leads to a speeding ticket, which is a countermeasure intended to compel the driver to remain within prescribed limits. So, unlike a recipe for home cooking, detailing the workflow (roughly, the speed limit signs) is only the beginning of building reliable process.

A more general picture of organizational process for the ground view can be seen in Fig. 15.2. This is a single instance of knowledge work such as an attorney crafting a contract or a salesperson calling on a customer. While the analogy to driving is close, a couple of steps need to be added to create a more general case. One is "Tools" that aid in the execution of the process, such as sales funnel management software aiding sales reps in selling processes or cloud-based project management tools that aid the project manager. Also, a block for resourcing is added: organizations will normally directly manage

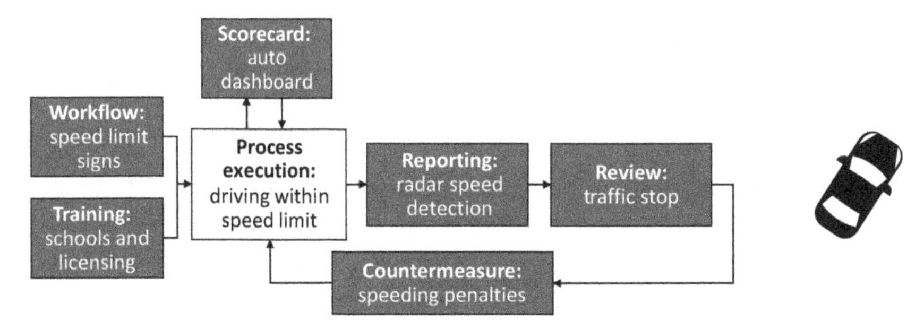

Fig. 15.1 The standard work of managing vehicle speed from the ground view.

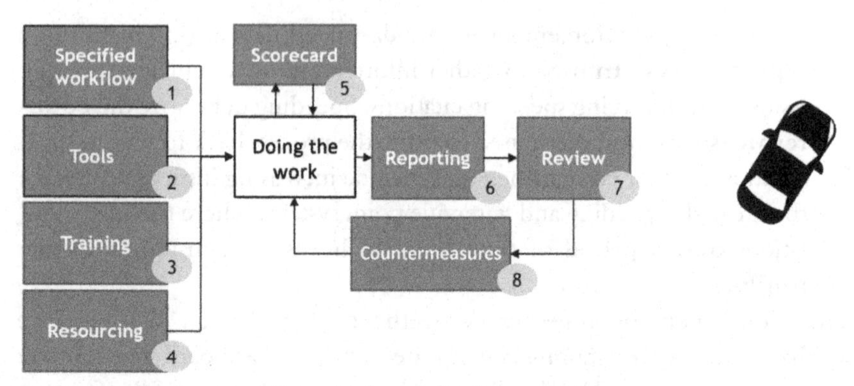

Fig. 15.2 Eight elements of standard work for the ground view.

how many resources are applied to a process, be that workers on an assembly line, sales representatives in a region, or engineers on a project.

The eight blocks of Fig. 15.2 are detailed in the next eight sections.

15.3.1 Step 1: Specified workflow

Ground view or "instance" workflow describes how a process should be executed one time. It is the recipe of the workflow. Documents can range from a single page to hundreds or even thousands of pages. They can be text, diagrams, videos, or a mix (such as assembly instructions). Process workflow that is documented but unused serves little purpose. You can tell if a workflow document is used by asking those who carry out the workflow how often they refer to the document system. Ask if those documents are easily available and if they are updated often. Check revision history to see how often the process workflow is updated. Workflow documents that are parked in an obscure SharePoint library or in a dusty binder are unlikely to bring much value.

15.3.2 Step 2: Tools

On the manufacturing floor, process instance tools include physical tools like a screwdriver. For knowledge work, tools are generally software such as those for financial calculations to evaluate a project or a physical display like a task list hanging outside the project manager's office. The more standardized the tool, the less variation the workflow will exhibit and the more fully it can be integrated into the process.

15.3.3 Step 3: Training

Training is required for any process. Sometimes the need to ensure a person is process-capable is so great, the training must be accompanied by certification such as a university degree, professional certification, or passing a bar exam. Other times, attending a 30-minute presentation is sufficient.

15.3.4 Step 4: Resourcing

Workflows are resourced by the organization's leadership. This may be direct—for example, when a management team assigns a sales rep and customer service rep to a territory or when the VP of R&D assigns a project team. Even when the resources are not formally assigned by the leader, they will at some point come under review: why do we have five people working on Project 1 when Project 2 is idle and so much more important?

15.3.5 Step 5: Scorecard

A scorecard (Section 8.5.2) is the display of key information used to monitor the quality and efficiency of a workflow. It's the primary information used by the team executing the process. For a team running a project, this might include a task board for the upcoming items or a set of milestones for the project. Robust workflows require information to ensure the process is being executed at the required cadence and that all the steps of the workflow are being executed properly including managing quality, efficiency, and safety.

15.3.6 Step 6: Reporting

Reporting is information that is made available for those outside the workflow such as managers who review performance, customers who are monitoring progress, and outside departments who must coordinate. Reporting is usually updated less regularly than the scorecard. Consider the project team we discussed above: the scorecard might include a task list for each person with a status that's updated daily and a list of all identified risks. By contrast, the reporting will likely be monthly or quarterly. Also, it will be summary in form and likely to address a broader range of areas:

- Performance to schedule during the last month
- Performance to budget during the last month
- Critical risks including status and likely effects
- Customer contacts this year
- Completed features since last month

The information in reporting is different from that in the scorecard because it supports different types of decisions. The scorecard is aimed at driving the stream of internal decisions made by the team: shifting focus from one task to another, requesting help on immediate problems, and ensuring customer needs (internal or external) are satisfied. Reporting is aimed at driving decisions by an outside team, most commonly the leadership, such as shifting resources and approving budget requests. Reporting is often done with a dashboard, which is discussed in Section 8.5.2.

15.3.7 Step 7: Review

The review is a leadership team examining the way a process was executed a single time. This is the forum used mainly by those outside the process team. It might be a weekly meeting dedicated to a large project. It could also be a monthly review of the six top sales opportunities. On the other hand, it could be informal reviews called when a problem is discovered. For example, if a key customer calls the company president saying they are so unhappy with a product function that they will take their business elsewhere, that's probably going to result in an impromptu review.

15.3.8 Step 8: Countermeasures

Countermeasures (CMs) are those steps taken after a review to address gaps found during the review. Examples of these CMs include putting a project on hold, or scheduling a customer visit to help untangle a problem. Like traffic citations for speeding, these are the levers the management uses to address shortcomings in the workflow.

Evidence of effective countermeasures shows the process loop is closed. For example, consider all the information we can glean from the following:

> *At the monthly review, the Hospital Director increased weekend nursing staff 20% due to delays in the Imaging Department.*

We know that:

- data on delays is collected and reported;
- the Hospital Director is reviewing the data; and
- the data is clear enough to drive an action.

All of this indicates that a process to measure delay and drive behavior is in place. It doesn't mean the process is perfected or even that the actions taken will all be effective. But it's a good sign; at the very least, it gives confidence that a foundation is in place from which continuous improvement can build.

15.3.9 Benefits of standard work

Perhaps the least-understood benefit of continuous improvement is how it becomes a force multiplier for managers. If you are a manager, consider how much time you must invest in "normal" work. Do you have to call a lab three times to expedite results? Do you have to write emails to get approval for something common? Do you frequently have to ask people on your team what they are working on or how long it will be before they are done? These are all examples of things that could be managed better through standard work. These activities waste your time and create tension: you might be annoyed that you have to ask for something people should know to do, and the recipient of that type of attention may be annoyed with what appears to be make-work or micromanagement. As the team moves to the model of Fig. 15.2, the efficiency of all involved goes up. This now frees the manager's time and energy to those things managers should be focused on, for example, resourcing decisions, dealing with exceptions through a review process, and, of course, finding the next area to improve.

15.3.10 Batch workflow creates waste

It is difficult for people to get rid of their misconceptions that it is cheaper or more efficient to do many parts at once rather than one piece at a time.

Taiichi Ohno [4]

In lean manufacturing, we strive to produce one piece at a time. In traditional manufacturing, people might weld 50 widgets, put them in a bin, send them to assembly where someone might bolt on a second piece to all 50 widgets, put them in another bin, send them to the inspection department, and so on. In lean manufacturing, we strive for something called "single piece flow": weld one widget, check it, bolt on one second piece, check it, and so on. This drives productivity up and defects down because if there's a mistake in the welding that's discovered when you bolt on the second piece, you just have 1 bad weld to rework, not 50.

In knowledge work, we face a similar issue in checking work. Consider a process to file patents: first we do project work, then we ask developers to identify inventions and write disclosures. Then a small group reviews the

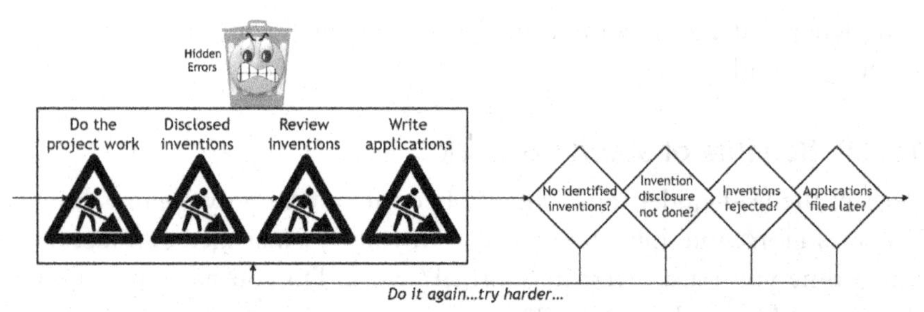

Fig. 15.3 Batch workflow to file patent applications from project work creates waste.

invention disclosure and decides if it should be sent to an attorney to write the application. This work is often done without intermediate checks, something like what's shown in Fig. 15.3. What can often happen is, at some point in the year, the boss might notice that too few patents are being filed, which can lead to blaming: "Why aren't more people identifying potential inventions or filling in disclosure forms or reviewing inventions?" We can think of this as errors hiding in the process. When we look deeper, we see the organization has created a large "batch" workflow to create patent applications.

The organization can create smooth workflow by placing frequent, simple checks—go/no-go decisions at each step, as shown in Fig. 15.4. For example, when doing projects, we could start with a plan from where we think the likelihood of a patentable invention is high and build the evaluation of those areas into the Action Plan/Success Map. Missing the plan becomes a Stop-Fix error. Similarly, we can create artifacts, like an invention disclosure template, where completion level is shown visually so that a person can tell at a glance if it's done. And so on.

Creating smooth flow in knowledge work is complex. Even when the requirements are unambiguous, our workflows usually require tacit knowledge to evaluate. One of my favorite examples of interplay of unambiguous standards and tacit knowledge is the strike zone in baseball, the imaginary box in front of a batter inside which the ball must pass to count for a strike. The strike zone is unambiguous, something anyone with an internet connection and 5 minutes can learn. But the ability to call a ball in that strike zone in a professional baseball game is a skill demanding tacit knowledge that takes decades to acquire. Removing ambiguity does not remove the need for expertise.

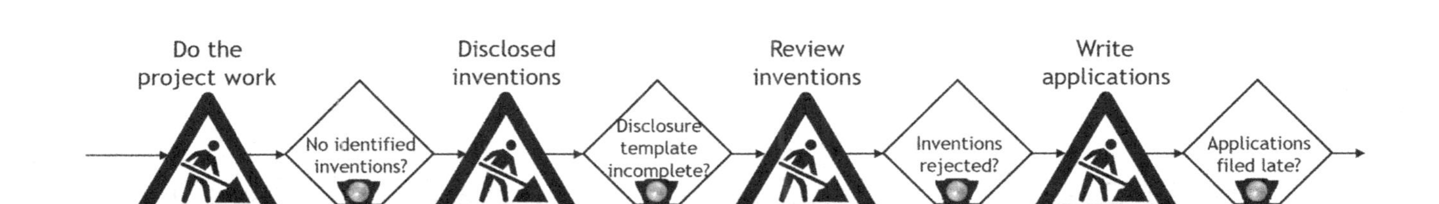

Fig. 15.4 Smooth workflow to file patent applications from project work.

15.3.11 Standards drive normal work. Managers review exceptions

Without review and feedback, every process will falter. Exceptions are the most effective way to understand the process: what it does well and where it is lacking. Managers can spend so much time greasing the wheels of normal work that they can't pay adequate attention to identifying and resolving exceptions. Consider a simple example: managing a team budget. In the absence of process, budgets are normally simply an extension of prior year spend. The team may not be engaged in the process to create the budget, and if not, when overspends occur, they are unlikely to engage fully in the correction. The manager can easily become immersed in what could be standard work: approving each spend, watching for month-end surprises, and calling staff in for meetings to deal with overspends. Cycles like this can continue for years. Standard work offers a better path.

Now imagine if each person who spends from the budget must at the start of the year provide an annual plan for their spend—not just scaling last year's spend, but thinking through year-over-year changes and laying in a plan for the major spends. Then, as the year proceeds, it becomes their task to stay within their plan. Spending within that plan can be approved easily. The job of a manager shifts to reviewing the budget plans for quality in January and looking for exceptions the rest of the year. The manager's focus can be applied to unexpected spending, something that takes much less time than detailed reviews of every spend throughout the year. It's challenging to quantify the waste this reduces, but my experience is that 10 × is a reasonable estimate. The waste a manager creates by overseeing things that don't need his or her oversight is replaced by laser focus on exceptions: those things the standard process missed. The manager will find problems earlier, understand the root causes better, and will be better able to guide the next step of process improvement. This positive cycle spirals up: as continuous improvement pushes more exceptions into standard work where the team can deal with them, the manager can focus fully on the smaller number of exceptions that remain. The beauty of continuous improvement is that everyone benefits: managers spend their energy doing what they are most interested in, employees are more autonomous, and the organization enjoys the benefits of everyone's improved performance.

In the next section, we will move to the next level of review: dealing with instances grouped together or aggregated in what we will call a "helicopter view." Aggregation review typically reaches higher up the organizational chart. Where reviewing a single project or patient experience may involve mainly the team doing the work, aggregate review is usually carried out by a management team.

15.4 The helicopter view

Returning to the process to manage vehicle speed, notice that Fig. 15.1 showed only the execution of the process for a single user, which is how the speed of one car is managed. There is another set of mechanisms used to manage the speeds of many vehicles through the use of aggregated results. For example, if it is discovered that many accidents occur along a section of roadway, the Department of Motor Vehicles may institute a study to measure average vehicle speed and see if it's being violated at a higher-than-normal rate, which likely would recommend more enforcement actions in that area. On the other hand, the data may show the vehicle speed is being complied with, but that accidents are still occurring; this might indicate that the speed limits for a section of roadway need to be lowered.

To ensure the speed limits are followed, the state might aggregate data such as traffic citations by time of day, day of week, and location. If so, there would need to be workflow defined and tools developed so that this data could be collected and analyzed consistently over time. Our process instance of Fig. 15.1 is augmented in Fig. 15.5 to contrast ground and helicopter views. You may have noticed that in Fig. 15.1 there was no means to address issues in process instance workflow or training. That doesn't appear there

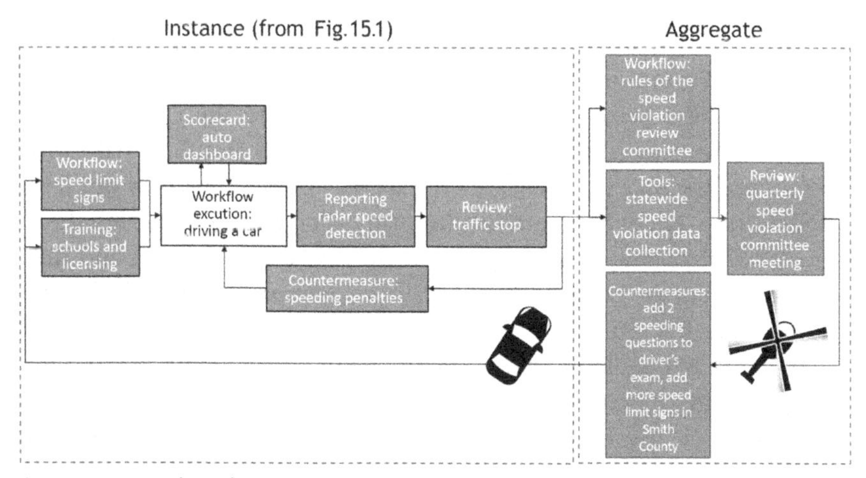

Fig. 15.5 Expanding the project/initiative view to the helicopter view.

because it's unlikely that a single traffic stop would result in changing the speed limit sign policy or the statewide driving test—those types of decisions can only be made by analyzing many instances.

Notice that there are many similarities between the instance (ground view) and aggregate (helicopter view): both have workflow definition, reviews, and countermeasures. However, the contents of these components are quite different: the workflow around driving a car has little in common with the workflow to analyze all the speeding citations in the state over the last year. The countermeasures are also different: how the driver's manuals are written, how new drivers are trained, and how speed limit signs are posted.

Almost every process an organization executes will have an instance/aggregation split: an annual review process from human resources will include the process to support a single performance review and the process for aggregation, such as measuring how many reviews are complete and the average ratings across the organization. A project management process will include the steps to manage one project and also steps to review the entire project portfolio. A call center will need process for each operator to answer one call and a means to measure average wait time for all callers. The instance and aggregate cases have very different workflows, tools, and countermeasure mechanisms. A project manager may see that his or her project is late and take steps to accelerate the project. The Director of R&D may see that 40% of his or her projects are late and look at the project management process for opportunities to improve.

A general workflow with both ground and helicopter views is shown in Fig. 15.6. Notice that the instance and aggregate sides have similar structure. The aggregate side shows training and resourcing in dashed blocks, but are not detailed here because those components are typically carried out by managers informally (they get themselves up to speed and they make time necessary to complete reviews); also, the aggregate side shows only one level of aggregation, though it could show many. For example, the Southeast Region might review all sales in that region (first aggregation), and then report that out to the US main office; they are aggregated again with all other regions of the US. So, while the aggregate side is shown here only once, understand that it could be repeated multiple times extending to the right, with each aggregate targeted at a higher level of review.

Details of the five blocks new in Fig. 15.6 are described in the sections that follow.

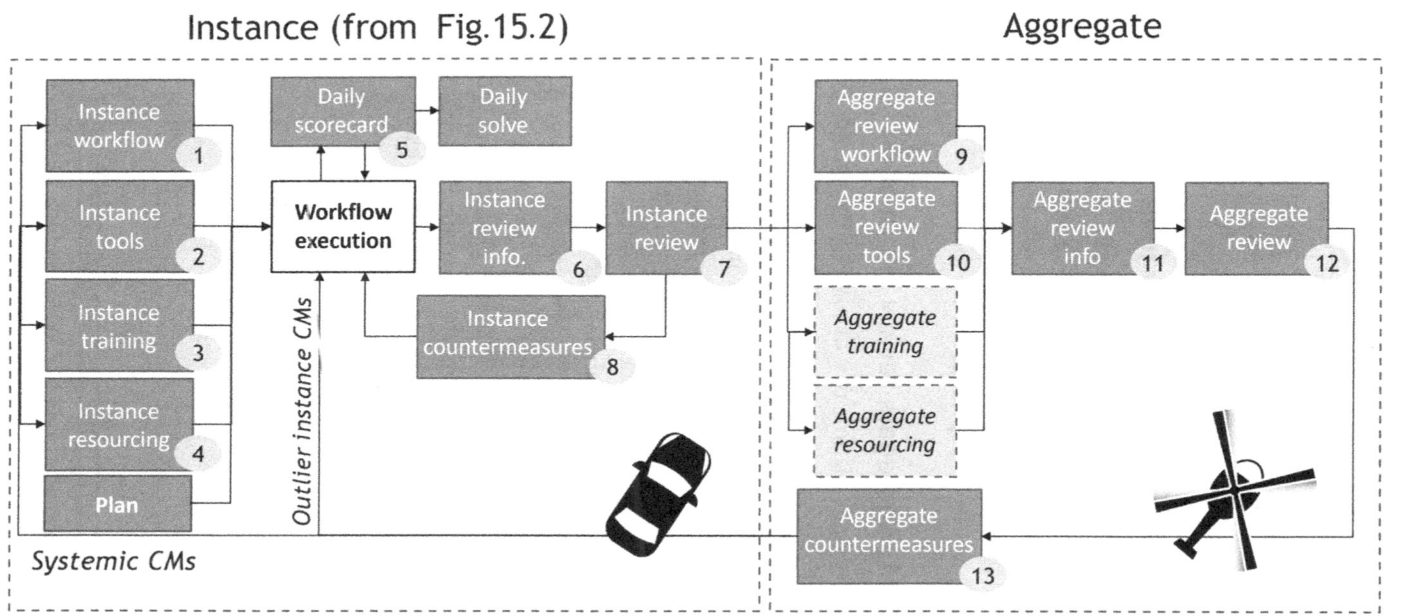

Fig. 15.6 Standard work: the ground view vs the helicopter view.

15.4.1 Step 9: Aggregate review workflow

Aggregate review workflow is the set of steps defined to review an accumulation of process execution data. This could be simply defining the team of people who will review the data, calling a monthly meeting for the review, and setting an agenda for that meeting.

15.4.2 Step 10: Aggregation review tools

The tools for aggregated review are normally software—spreadsheets or online tools that generate data in the required format. For a project portfolio, this might be planned vs actual completion dates, planned vs actual expenses, and so on for all projects. In a hospital, aggregate tools might provide all the wrong-site surgeries in a hospital chain by location.

15.4.3 Step 11: Aggregate review information

Aggregate review information is a display of data needed to support the review process. For example, consider the simple display in Fig. 15.7. Here, key information about the top six sales opportunities is displayed along with an "all other" row. As is typical, key data from the most important instances (how late each opportunity is) can be displayed along with aggregated data (average late time for all opportunities). Other examples include a monthly review of hospital admissions, which might look at average time to admit and then look in detail at the outliers (if the average time is 35 min, but in three cases admission took >180 min, those three might be looked at in detail). A construction company might review average budget and performance to schedule for all its 35 active projects, but spend a great deal of time reviewing the top five projects in detail. Reviewing key instances and aggregate data brings balance, allowing the leadership to address the most important instances with immediate action while improving the entire process over time.

January sales report, Northeast US (all opportunities >= $100k)
Date: Jan 15

Customer	Sales rep	Value	Plan close date	Forecast close date	Days late
RG Samples	Smith	$400k	13-Mar	10-Jun	89
EHG Works	Douglas	$320k	1-Mar	17-Mar	16
Whistle Wyle Works	Smith	$250k	23-Feb	12-Mar	18
Am Ritterskamp	Raney	$250k	11-Feb	28-Feb	17
O. Pen Sesame, Inc.	Daze	$170k	31-Jan	7-Feb	7
Conduits and Cakes of NYC	Daze	$120k	30-Jan	18-Feb	19
All other (26 Opportunities)		$660k			
Total		**$2170k**	Avg days late		**28**

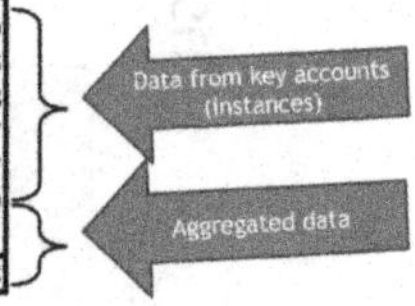

Data from key accounts (instances)

Aggregated data

Fig. 15.7 Example of aggregate review information for a sales process.

15.4.4 Step 12: Aggregate review

Aggregate review is typically a management team looking over the aggregate review information of Step 11.

15.4.5 Step 13: Aggregate countermeasures

Aggregate review countermeasures are artifacts of the aggregate review, two of the most common being as follows:

1. Instance CMs: CMs equivalent to Step 9 that happen to be discovered in the aggregate review.
2. Systemic CMs: Actions that affect the process as a whole such as changing the driver's manual in our example of managing vehicle speeds.

Instance CMs, which would normally be discovered in the instance review of Step 7, may only come to light during an aggregation review for a variety of reasons. For one, the review team is typically more senior, so they will naturally see issues that might be missed at the instance review. Another reason is that the aggregate review allows for projects to be compared so outliers will stand out. This is information usually unavailable at the instance review of Step 7. As an example, consider the aggregation data simulated in Fig. 15.7. This table shows all sales opportunities in the Northeastern US region above $100k value, ordered from largest to smallest. This causes *RG Samples* (first entry) to stand out because it is both very large and very late ($400k, 89 days). Even if the primary purpose of the aggregation review is to review summary data, such an outlier would likely generate instance-based CMs.

Systemic CMs are those driven by summary data such as average days late in Fig. 15.7 (bottom right). Let's say that in this process, the expected value of average days late for large opportunities is <10 days; having an average of 28 days (as in Fig. 15.7) could therefore drive systemic action—modifying workflow, tools, training, or resourcing.

A well-known systemic CM applied in software and IT organizations over the last 10–15 years occurred after aggregated data showed that software project managed with Gantt charts and related methods provided poor results. A CM applied across many industries has been to change the project management method to Agile Scrum, which is radically different from the traditional critical path management methods. Changes of this magnitude are rarely driven by results of a single instance—results of many process instances aggregated together are required to justify such broad action.

This section has reviewed 13 components of workflow definition. The first goal was to make a clear distinction between workflow definition (what many people mistakenly think of as *process*) and the many other components needed to have a fully functioning process. But the deeper purpose is to clarify what's needed as a foundation for continuous improvement. Certainly, it's not necessary for every one of these 13 components to be mature just to get started. And this list of 13 is not meant to be comprehensive—different organizations will have different needs.

15.5 Process robustness

I can recall the time when automobiles had few safety features. There were no antilock brakes nor back-up cameras. There were no shoulder belts and I can recall riding in cars old enough to lack even lap belts; it didn't matter because no one—at least not a person I knew—used seat belts. Small children commonly stood up in the back seat or rode in the bed of a pickup truck. Dashboards were steel; low seatbacks provided no neck support in the event of an accident. There were no airbags. In those days, injury had one means of avoidance: try harder or, in the language of vehicle safety, drive defensively.

Over the years, many safety devices were added to cars. Some were low-tech solutions like shoulder belts, child car seats, and alarms when seat belts were disconnected. Also, from the 1980s forward, seat belt usage increased dramatically due to new laws. Other solutions needed technical development such as antilock/antiskid brakes and airbags. Progress began when governments and the auto industry rejected the mindset that drivers just needed to try harder to avoid accidents; they created new devices, laws, and awareness so that accidents were less likely, and when they did occur, they caused less harm. The results were outstanding. In 1980, the motor vehicle crash death rate was about 25/100,000 people; in 2017, it was less than half that [6]. Today's cars are equipped with standard devices that make driving safer, understanding that drivers will make some level of errors (see Fig. 15.8).

We can apply similar thinking to knowledge workflow. First, adopt business humility: "There are many things we don't know, so we must install checks along the way." For example, an attorney filing a patent application is likely to spend a great deal on an invention—estimates for the average lifetime costs of an internationally filed patent exceed $100k. How do we guide attorneys who are generally at arm's length from the business they serve so that they don't follow a patent-filing process, blindly spending great

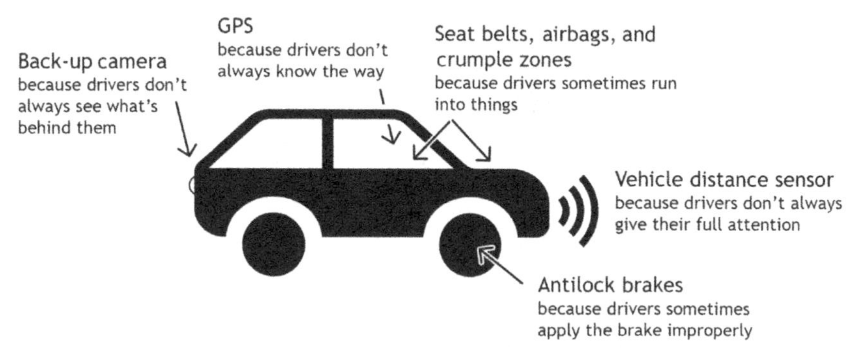

Fig. 15.8 Modern safety devices help mistake-proof driving.

Table 15.2 Increasing robustness through validation.

For a person working in...	**...with a workflow for...**	**...a validation could be...**
Product Design	Selecting a new component	• Ensure supplier is approved by Operations • Point to standard tests for new components
Law	Writing a patent application	• Formal sign-off with Chief Technical Officer (CTO) before creating application • Standard business case for all inventions before writing application
Sales	Developing a new account	• Standard account planning template • Standard revenue projection method to evaluate account

sums for small value? We can build in checks along the way to use expert evaluation. These steps of validation assume some level of "driver error" such as an overeager engineer submitting a misguided invention disclosure, that, if acted upon directly, could waste many thousands of dollars. This and two other examples are shown in Table 15.2.

15.6 Conclusion

Standardizing workflow provides a rich set of opportunities to cut waste, as shown in Fig. 15.9. It **simplifies** by bringing smooth, constantly validated workflow. It creates robustness, the ability for the workflow to respond well when people make ordinary mistakes. It also provides *ground* and *helicopter*

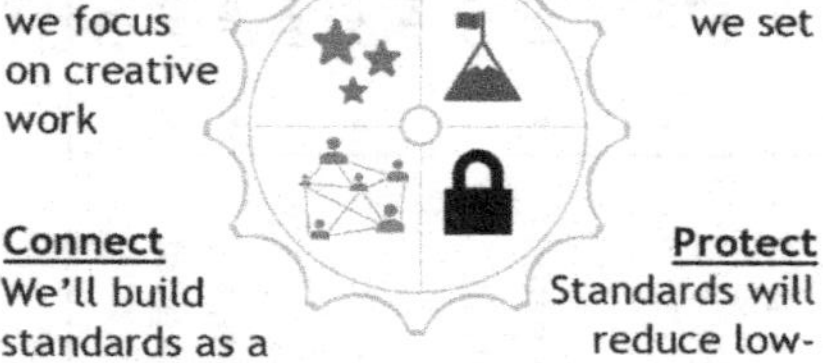
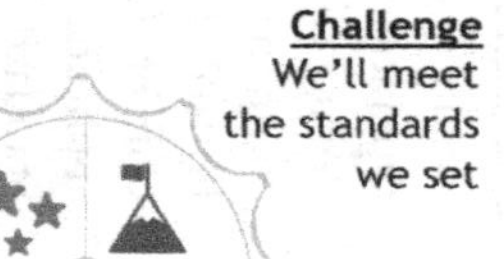

Fig. 15.9 Standard work simplifies, engages, and encourages experimentation.

views to guide broad review and countermeasures. It **engages**, bringing the promise of executing tedious work rapidly and with high quality. It builds team connection by drawing in the people who do the work to create the standards by which their work should be judged. It creates achievable measures to replace the arbitrary and vague requirements too common in knowledge work, and it provides a floor that can be applied to all work to "protect the brand." And standard work creates a large **experiment**: we create standards to respond to gaps in results. The measurement of those gaps drives action: if the gaps are filled, we move on to the next area; if not, we must dig in to discover why our standards are not working and correct them.

References

[1] T. Ohno, Taiichi Ohnos Workplace Management, Special 100th Birthday Edition, McGraw-Hill Education, Kindle edition, 2012, 175.
[2] T. Ohno, Taiichi Ohnos Workplace Management, Special 100th Birthday Edition, McGraw-Hill Education, Kindle edition, 2012, 142.
[3] T. Ohno, Taiichi Ohnos Workplace Management, Special 100th Birthday Edition, McGraw-Hill Education, Kindle edition, 2012, 141.
[4] T. Ohno, Taiichi Ohnos Workplace Management, Special 100th Birthday Edition, McGraw-Hill Education, Kindle edition, 2012, 18.
[5] L. Lupo, The Quality C's of McDonald's, Quality Assurance & Food Safety, (June 2007)Available at https://www.qualityassurancemag.com/article/-best-practices–the-quality–cs–of-mcdonald-s/.
[6] https://www.vox.com/2014/4/2/5572648/why-are-fewer-people-dying-in-car-crashes.

CHAPTER 16
Workflow improvement cycle

> *I think it ruins people when there is no race to get each person to add their good ideas to the work they do within a company...the ability to add your creative ideas and changes to your own work is what makes it possible to do work that is worthy of humans.*
>
> ***Taiichi Ohno [1]***

16.1 Introduction

This chapter will build an example of a workflow improvement cycle using the Knowledge Work Improvement Canvas (KWIC), which was briefly introduced as Fig. 9.21 as part of the presentation of the Success Map. Here we will use the KWIC as a tool for general workflow improvement.

16.2 The 8 Wastes

The concept of improving is engaging. Who doesn't want to get better at what they do? However, without a worthy problem to improve or a reliable means to manage improvements over time, the effort produces little value. This leads to the examples of the 8 Wastes as shown in Fig. 16.1.

16.3 A poorly managed cycle of improvement

The story of a poorly executed cycle of improvement is told in Fig. 16.2. Hannah, Greg's supervisor, wants the group to get better at virtual meetings because the interactions with remote team members have been poor. Certainly, that's a worthy goal and something of which Greg is capable. So, Hannah does what good leaders do: she delegates, asking Greg to "improve virtual meetings." As good as her intentions are, the problem is poorly defined. Also, Hannah doesn't coordinate this work outside her

Improve
https://doi.org/10.1016/B978-0-12-809519-5.00016-8

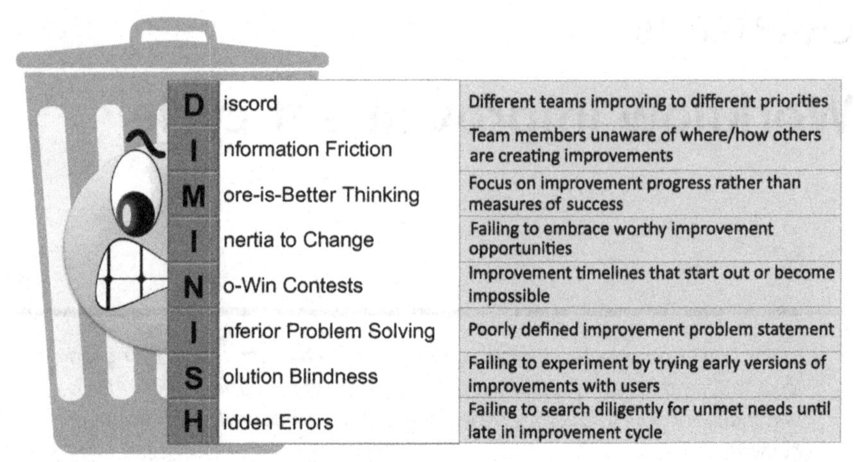

Fig. 16.1 Examples of the 8 Wastes applied to a poorly managed cycle of improvement.

group, which will eventually create problems for Greg when he tries to work with IT.

The result is predictable. Greg doesn't completely understand the problem from Hannah's viewpoint and IT is on another path. Other work in Greg's day crowds out this project, which seemed to him to be of lower priority. Hannah loses touch, coming into contact with the initiative occasionally, only when she's frustrated because the improvement cycle doesn't seem to be moving. In the end, IT has obscure rules that prevent the whole effort from going forward. The project withers, Greg's effort goes for naught, and Hannah feels nothing goes forward if she doesn't do it herself.

Many of the steps in Fig. 16.2 bring one of the 8 Wastes, as shown in Fig. 16.3. It starts with **Inferior Problem Solving** because the problem isn't defined in a measurable way. Partway through the project, Hannah reveals a new requirement, held back 3 weeks by **Information Friction**. IT is not so sure they want someone from Hannah's group leading an effort they feel is in their area, so they don't actively cooperate, which causes **Discord** that could have been avoided had Hannah aligned with the head of IT. Hannah checks in on the project once every month or two and, frustrated by the lack of progress, chides Greg to try harder, creating **More-is-Better Thinking**. Later, IT slows the whole process by forcing Greg to wait past the project deadline because of **Information Friction**, putting him in a **No-Win Contest**. The story continues adding waste upon waste, accomplishing nothing of value.

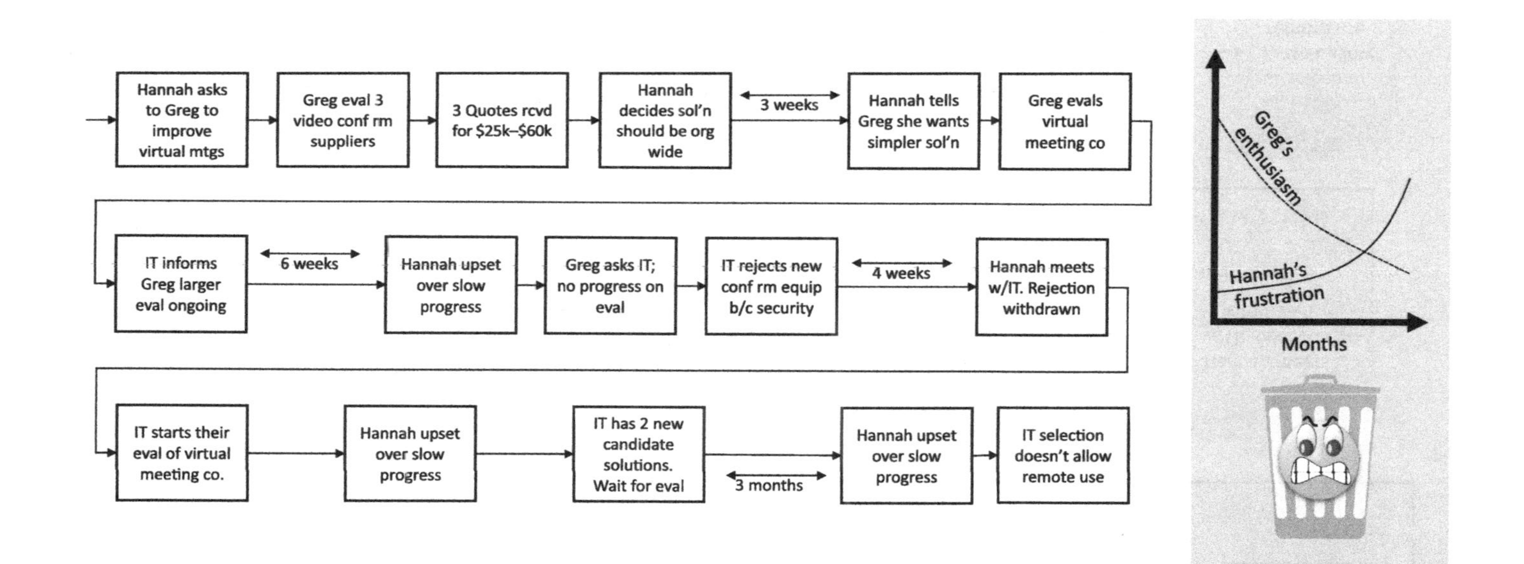

Fig. 16.2 Common outcomes of poorly focused improvement cycles: Rising frustration and declining enthusiasm.

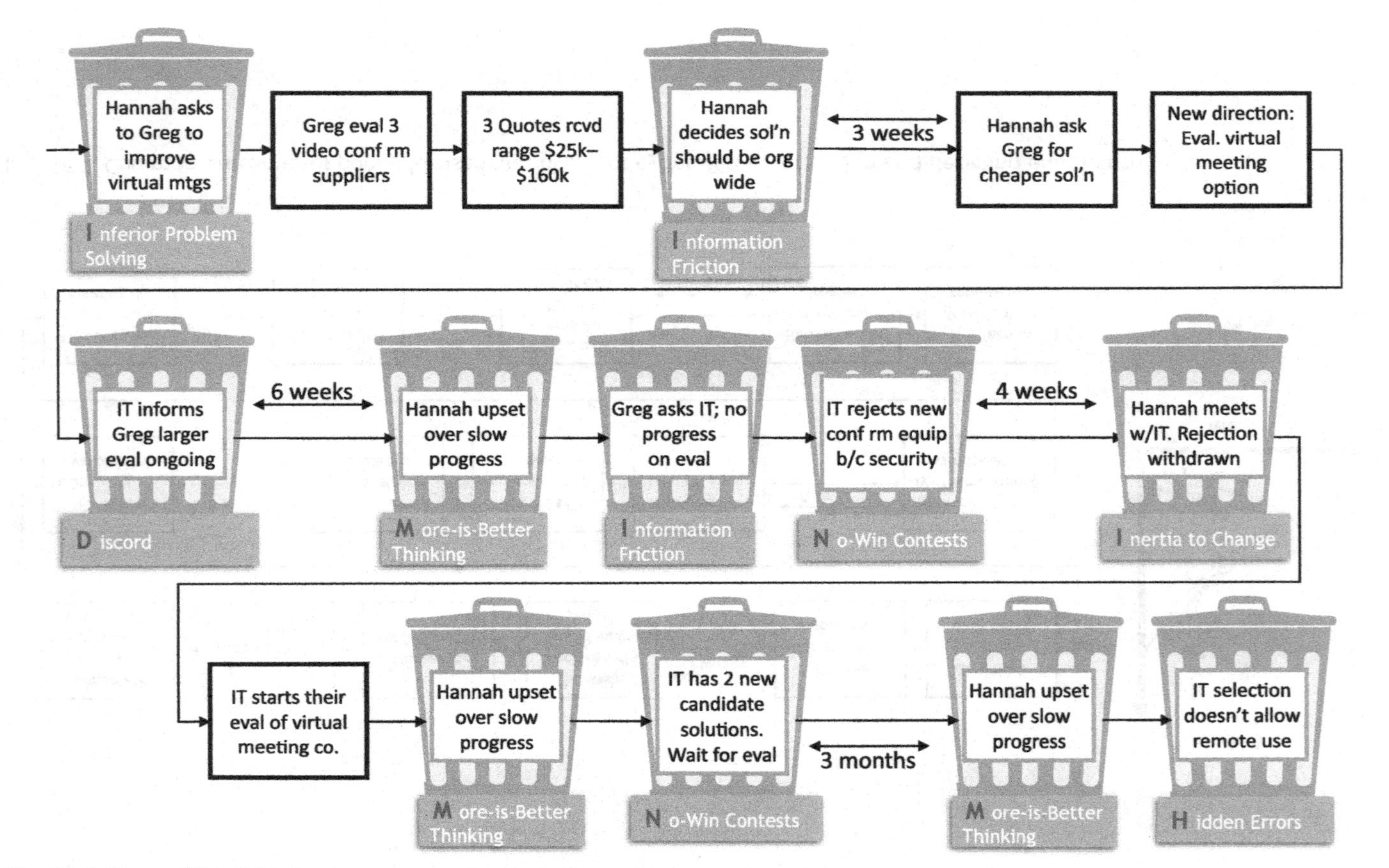

Fig. 16.3 Many of the 8 Wastes are common in poorly focused improvement cycles.

16.4 Leading improvement cycles

Mr. Ohno was passionate...He said you must clean up everything so you can see problems. He would complain if he could not look and see and tell if there is a problem.

Fujio Cho, President, Toyota Motor Corporation [2]

There is a better way to lead improvement. To start with, we will standardize how we manage a cycle of improvement, which will immediately reduce waste. We begin with the step of "Guide" so that Hannah scopes the effort, clearly stating the need so Greg doesn't learn of a major requirement partway in; she then sets out the needs by date and defines a small cross-functional team. This cross-functional team is just Greg with Lisa from IT, but it requires that Hannah has talked with Lisa's supervisor and obtained alignment as the start. In our example, Hannah would meet with Greg and Lisa for a few minutes to review the guidance, giving them a chance to negotiate their priorities. This step is simple; it requires perhaps 30 minutes of Hannah's time, but even in that short time she eliminates paths for **Discord**, **No-Win Contests**, and **Information Friction**.

The next step is Focus, where Greg grooms the Guidance from Fig. 16.4 into a problem statement as shown in Fig. 16.5. Key metrics and goals are chosen, which here involve the portion of meetings using video information going to 85%. Greg lists the problem solving methods, which might sometimes be a

Guide (Sponsor)	
Sponsor	Hannah
Explain need	We need to be able to hold virtual meetings so we can fully include people remotely in meetings as standard
Need-by date	6 weeks
Owner name	Greg
Team members	Greg, Lisa (IT)

Fig. 16.4 In the Guide step, the leader clarifies the needs.

Focus	
Problem or opportunity	We cannot communicate video information in most of our meetings with remote people
Metric	% of meetings that use video information
Current perf.	~10%
Target perf.	85%
Solve method	Informal problem solving in team

Fig. 16.5 In the Focus step, the improvement cycles owner creates the problem statement.

3-day team meeting (a *Kaizen* event), but for this simple problem is just an informal solve. Again, these steps require simple quantification and take 30 or 60 minutes; as simple as this is, it helps Greg, Lisa, and Hannah confirm they are working to the same scope and goals before any real effort is expended.

Then Greg and Lisa set out to solve the problem, which here has two countermeasures: choosing a supporting software set and IT working to ensure secure connections are possible. These steps are simple, as they usually are for modest improvement cycles like this. But they are still needed to create even flow from the need to the countermeasures, engaging the stakeholders along the way (Fig. 16.6).

Finally, we add a Success Map, a combination of an Action Plan, a Test Track, and a Bowler, as shown on the right-hand side of the canvas in Fig. 16.7. If we place the canvas where Hannah can retrieve it on demand, there is no need for her to reach out to Greg every time she wants an update. The Success Map shows the current state of the work at a glance so that Hannah can get a sense of weekly progress in a few seconds. No emails. No phone calls gone to voice mail. And, when a deeper review is needed, the two can meet in person. Here the Success Map speeds the discussion to meaty topics because there's no need to spend time reminding each other of what each thought the other was doing, a common source of waste. The Success Map cuts the frustration of the **More-is-Better Thinking** because Hannah and Greg share one definition of success.

Solve		
Root causes		Video information cannot be added outside of main site due to security issues
Solution		Choose application that meets IT security requirements that can be installed on all R&D mobile/desktop systems and those of closely related functions
Counter-measures	1	Select SW that is practical to install on a large number of machines
	2	IT to support connection from inside and outside of main site
	3	

Fig. 16.6 In the Solve step, the team creates the countermeasures.

16.5 Conclusion

As shown in Fig. 16.8, creating a standard, unambiguous means to drive improvement **simplifies**, **engages**, and **experiments**. It simplifies by applying formal problem solving to the improvement process, specifying a Guide-Focus-Solve flow to ensure that the right areas are chosen and, through SPoT, that the whole team understands why this area is in need of improvement. And the Knowledge Work Improvement Canvas (KWIC) standardizes the visualization. The Engagement Wheel provides a balanced view of how creating an efficient workflow for an improvement cycle can raise engagement. We will inspire the team by providing confidence that we will get better at the things we do. We connect the team, who will own change together, as Greg and Lisa did. Improvement will bring a challenge through the Success Map. We'll protect the team by clearly defining what need we are filling and how we will fill it. Finally, each cycle of improvement becomes an experiment as we measure results, compare to goals, and adjust as gaps emerge.

Initiative	Upgrade Video Conferencing Capability	Status	4-Success Map tracking	Last update	1-May	Update due	15-May

Guide (Sponsor)

Sponsor	Hannah
Explain need	We need to be able to hold virtual meetings so we can fully include people remotely in meetings as standard
Need-by date	6 weeks
Owner name	Greg
Team members	Greg, Lisa (IT)

Focus

Problem or opportunity	We cannot communicate video information in most of our meetings with remote people
Metric	% of meetings that use video information
Current perf.	~10%
Target perf.	85%
Solve method	Informal problem solving in team

Solve

Root causes		Video information cannot be added outside of main site due to security issues
Solution		Choose application that meets IT security requirements that can be installed on all R&D mobile/desktop systems and those of closely related functions
Counter-measures	1	Select SW that is practical to install on a large number of machines
	2	IT to support connection from inside and outside of main site
	3	

Action Plan

Deliverable	Next Action	Date	Owner	Plan	Replan	Fcst	Actual
Select application	Define criteria for selection	15-Mar	Greg	16-May		16-May	
Prototype meeting	Install on 8 machines (remote and	15-May	Lisa	30-May		30-May	
Develop training material			Greg	6-Jun		6-Jun	
Hold 2 training sessions, record 1			Greg	13-Jun		13-Jun	
Ensure install of new SW on >85%			Greg	4-Jul		4-Jul	

Test Track

Case	Dates	Lap 1	Lap 2	Lap 3	Lap 4	Lap 5
Design review, project XYZZY (prototype meeting)	Plan	1-May	15-May	22-May		
	Replan					
	Forecast	1-May	15-May	22-May		
	Actual					
Patent applicatoin meeting	Plan	11-May	25-May	1-Jun		
	Replan					
	Forecast	11-May	25-May	1-Jun		
	Actual					
Patent applicatoin meeting	Plan	19-May	2-Jun	9-Jun		
	Replan					
	Forecast	19-May	2-Jun	9-Jun		
	Actual					
	Plan					
	Replan					
	Forecast					
	Actual					

Bowlers

Goal 1	Meetings held with video											
Date	1-May	1-Jun	1-Jul	1-Aug	1-Sep	1-Oct	1-Nov	1-Dec	1-Jan	1-Feb	1-Mar	1-Apr
Plan	2	5	10									
Replan												
Fcst	2	5	10									
Act												

Goal 2												
Date	1-May	1-Jun	1-Jul	1-Aug	1-Sep	1-Oct	1-Nov	1-Dec	1-Jan	1-Feb	1-Mar	1-Apr
Plan												
Replan												
Forecast												
Actual												

Fig. 16.7 The Success Map (Action Plan, Test Track, and Bowler) completes the Knowledge Work Improvement Canvas (KWIC).

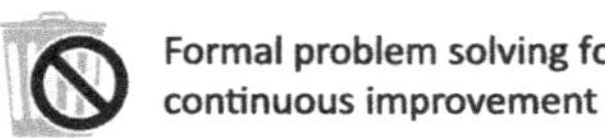
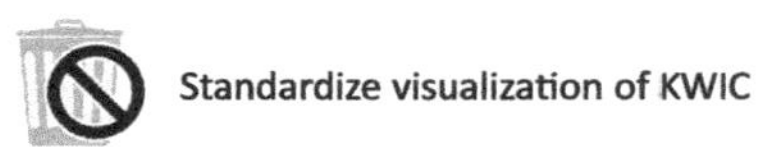

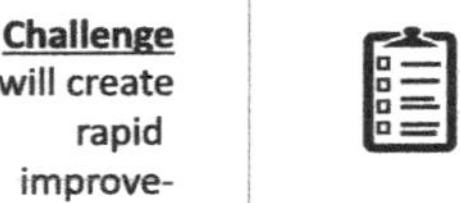

Fig. 16.8 Creating effective improvement cycles simplifies, engages, and encourages experimentation.

References

[1] T. Ohno, Taiichi Ohnos Workplace Management, Special 100th Birthday Edition, McGraw-Hill Education, Kindle Edition, 2012, p. 178.
[2] J. Liker, The Toyota Way: 14 Management Principles From the World's Greatest Manufacturer, McGraw-Hill, 2004, p. 149.

CHAPTER 17

Workflow—Checklists and expert rule sets

Avoidable failures are common and persistent, not to mention demoralizing and frustrating, across many fields—from medicine to finance, business to government. And the reason is increasingly evident: the volume and complexity of what we know has exceeded our individual ability to deliver its benefits correctly, safely, or reliably.

Atul Gawande [1]

17.1 Introduction

This chapter presents checklists and expert rule sets, both of which build up organizational memory for complex workflows. Knowledge work relies on experts who each bring a unique set of experiences and expertise to every situation. Knowledge work depends on the tacit knowledge of those experts, knowledge that cannot be captured in step-by-step processes. However, even in the most complex workflows, there are guiding principles, what we will call *rules*. These rules define a common way to approach complex problems. Some of those rule sets are codified, especially in heavily regulated industries like pharmaceuticals, medical devices, defense, and aviation. More often, they are informal, built up from years of work—for example, a special test when a device will be used at a high temperature, unique requirements for an eComm site in China, and how to publish professional recommendations for medical devices in the EU. These rules are a sort of organizational memory: *how we do things here*. Perhaps we were happy with a new technique and we want to use it again. Or maybe we made a painful mistake that we don't want to repeat.

Every organization has these rules, but they are usually unstated, carried in the heads of the people who do the work, and that can cause a great

https://doi.org/10.1016/B978-0-12-809519-5.00017-X

deal of waste. When the people who know the work best leave the organization or transition into a new role, a great deal of that knowledge is lost. Even when people don't leave a role, there is waste. When a person takes on an unfamiliar job, they will make many mistakes until they learn these rules. The result is that quality of work is unnecessarily dependent on the person doing the work. The core of these wastes is **Information Friction**: communicating the "rules" is unnecessarily slow and error-prone.

Waste can be cut by enhancing organizational memory with checklists and expert rule sets. Start by writing down the primary rules in a defined workflow, perhaps on just a page or two. Formal expert rules break down barriers between the people who know and those who don't. Thus, it builds the group's capability by refining the primary guidelines and also provides a natural mechanism to teach people new to the organization.

Over time, checklists and rule sets grow. When major mistakes occur, the rules can be augmented so that mistake is unlikely to happen again. When a new method works well enough that it should be widely used, add it to the rule set. Every time we take a team through a 2-hour review via a rule set, we will discover small ways to bring improvement as well as a few larger ones on occasion. That's the moment to update the rule set. Aspire to update these documents each time you use them. Checklists and rule sets that don't get updated will wither.

Each checklist or expert rule set needs an owner. That person should be a leading expert for the organization in the domain where the checklist or rule set is focused. Ideally, that person would be active with the checklist/rule set in the department, attending a wide range of reviews. This puts that person in an ideal place to give life to the rule sets. If the organization supports multiple domains, as most do, there will be multiple rule sets, each with an owner. For example, one engineering department might support electrical design, mechanical design, and magnetic material science. In such a case, there could be three expert rule sets, each with different owners. Owners go to reviews to see the rule sets in action, ensuring that they are used properly and changing behavior for the better. Go and "Watch it Work" (Section 12.2) rather than assuming. Rule set owners also set long-term direction, for example, planning expansion in key areas or getting outside experts to consult on details. Done well, these lists hold some of the organization's most precious knowledge. They are worth having the strongest experts creating and sustaining them (Fig. 17.1).

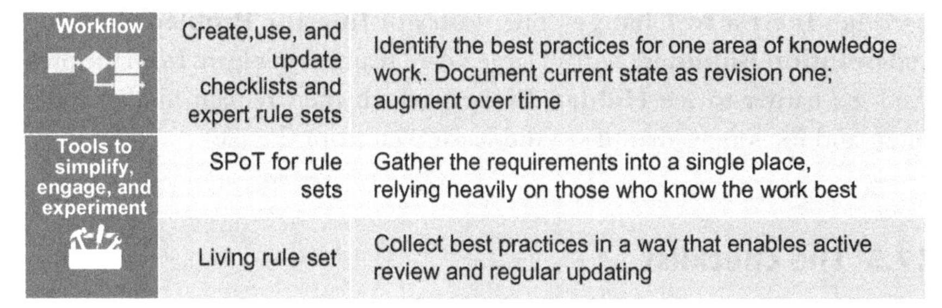

Workflow	Create,use, and update checklists and expert rule sets	Identify the best practices for one area of knowledge work. Document current state as revision one; augment over time
Tools to simplify, engage, and experiment	SPoT for rule sets	Gather the requirements into a single place, relying heavily on those who know the work best
	Living rule set	Collect best practices in a way that enables active review and regular updating

Fig. 17.1 The workflow and tools for an improvement cycle.

17.2 The 8 Wastes

The waste endured by not having a rule set is varied, as shown in Fig. 17.2. It can create **Discord** due to experts being misaligned or even competing with each other. Expert rules directly cut **Information Friction** by removing barriers to the flow of knowledge from the organization's key expert to others. Without expert rules, evaluating work can become a **More-is-Better Thinking** exercise; there is no standard to review by and thus no way to understand when you're done, so it can seem that spending more time makes a better review. For those doing work, the lack of the standard can become a **No-Win Contest** because there is nothing concrete to work toward—whatever you do seems to be wrong, especially for those who are junior. Not having a standard also creates barriers to improving, and so

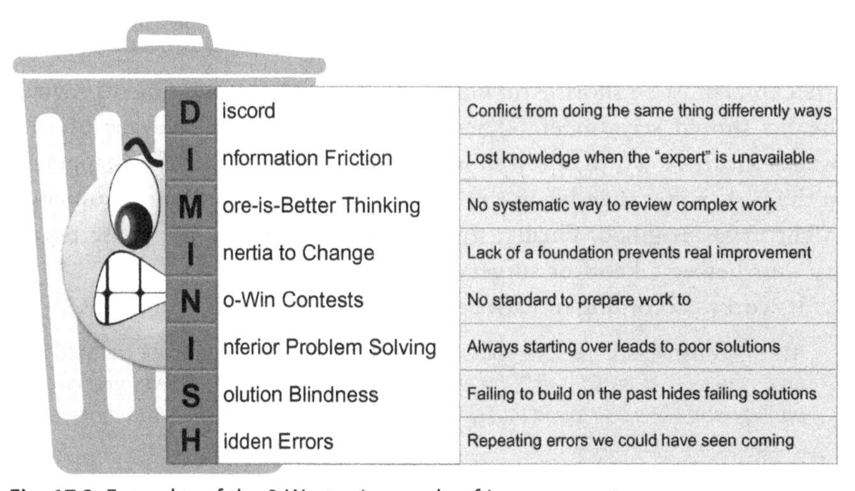

Discord	Conflict from doing the same thing differently ways
Information Friction	Lost knowledge when the "expert" is unavailable
More-is-Better Thinking	No systematic way to review complex work
Inertia to Change	Lack of a foundation prevents real improvement
No-Win Contests	No standard to prepare work to
Inferior Problem Solving	Always starting over leads to poor solutions
Solution Blindness	Failing to build on the past hides failing solutions
Hidden Errors	Repeating errors we could have seen coming

Fig. 17.2 Examples of the 8 Wastes in a cycle of improvement.

becomes **Inertia to Change**. The wastes of **Inferior Problem Solving** and **Solution Blindness** remain large when teams don't learn from the past. And it's harder to see **Hidden Errors**, which then remain hidden more often and for longer than if we did learn well from the past.

17.3 The checklist

> *Four generations after the first aviation checklists went into use, a lesson is emerging: checklists seem able to defend anyone, even the experienced, in many more tasks than we realized.*
>
> ***Atul Gawande [2]***

Checklists are common in expert endeavors, for example, construction, pre-surgery, and airlines. None of these checklists seeks to make an expert task appropriate for a nonexpert. For example, only a trained pilot can walk through a pre-takeoff checklist. Only trained medical staff can execute a pre-surgery checklist. The checklist doesn't rob creativity, nor does it make a complex task simple. It does prevent ordinary flaws in memory and attention from becoming serious problems.

> *Knowing that we are prone to this type of forgetting prompts us to use checklists. That way, even if we forget something, the checklist will remind us and keep us, in the end, from leaving anything out.*
>
> ***Shigeo Shingo [3]***

Checklists must be short. Atul Gawande recommends the time to review the list should stay under 60 seconds. They must be relevant for the users—in other words, they should prevent errors that have happened in the past, rather than being a collection of rare "corner cases" someone might have imagined. Finally, they need to fit in a natural pause point, say, just before takeoff or surgery.

It works. According to Gawande in a 2010 interview with NPR [4]:

> *"We get better results," he says. "Massively better results. We caught basic mistakes and some of that stupid stuff," Gawande reports. But the study returned some surprising results: "We also found that good teamwork required certain things that we missed very frequently."*

Back in Section 9.3.7, we discussed a checklist to help with approval processes. An example of such a checklist is shown in Fig. 17.3.

Large Requisition Approval Checklist

Note: Having all items complete and answering "Yes" to all Yes/No questions speeds approval.

- ❑ Team Review
  - Did project team review this expense? (Yes)/(No) Comment ________
  - Does team agree this is necessary? (Yes)/(No) Comment ________
  - Does team agree this is the lowest cost? (Yes)/(No) Comment ________
  - List team ________ ________ ________ ________ ________ ________
- ❑ Quotations
  - Were two quotes obtained from approved vendors? (Yes)/(No)
  - Are quotes attached to this form? (Yes)/(No)
  - If no to either, please explain why not:
  - ________
  - ________
  - ________
- ❑ Approved project
  - Project name ________
  - Project manager ________
  - Total project spend ________
  - Brief explanation of why this item is needed for this project:
  - ________
  - ________
- ❑ Budget line item
  - Budget number ________
  - Budget owner ________ I confirm this is in my budget ____(Initials)

Fig. 17.3 The Large-Requisition Approval checklist.

17.4 The expert rule set

That's the purpose of memory. It's not to remember the past. It's to stop the same damn thing from happening over and over.

Jordan Peterson [5]

The expert rule set, sometimes called design standards, is a level above a checklist. Where the checklist is inserted in daily work and targeted for a couple of minutes, the expert rule set will take longer but it is applied only occasionally, say at two or three key points in a project, so the time it requires is less of an issue. Expert rule sets also create a large, continuous experiment. Every time knowledge staff discover a serious error, it's a candidate for augmenting the list. For example, Toyota uses them extensively:

> *Toyota Engineers...make extensive use of design standards that go back to when Toyota first started to engineer cars. Within each section—[for example] door latches...—engineering checklists have evolved from what has been learned as good and bad design practice. The engineer uses these checklist books...and develops them further...[6].*

Website: Development design rule set (Rev 7.3)

1. Test at 2 x full user design load
2. Translate website to Mandarin and have 2–3 local people evaluate
3. E-commerce site provides live order when accessed from China, Germany/France, Brazil, and United States
4. New user registration <2 minutes with new user

Fig. 17.4 A simple expert rule set for website development.

A simple expert rule set for website development is shown in Fig. 17.4. The workflow to create and sustain an expert rule or checklist set is shown in Fig. 17.5. At the outset, the goal is to capture the current state of critical rules with no need to improve it. Simply writing down what the most capable experts are doing on a regular basis is enough. In fact, you probably want to avoid getting too much detail at the start, because people left to their own often gravitate toward overly complex cases. Instead, focus on capturing the rules that will help the team avoid recurrence of painful mistakes from the recent past—the clause in a contract that didn't mean what we thought it would, the software test that is often forgotten, or the power condition that we forgot to test for and that therefore disrupted business in a key region of the world. You may want to limit the number of entries in the beginning, for example, asking key experts to write down the top 10 or 20 mistakes from the past—things that we never want to repeat—then capture the rules that will prevent those mistakes. Over time, the rule set will grow and improve. But it's much easier to add a rule than take one away, so in the beginning, keep the list to a manageable number.

Expert rule sets can set the agenda for work reviews where, without a rule set, the team may simply pull out a file history or scroll through a set of drawings. Expert rule sets provide an organized way to review work and, in doing so, offer an organized way to prepare for reviews. Earlier in my career, I watched the frustration of people who arrived at reviews where the "rule set" resided mostly in the boss's head. Preparing for the review was an unwinnable contest; people almost always had significant errors exposed. For some people, this was just part of the job, but for others it seemed unfair; having a written, living expert rule set that captured common problems would have helped everyone prepare better.

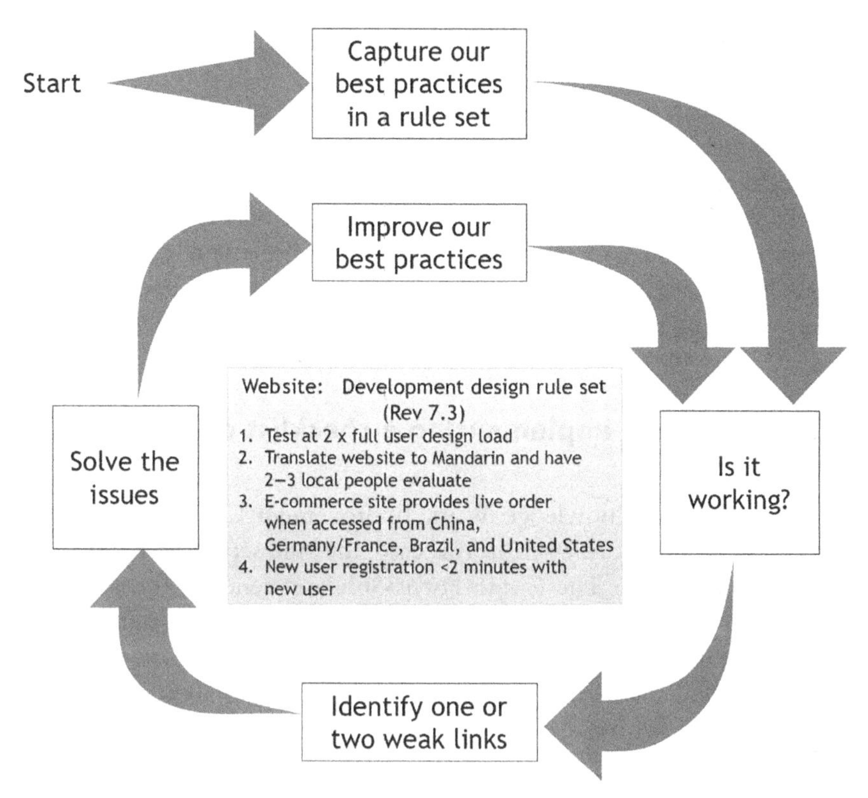

Fig. 17.5 Expert rule sets are living documents that capture the wisdom of the team over years.

For expert rule sets to be effective, they must be alive; the process of Fig. 17.5 is a never-ending cycle. For them to live, there must be an owner who actively manages the set, personally observing their use within the organization and updating them often. Almost all organizations support multiple functions and so will have multiple sets of experts. Each of these functions can create and maintain its own expert rule sets. Each should be maintained by a person who is recognized within the organization for expertise in the relevant area. If it's too large a burden for one person, the responsibility can be rotated through several.

US companies have tried to imitate Toyota's approach [with] large data bases... without success...The reason is they have not trained their engineers to have the discipline to use the standards and improve on them. Capturing knowledge is not difficult [6].

In The Toyota Way, Jeffrey Liker provides several rules. The two most important for our purposes are as follows [6]:

1. Expert rule sets must be specific enough to be useful guides, but retain flexibility. For knowledge work, standards become more variable than they would be for manufacturing work.
2. The people doing the work must improve the standards. Standards must not be used as one group policing or controlling another; standards must evolve from within the team using them. This reduces friction and it is empowering to see others using your improvements.

17.5 A KWIC for implementing a checklist or expert rule set

Fig. 17.6 shows a Knowledge Work Improvement Canvas (KWIC) to implement an expert rule set, in this case, the rule set for web development from Fig. 17.4. The Guide-Focus-Solve sequence of formal problem solving is shown on the left and a Success Map is on the right. If you want to create several expert rule sets, consider focusing on the most urgent one and then, after gaining experience in that area, expand the thinking into other areas.

17.6 Conclusion

Checklists and expert rule sets **simplify**, **engage**, and **experiment** as shown in Fig. 17.7. They **simplify** by collecting the wisdom of the team and providing an agenda for peer review. Also, standardizing the ownership and methods to use and update these lists guarantees sustained relevance and improved content over time. They **engage** as shown by the Engagement Wheel. They can inspire an expert team by having a structured way to gather and improve the team's expertise over time. They increase connection by helping everyone benefit from the knowledge of others. They provide challenge by giving a concrete standard to prepare to and protect the "brand" of the entire team by preventing repeated mistakes. Finally, checklists and expert rule sets create **experiments** where the team monitors itself for mistakes so that it can install rules to prevent them from happening again.

Initiative	Design Rule Set for Web Page Development	Status	4-Success Map tracking	Last update	1-Oct	Update due

Guide (Sponsor)

Sponsor	Torben
Explain need	We often make the same errors when releasing a new web page. We need to identify errors and weak design from the past
Need-by date	By year end
Owner name	Kelly
Team members	Kelly, Yex, Eliana

Focus

Problem or opportunity	We need to capture the rules our internal community of experts use so we can 1) apply them to every web page we develop and 2) strengthen the rule set over time
Metric	#errors found with design rule set in formal reviews %projects running formal reviews with design rule set
Current perf.	0
Target perf.	5 errors per project 10% projects using design rule set in formal reviews
Solve method	2-day event with international team of web development experts

Solve

Root causes		We have no means to capture design rules from different experts
Solution		Create a SPoT (Single Point of Truth) for web design rules. Add workflow measurement to ensure all projects use formal review process
Counter-measures	1	Create a SPoT (Single Point of Truth) for web design rules
	2	Select owner for design list for 12-month rotating period
	3	Measure projects that apply design rules in active web-page development projects

Action Plan

Deliverable	Next action	Date	Owner	Plan	Replan	Fcst	Actual
2-day event in St. Louis with all web experts	Plan meeting; send invitations	1-Sep	Kelly	2-Oct		2-Oct	
1st-pass design rule set			Kelly	9-Oct		9-Oct	
Select owner for next 12 months	Meet with Stan, Robin, Prikosh	15-Sep	Yex	1-Oct		1-Oct	
Present rule set to all designers for comment			Eliana	1-Nov		1-Nov	
Publish Rev 1.0 to SPoT			Eliana	1-Dec		1-Dec	
Review use on >= 3 projects with dev teams	Select 3 projects	1-Dec	Kelly	15-Jan		15-Jan	

Test Track

Case	Dates	Review	Report found errors	Update rule set	Lap 4	Lap 5
Project Chilstrta: Design rule review	Plan	2-Feb	7-Feb	15-Feb		
	Replan					
	Forecast					
	Actual					
Project #2 (TBD)	Plan	27-Feb	4-Mar	12-Mar		
	Replan					
	Forecast					
	Actual					
Project #3 (TBD)	Plan	24-Mar	29-Mar	6-Apr		
	Replan					
	Forecast					
	Actual					
	Plan					
	Replan					
	Forecast					
	Actual					

Bowlers

Goal 1	# errors found per design review in formal design reviews using rule set											
Date	1-Jan	1-Feb	1-Mar	1-Apr	1-May	1-Jun	1-Jul	1-Aug	1-Sep	1-Oct	1-Nov	1-Dec
Plan	5	5	5	5	5	5	5	5	5	5	5	5
Replan												
Fcst												
Act												

Goal 2	%projects using formal review with design set											
Date	1-Jan	1-Feb	1-Mar	1-Apr	1-May	1-Jun	1-Jul	1-Aug	1-Sep	1-Oct	1-Nov	1-Dec
Plan	50%	60%	70%	80%	90%	100%	100%	100%	100%	100%	100%	100%
Replan												
Forecast												
Actual												

Fig. 17.6 A Knowledge Work Improvement Canvas (KWIC) for installing an expert rule set.

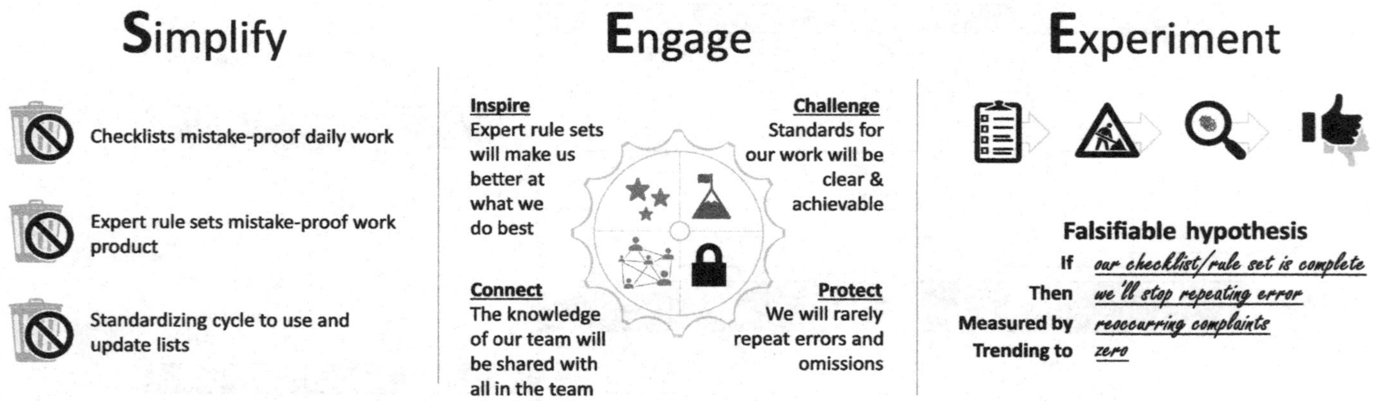

Fig. 17.7 Expert rule sets increase engagement with a combination of several factors.

References

[1] A. Gawande, The Checklist Manifesto, Henry Holt and Co., Kindle Edition, Location 197, 2011.
[2] A. Gawande, The Checklist Manifesto, Henry Holt and Co., Kindle Edition, 2011, p. 48.
[3] S. Shingo, The Sayings of Shigeo Shingo: Key Strategies for Plant Improvement, English translation, Productivity Press, 1987, p. 168.
[4] A. Gawande, Atul Gawande's 'Checklist' For Surgery Success, NPR Morning Edition, Jan 5, 2010. https://www.npr.org/templates/story/story.php?storyId=122226184.
[5] J. Peterson, 12 Rules for Life, Random House, An Antidote to Chaos, 2018, p. 239.
[6] J. Liker, The Toyota Way: 14 Management Principles From the World's Greatest Manufacturer, McGraw-Hill, 2004, p. 147.

References

CHAPTER 18

Workflow—Problem Solve-Select

...decision making is a two-step process: first learning and then deciding.

Ray Dalio [1]

18.1 Introduction

This chapter will review the problem Solve-Select workflow, a variant of the formal problem-solving method that was presented in Chapter 12. Many problems are so complex that they cannot be resolved by a team of domain experts working independently; these problems require a business and/or leadership team to stay connected with the domain experts throughout the process. In these cases, the method presented in Chapter 12 works poorly. Recall that process had a single handoff between decision-makers and the "solve team"; the two groups worked almost in isolation.

Ray Dalio said that decision-making is first learning and then deciding. The problem Solve-Select workflow is this principle applied to the organization, which must learn as a whole by evaluating alternatives that were created or discovered by the experts. Then the organization can decide as a whole upon those alternatives, led by those responsible for the organization.

In these types of problems, technical/domain issues need to be addressed even when business issues cannot be fully defined before the solve process starts. In such a case, the single handoff between the "Guide" and "Solve" steps of Chapter 12 becomes a continuum with the domain experts as the solve team working hand-in-glove with organizational leadership as the select team. An example is devising a response to a new competitive product or service: the leadership (select team) needs to see a range of options, here called a "solution menu." So, the formal problem-solving workflow of Chapter 12 is extended to a "Solve-Select" workflow. In this chapter, we will discuss the Solve-Select workflow including a presentation of the solution menu and a Solve-Select canvas, a canvas adapted for this workflow (Fig. 18.1).

Improve
https://doi.org/10.1016/B978-0-12-809519-5.00018-1

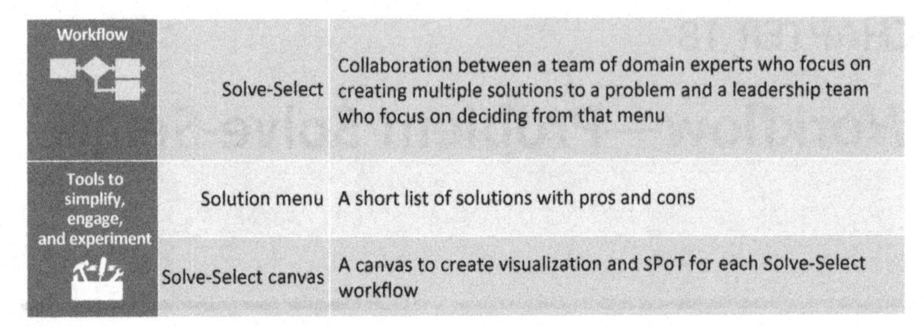

Workflow	Solve-Select	Collaboration between a team of domain experts who focus on creating multiple solutions to a problem and a leadership team who focus on deciding from that menu
Tools to simplify, engage, and experiment	Solution menu	A short list of solutions with pros and cons
	Solve-Select canvas	A canvas to create visualization and SPoT for each Solve-Select workflow

Fig. 18.1 The workflow and tools for an improvement cycle in the Solve-Select workflow.

18.2 The 8 Wastes

Waste from the lack of a Solve-Select process can come from each of the 8 Wastes of Knowledge Work, as shown in Fig. 18.2. **Discord** can be created when management teams feel that the technical team has failed to be transparent, or when technical teams feel managers are excluding them from decisions where their views should be considered. **Information Friction** occurs when the leadership team needs multiple solutions to be considered, but the technical team provides information on only the option they favor. **More-is-Better Thinking** can result when, lacking a means to drive rapidly to an effective solution, the organization debates indefinitely without reaching a conclusion. **Inertia to Change** can prevent the organization from improving

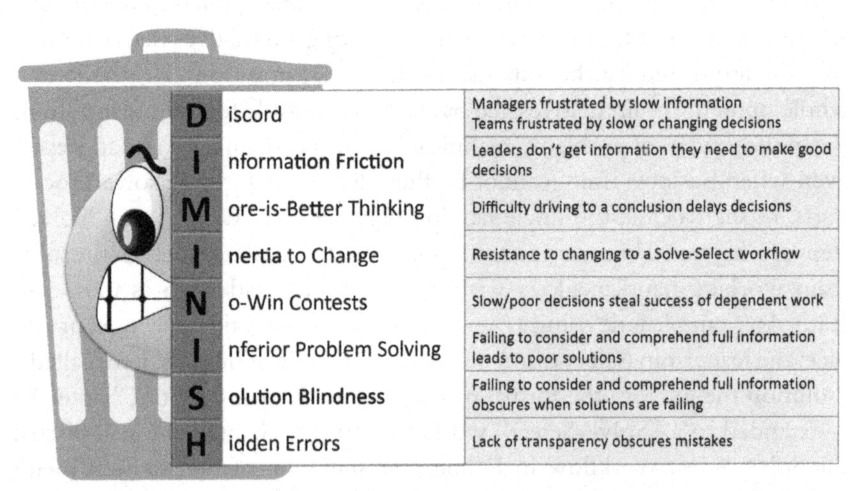

Discord	Managers frustrated by slow information Teams frustrated by slow or changing decisions
Information Friction	Leaders don't get information they need to make good decisions
More-is-Better Thinking	Difficulty driving to a conclusion delays decisions
Inertia to Change	Resistance to changing to a Solve-Select workflow
No-Win Contests	Slow/poor decisions steal success of dependent work
Inferior Problem Solving	Failing to consider and comprehend full information leads to poor solutions
Solution Blindness	Failing to consider and comprehend full information obscures when solutions are failing
Hidden Errors	Lack of transparency obscures mistakes

Fig. 18.2 Examples of the 8 Wastes of Knowledge Work from a poor Solve-Select process.

how solutions to how complex problems are chosen. **No-Win Contests** can be created in work that depends on rapid decisions being delayed by slow or poor decision-making. **Inferior Problem Solving** and **Solution Blindness** both occur when we fail to visualize the full breadth of a problem, for example, neglecting relevant solution alternatives. **Hidden Errors** can occur when either team misunderstands the problem.

18.3 The Solve-Select workflow

The Solve-Select workflow is based on the need for a combination of solid problem solving married to diligent knowledge transfer. Unlike traditional formal problem solving where a problem is handed off by a leadership team to await a solution, in the Solve-Select process, the two teams work together at many steps along the way to a solution. Rather than having a single point at which the problem is defined, the problem is spelled out as well as it can be at the outset, allowing the solve team to build out a few options. This then sheds light allowing the problem to be refined (Fig. 18.3).

Consider a common example: a competitor just released a new product or service and we need to react. But so little is known at the beginning that a leadership team needs to understand better the options to lay out clear goals. If we can respond well by making a simple modification to something we already provide, that might be appropriate. But we probably will want to look at other options such as a wholly new product or service, or perhaps, if the threat turns out to be small, we may do nothing at all. But without domain experts to bound the cost and effectiveness of those alternatives, a management team probably lacks the information to define the problem further. This is where the solution menu of Fig. 18.4 fits.

The solution menu describes a handful of competing alternatives, perhaps two to four, that have enough information for any stakeholder to understand better the value and cost of each. As an R&D leader, I found solution menus so valuable that virtually every problem solve we presented included a solution menu for organizational leaders to choose from. At the very least, we compared a favored solution to doing nothing, which resulted in a two-item menu. But in most cases, there were one or two more options we could build out. Building solution menus helped everyone: technical people could consider multiple business angles, and it gave confidence to the leadership that a broad range of solutions was being considered by the technical team. The Solve-Select workflow creates information flow between the solve team of domain experts and the select team of business leaders. As shown in Fig. 18.5, the conversation between the two teams continues through the process.

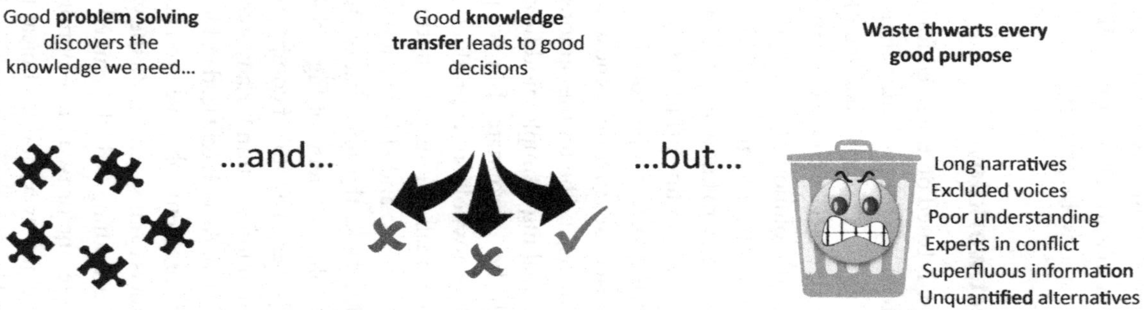

Good **problem solving** discovers the knowledge we need...

...and...

Good **knowledge transfer** leads to good decisions

...but...

Waste thwarts every good purpose

Long narratives
Excluded voices
Poor understanding
Experts in conflict
Superfluous information
Unquantified alternatives

Fig. 18.3 The Solve-Select workflow integrates formal problem solving with organizational decision-making.

Menu items	Pros and cons	Cost
Replace display with touchscreen	**Pros**: More modern look and feel, full functionality **Cons**: Parts cost goes up $78, will take 12 months	4000 hours design time
Move user interface to mobile device	**Pros**: Reduce parts cost $60, high differentiation from older products **Cons**: Some customers unwilling to use mobile device, high cost project	7500 hours design time
Leave hardware as is; upgrade interface software	**Pros**: Fast to market (3–5 months) **Cons**: Looks older, does not support two needed features	300 hours design time

Fig. 18.4 The solution menu: the problem-solve team provides multiple solutions to the organizational decision-makers.

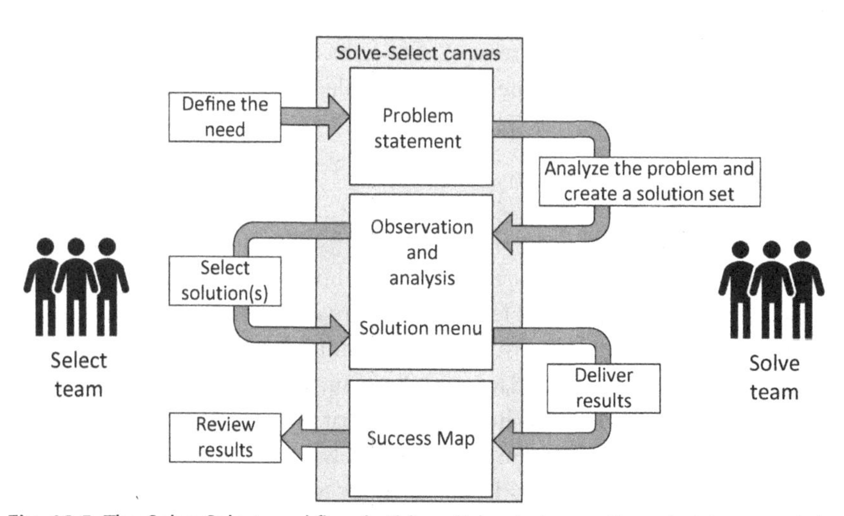

Fig. 18.5 The Solve-Select workflow builds a dialog between the select team and the solve team.

18.4 The Solve-Select canvas

A schematic for the Solve-Select canvas is shown in Fig. 18.6. Instead of a set of countermeasures for a single approach, the solution menu describes alternative solutions at a level a leadership team can use to make decisions. There is a simple Success Map, perhaps just an Action Plan that drives alternatives down to a single solution over time.

A sample Solve-Select canvas is shown in Fig. 18.7. This example is to provide a solution to a problem where remote employees cannot access information adequately. Notice that in the solution menu, the leadership has chosen Countermeasures #1 and #3, as indicated with checkmarks in the left-hand column. The bottom right holds an Action Plan; depending on the scale of the problem, a full Success Map (with Test Track and Bowlers) may be warranted.

18.5 Knowledge Work Improvement Canvas (KWIC) to adopt Solve-Select workflow

Fig. 18.8 shows a Knowledge Work Improvement Canvas (KWIC) to deploy the Solve-Select workflow in an organization. Note that the KWIC addresses how to deploy, so there would be one KWIC for an organization. Once deployed, there will be many Solve-Select canvases such as that shown in Fig. 18.7, one for each time the Solve-Select process is applied.

18.6 Conclusion

The Solve-Select workflow **simplifies**, **engages**, and **experiments**, as shown in Fig. 18.9. Three examples of simplification are structuring communication between business leaders and domain expert with many touchpoints, using the solution menu to present multiple approaches and using a Solve-Select canvas as a standard visualization. Engagement is increased as shown by the Engagement Wheel. A team that consistently receives timely, well-made decisions will be inspired. The Solve-Select workflow connects leadership and knowledge work staff to work together to solve some of the most important problems that face an organization. Moving the organization to reliable decisions will enable teams to succeed, completing dependent work on time and on budget. Confidence in solid decisions for the future protects the organization. And the Solve-Select process brings

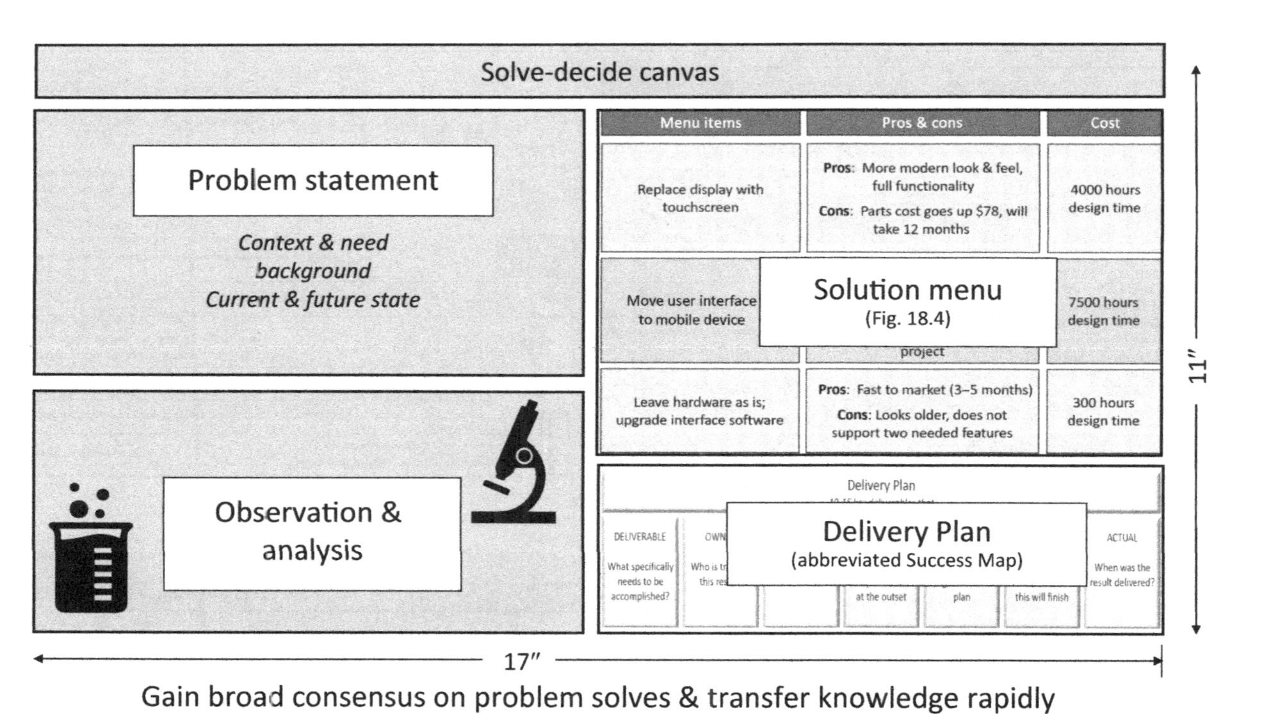

Fig. 18.6 A schematic of the Solve-Select canvas.

Solve-Select Canvas

Problem	Shared directories are not effective for providing access for all employees	Owner	Mgr, Web Svcs	Sponsor	VP IT	Last update	3-Oct	Rev	3

1. Background

For the last 2 years, employee access to shared data through shared directories has not worked well. Data transfers during business hours are so slow they are unusable by people in remote offices. This has made standard work, esp. in Sales, harder to maintain. Sales faces special challenges because they are often traveling and many salespeople work from home. As a result, most of their use is remote from our server where the delays are the largest.

2. Problem statement

Symptoms	Up to 30-second delay to access files for people in remote offices during peak hours 5–10-second delays are common at all times	
Goals	Reduce average time to access files	From up to 30 sec to 1 sec
	Maximum cost per employee per month	$10–$20/person for subscriptions (if we use a subscription service)
	Maximum cost install/purchase including IT hours	$60k
	Time to complete work	Installed by 31-Dec
Solve Team	Lead: Manager Web Services with 2–3 IT experts; must be 2–3 remote users on team, and 1 from sales. R&D is a large user of shared directories and so should also join team.	Select team: Dir IT, VP Marketing, VP R&D, VP Ops, CFO.

Measures

3. Analysis

The IT team ran tests on access delay through the VPN between September 1 and September 22. During that time, about 30,000 remote file accesses were executed. During peak hours 60% were from R&D who were mostly on-site. These files were placed on a shared server 3 years ago before this issue was being experienced. There is little need for R&D to have fast remote access. Moving these to another server will reduce delay by about half with few negative effects.

The remainder of the delay is due to IT infrastructure. We can replace the server bank used for VPN access for about $250k. Alternatively, we have evaluated two blue-chip suppliers of web-based shared folders. We believe any of these three options meets all security standards.

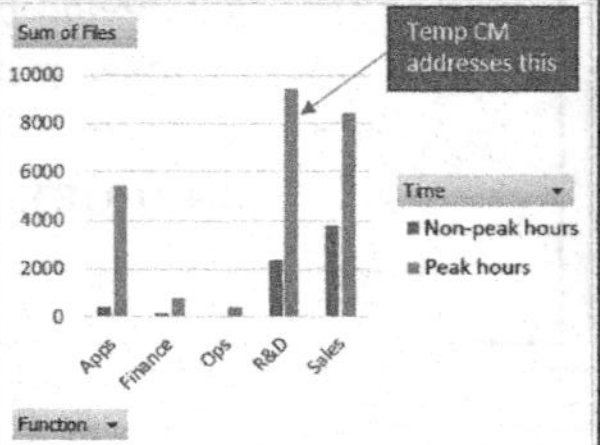

4. Root cause

	Symptom	Root cause
A	30-second delays during peak hours	High engineering use of shared drive. Engineering data can be moved to another server without significantly affecting their work since they are on site.
B	5–10-second delay during all times	Our server bank is under powered for the amount of file loading across the VPN.
C		
D		

5. Solution menu

	Countermeasure	How does this meet the goals?
✓ 1	(Temp) Move R&D data to secondary server (will increase delays for R&D, but this is acceptable b/c team is almost always on-site)	Will reduce overall access delay to 10–15 seconds. Can be done in 1–2 weeks
2	Add bank of high speed servers & upgrade VPN	Reduces access time to 3–4 seconds. No subscription. About $250k/8 months to install
✓ 3	Subscribe to web-based shared memory (Supplier 1)	Reduces access time to 1 second, subscription of $25/month, but $30k upgrades required
4	Subscribe to web-based shared memory (Supplier 2)	Reduces access time to 1 second, subscription of $40/month, but no upgrades required
5		

6. Solve-Select Action Plan

Expected result	Next Action	Next Date	Owner	Plan	Replan	Fcst	Act
Decision on CMs (team proposes: 1 & 3)	Schedule decision meeting	10-Oct	Kim	17-Oct		17-Oct	
Implement CM1			Kim	24-Oct		24-Oct	

Fig. 18.7 An example Solve-Select canvas.

Initiative	Implement Solve-Select Process	Status	4-Success Map tracking	Last Update	9-Sep	Update Due	

Guide (Sponsor)

Sponsor	Olivia
Explain need	We need a means to provide rich, consensus-driven solutions in tandem with efficient knowledge transfer between functional groups and our leadership so we can make high-quality decisions rapidly
Need-by date	Full implementation within 16 weeks
Owner name	Kim
Team members	Kim, Lisa, Sunnel, Justin, Grainne

Focus

Problem or opportunity	Improve our schedule and efficiency performance by making high-quality decisions rapidly
Metric	% of Decisions made in first meeting
Current perf.	<20%
Target perf.	85%
Solve method	2-day off-site Improvement Meeting with external team-based problem solve expert

Solve

Root causes	Information friction both to the technical/commerical team (esp. understanding the precise problem) and to leadership team (to help select the best overall solution)
Solution	Standardize use of canvas view of problem solve resulting in solution menus.

Countermeasures		
	1	Create a solve-decide canvas template
	2	Allow narration, slide shows, and other "linear" information transfer only as appendices
	3	

Action Plan

Deliverable	Next Action	Date	Owner	Plan	Replan	Fcst	Actual
Select expert to lead 2-day event			Grainne	6-Oct		6-Oct	
Schedule event			Kim	27-Oct		27-Oct	
Hold event with 1-2 live solve/decide cases			Justin	29-Oct		29-Oct	
Release V1.0 of canvas/announce new stds			Kim	12-Nov		12-Nov	
Training for team			Kim	26-Nov		26-Nov	

Test Track

Decision	Dates	Canvas Complete	Solve Decide Mtg	Decision Finalized	--	--
Select mechanical design tool	Plan	19-Oct	20-Oct	27-Oct		
	Plan	19-Oct	20-Oct	27-Oct		
	Forecast					
	Actual					
New distribution contract for North Amer.	Plan	9-Nov	16-Nov	23-Nov		
	Plan	19-Oct	20-Oct	27-Oct		
	Forecast					
	Actual					
--	Plan					
	Plan					
	Forecast					
	Actual					
--	Plan					
	Plan					
	Forecast					
	Actual					

Bowlers

Goal 1	Decisions made with one meeting between leadership and technical/commercial teams											
Date	1-Dec	1-Jan	1-Feb	1-Mar	1-Apr	1-May	1-Jun	1-Jul	1-Aug	1-Sep	1-Oct	1-Nov
Plan	90%	90%	90%	90%	90%	90%						
Replan												
Fcst												
Act												

Goal 2	--											
Date	1-Dec	1-Jan	1-Feb	1-Mar	1-Apr	1-May	1-Jun	1-Jul	1-Aug	1-Sep	1-Oct	1-Nov
Plan												
Replan												
Forecast												
Actual												

Fig. 18.8 A Knowledge Work Improvement Canvas (KWIC) to adopt the Solve-Select workflow.

Simplify

Structured communication business from/to domain

Visualization options in a solution menu

Standard visualization of solve-decide canvas

Engage

Inspire
We'll solve problems right the first time

Challenge
We'll eliminate bad decisions that hold us back

Connect
Leadership and domain experts will work together

Protect
The right team diligently solving critical problems

Experiment

Falsifiable Hypothesis

If | **Then** | **Measured by** | **Trending to**

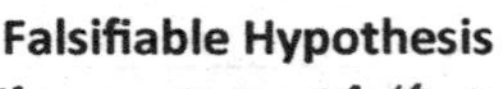

Fig. 18.9 The Solve-Select workflow **simplifies, engages**, and **experiments**.

experimentation in gaining alignment—here the goal is to get good decisions rapidly. The falsifiable hypothesis then is built around the solution menu as a mechanism to gain consensus.

Reference

[1] R. Dalio, Principles, Simon & Schuster, 2017, 236.

CHAPTER 19

Workflow—Visual management for initiatives and projects

Time is the scarcest resource and unless it is managed nothing else can be managed.
Peter Drucker [1]

Project and initiative management is one of the most commonly identified areas of dissatisfaction with knowledge work stakeholders. It's commonly stated that most projects are late or fail to deliver to expectations, or both [2]. A project or initiative has a group of people working for a limited time to meet specific requirements. The teams may be permanent—a group of people that move together from one project to another—or they may be joined together only until the project is complete. The word "project" is commonly used in technical areas and "initiative" is often used for the commercial equivalent; here the two words are synonymous.

This chapter will discuss techniques that allow management of more complex projects than can be managed with the traditional Action Plan of Section 9.3. The traditional Action Plan works well for simple initiatives, those that can be adequately captured on about a single page. As initiatives get more complex, the Action Plans must get longer and longer. It's not unusual to see people attempt to use Action Plans with 200+ lines, something unlikely to work because of the high effort to maintain the Action Plan in the face of change and the difficulty of seeing the principle issues spread among the many smaller action items. Initiatives in knowledge work regularly encounter unexpected events. Accommodating those changes in a 4- or 8-page Action Plan is exceedingly difficult because of task interdependences. The result is that long Action Plans fall quickly out of date; if that happens, the results are almost always disappointing. Our solution here is the Visual Action Plan, a visualization that displays the same task parameters as a traditional Action Plan but much more efficiently. The visualization allows you to represent perhaps 5 × the complexity for the same effort and in the same amount of space.

Improve
https://doi.org/10.1016/B978-0-12-809519-5.00019-3

This chapter teaches two primary techniques: The Visual Action Plan itself and the regular team stand-up, a short meeting held at a cadence of weekly to daily, as shown in Fig. 19.1. Both are proven techniques in knowledge work, combining the simplification of improved visualization with the increased engagement of a coordinated team.

The workflows here will be focused on managing knowledge work projects, fields with a great deal of variation from one project to the next. By contrast, projects outside knowledge work are more constant—for example, building one house is a project, even though that team may have built the same model house many times. No project in any field happens exactly the same way twice, but knowledge work projects—for example, business acquisition, product development, and expanding into new geographic regions—have more variation than other fields.

Project management suffers from each of the 8 Wastes of Knowledge Work, and frequently in large measure, as shown in Fig. 19.2. This list is focused on the waste created by long and rambling Action Plans and where the team fails to stay coordinated. The Visual Action Plan and regular team stand-ups are two tools that combat this.

Practical examples of waste in project management are shown in Fig. 19.3. Amber, a well-meaning, hardworking project manager is given a small project. Over a few weeks, she plans it out in great detail, creating a 250-line Action Plan. Two weeks later, the project starts in earnest and Amber learns that two of the people she was told would be on the team are now booked on other projects. When the three of them get together weeks after initial commitments are made, the two tell Amber they will do their best to work on her project too, but they stop short of firm commitments.

Workflow	Visual management	Manage a project or initiative using Visual Action Plans to improve team performance
Tools to simplify, engage, and experiment	Visual Action Plan	A canvas-view project plan where tasks are shown as blocks flowing left to right with block width representing task time. Each row represents the contribution of one team member. Optional leader lines between tasks can highlight task dependency
	Regular team stand-up	Short, high-cadence (daily to weekly) meetings to measure schedule performance and to react to issues as a team

Fig. 19.1 Workflow overview for visual management of projects and initiatives.

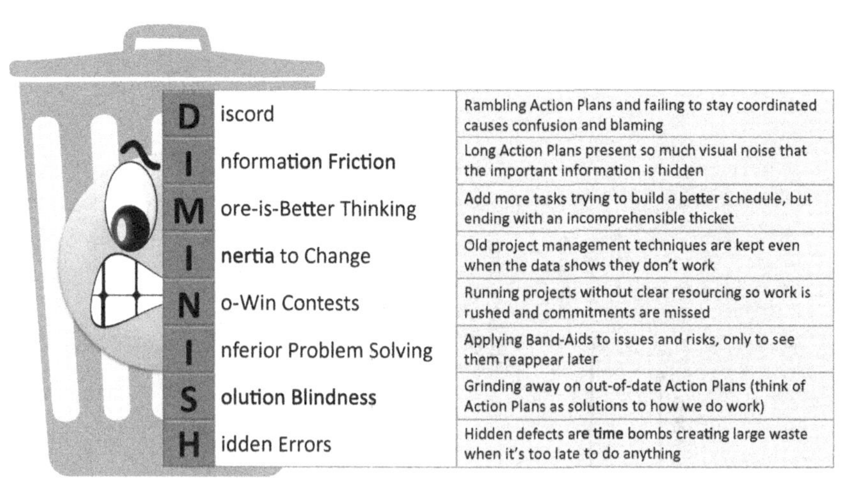

Fig. 19.2 Each of the 8 Wastes of Knowledge Work affects project/initiative management.

A couple of weeks later, a 3-week schedule slip comes to light, much of it caused by the two overbooked people not being able to get their tasks on Amber's project done. The project now seems hopelessly over schedule. Amber recognizes the Action Plan needs to be reworked, something she now has no time for. Amber works hard trying to get more resources to get back on schedule: telling the boss, sending lots of emails, and scheduling meetings to address the issue. In her mind, Amber is "trying" to hold the schedule, but there is no consensus on a plan to make up time. The rambling, out of date Action Plan is of little use now, but it has details recorded in it so Amber keeps referring to it. Confusion keeps increasing. Some weeks later at a quarterly project review, the slip, which had been known by only a few, comes into the view of the leadership, who make no effort to hide their disappointment. As a follow-up, the department leadership wants a full review of the project by the end of the week, creating extra hours of work for Amber to explain the previously hidden problems. The project from that point is permanently late because there is no practical way to make up the gap; the team now faces a months-long **No-Win Contest**. Trust from the leadership is the first casualty of this delay; engagement of the team is the second. It doesn't have to be this way.

For most organizations, the tools of choice for project management are almost certainly the Action Plan from Section 9.3 and the Gantt chart. As discussed, the Action Plan works well for small initiatives, say 20 or 30 tasks.

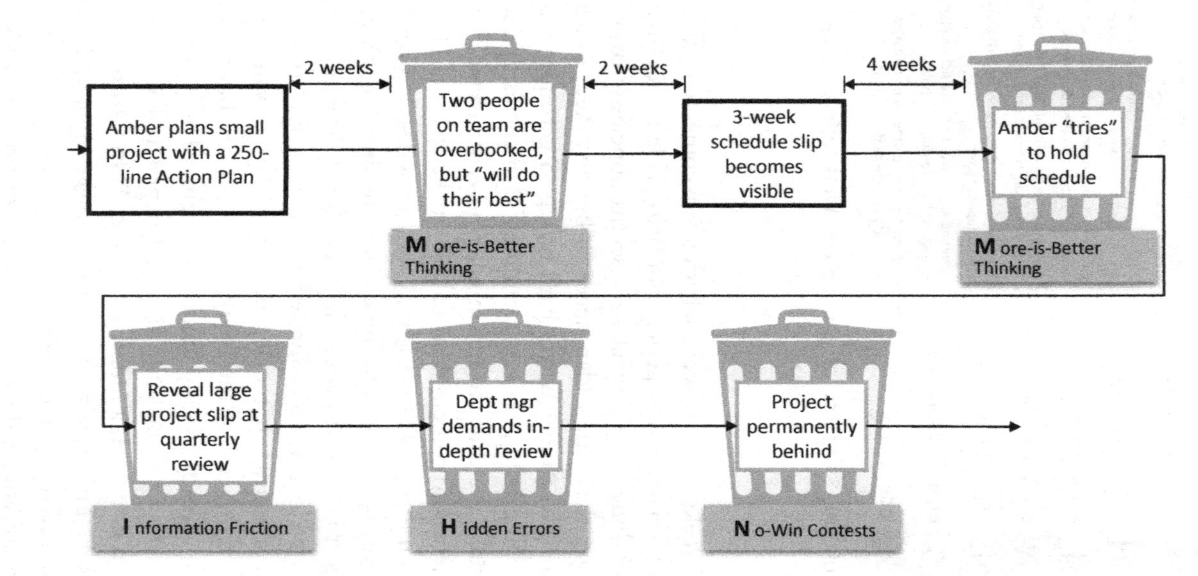

Fig. 19.3 Typical pattern of waste in project management.

Beyond that, the Action Plan is too low density owing to its limited use of visualization—essentially that of an ordinary table. A partial example of a >100 line Action Plan is shown in Fig. 19.4; Action Plans quickly grow to long lengths even with relatively simple projects.

19.1 Why avoid the Gantt chart?

The Gantt chart is a visualization made a staple of project management for decades in part by the popular application Microsoft Project. It visualizes projects as a set of interdependent tasks where the horizontal axis is calendar time and each task occupies one row. The task duration is shown as a rectangle

Expected result	Owner	Start	Plan	Replan	Forecast	Actual
Web development						
Feasibility study	Rachel	Jan 1	Feb 10			
Revise spec	Rachel	Feb 10	Mar 3			
Customer first use	Rachel	Mar 3	Mar 25			
All functions implemented	Rachel	Mar 25	May 9			
Web Category 0 bugs fixed	Rachel	May 9	Jun 1			
Web Category 1 bugs fixed	Rachel	Jun 1	Jul 2			
Version 1 released	Rachel	Jul 2	Jul 10			
SW tests running and reviewed						
Test plan complete	Ethan	Jan 1	Feb 1			
All critical SW tests automated	Ethan	Feb 11	Apr 8			
SW daily tests running	Ethan	Apr 8	April 30			
SW ready for spring show	Ethan	Apr 8	April 30			
All SW crashes fixed	Ethan	April 30	May 30			
All SW customer-down bugs fixed	Ethan	May 30	June 30			
Manage key accounts						
Customer coop. agreement visits	Kim	Jan 7	Feb 1			
2 Coop. agreements completed	Kim	Feb 1	Mar 9			
5 Account plans						

>100 more lines

Fig. 19.4 Action Plans are often a poor choice to visualize projects of size. Here the first dozen or so tasks of a 250-task Action Plan are shown.

stretching left from the start date to the end date; task interdependencies are optionally illustrated by a thin leader line interconnecting one task's end to the next task's start.

The Gantt chart is popular and, in the hands of trained experts, is a powerful tool [3]. Despite that, it's avoided here owing to the inherent difficulty of most users to control granularity. This is demonstrated by the all-too-common problem where users add tasks of ever-decreasing granularity to form an incomprehensible tangle of a project plan. It seems people feel obligated to add tasks of the smallest consequence so that Gantt charts can grow to many pages—so long that the structure of the project is lost in the detail of hundreds of small tasks. Then, when something mid-project changes that demands that structure change, as inevitably happens in knowledge work, updating the Gantt chart becomes a Herculean task, which of course, no one has time for in the busy days after problems are discovered. The Gantt chart can then become quickly outdated; project management becomes chaotic as the Gantt chart becomes less and less relevant over time. These problems are similar to those of the long Action Plan described in Section 9.3.2.

As stated above, the Gantt chart is a popular tool and many people are successful with it. My experience is that people who have formal project management training such as that necessary to be certified as a Project Management Professional [4] (PMP) by the Project Management Institute (PMI) often do well. So, if you and your organization use Gantt charts and get good results, there may be no need to change. But most organizations need people to lead projects that are not certified experts, especially for projects of modest size. This is where the Gantt chart is problematic. I've experienced many times in multiple organizations and in different regions of the world how difficult it is for a leader or anyone outside the core team to tell whether or not the Gantt chart-managed project is working. As a leader in an organization, it's not enough to know the Gantt chart *can work*; a leader must be able to recognize rapidly when it *is not working*. It's not that the project managers are deceitful, but when the project is out of control, they are unlikely to escalate because they don't properly understand the scale of the problem. Of course they don't—if they did, they wouldn't have let it get out of control in the first place. The power of Visual Action Plan is the transparency it creates, making it easier for everyone to see issues—the project manager, the project team, and the organization's leaders.

19.2 The Visual Action Plan

> *It's amazing how much the computer software available today simplifies clerical work. It's also amazing to what extent this sophisticated software doesn't help us solve the real problems.*
>
> ***Eliyahu Goldratt [5]***

The Visual Action Plan is based on the Critical Path Method (CPM) [6], which became popular in the middle of the last century. CPM is often used in conjunction with the Gantt chart. Usually, each task in a Gantt chart occupies a full row, something that can cause plans to quickly outpace the height of a Canvas View. To increase information density, we will borrow the swim lane diagram [7, 8], sometimes called a cross-functional flow chart [9], where each row is one function or person and the many tasks owned by that person are shown one after the other in the one row.

An example is shown in Fig. 19.5. This is a paper schedule in the sense that the task boxes represent what might be written on sticky notes; the critical path could be laid out with thin red painters' tape, here in most tasks of Rows 1 and 3 and the middle task of Row 2. Here, there are six lanes, each corresponding to a functional subgoal: web development, software testing, software test lead, key accounts, product marketing, and eComm development. The paper background could be a few poster size (20 in. × 30 in., 50 cm × 75 cm) sticky notes or a permanently mounted whiteboard. All this can be mimicked in a software presentation application like Microsoft PowerPoint or Microsoft Visio, which are effective for distributed projects teams.

Our example has a critical path (in fact, two) indicated by a thick, dark line across the top of chains of tasks. A task is said to be critical if "delaying that task by one day will delay the project end by one day." All the other tasks are said to be off the critical path, and small delays in those tasks are of less concern. One of the complexities of critical paths is that many things must happen for a successful project; demanding there be one and only one critical path creates difficult conversations. It's something like asking a group of people which is critical for a car on a 300-mile drive: gasoline, oil, or water? You could probably arrive at an answer, but whether the conversation adds much value isn't clear because all three are required to go 300 miles. My experience is that some project managers feel specifying critical path ahead is necessary for good results and others work out the critical path

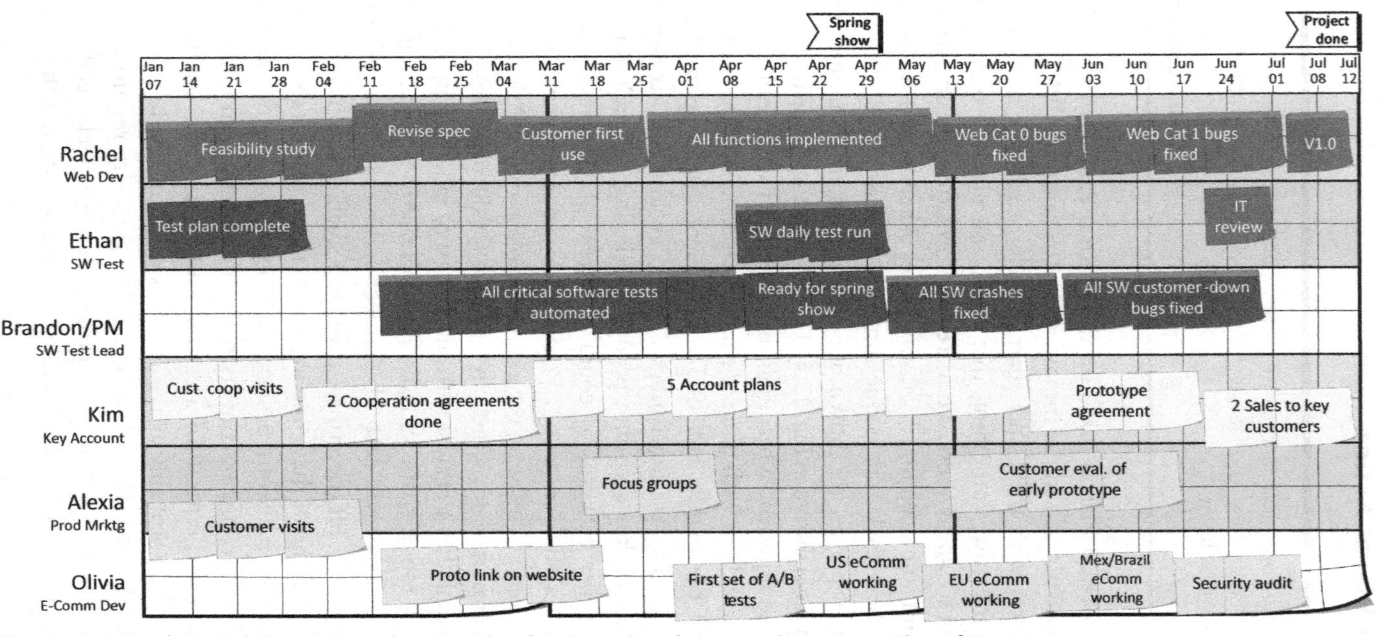

Fig. 19.5 Example: Initial "paper" Visual Action Plan for the web/software update Action Plan of Fig. 19.4.

when the team faces a delay. I've seen both approaches work well, so I recommend a formal critical path as a norm, but remain flexible on the topic when experienced project managers elect to avoid it.

A critical task is one where delaying that task by one day will delay the project completion by one day.

The Visual Action Plan is based on the swim lane diagram, which is commonly used in manufacturing. The primary difference for project planning is placing calendar time on the horizontal axis [10, 11]: the left and right sides of the task represent its start and end date (respectively), so that the width of a task represents its duration. Dependencies can optionally be shown with leader lines, something that is particularly helpful for the most important connections. Trying to show leader lines for every connection can create a spider's web that adds little to understanding because of the many, ambiguous dependencies common in knowledge work. Notice that Fig. 19.5 has no leader lines, which is an alternative I have also seen work: the dependencies are simple enough that the team can keep track of them from memory.

19.2.1 Diligently defining "done"

The demands for defining tasks in Visual Action Plans are equivalent to those for strong Action Plans in Section 9.3: define deliverables instead of effort, list clear ownership, and define the dates of the task. It's often challenging to define "done." Part of the problem is that almost every project in knowledge work has something different from what the team has done in the past. Diligently defining "done" is necessary to ensure high-quality knowledge work; the lack of diligence is disengaging to everyone, especially the Guardians of Section 7.2 who weigh "protecting the brand" heavily. Organizations running complex projects will need to define what's required to be done in detail separately because a phrase on a sticky note is insufficient to capture that. For example, the sticky note at the right side of Ethan's lane in Fig. 19.5 reads "IT review." But what is an IT review? Is it a meeting or an email? Who participates? What topics does it cover? Who decides if the review is "passed"? For most tasks in knowledge work, this level of detail is best captured in permanent documents outside the project schedule, but readily available to the project team and stakeholders.

19.3 The regular team stand-up

The regular team stand-up is a short, high-cadence meeting used to update the Visual Action Plan, identify delays, and continuously reconfirm the project schedule forecasts. The meeting can be held daily to weekly and can be as short as 15 minutes. It should normally include the main team members, basically the people who work in the project enough to own a lane in the Visual Action Plan. There are often others whose role is small enough that their work is organized by main team members; those people normally do not need to attend the stand-up meeting.

One important characteristic about stand-up meetings is they need to provide the opportunity for everyone to contribute equally. If the venue diminishes someone's place in the meeting, they won't contribute reliably. That's often unintentionally done to remote team members by having them join team meetings by phone where the majority of the team is in a conference room; this reduces greatly the remote members' ability to comprehend and join in. It's more effective to have the meeting either all physical (in which case all team members review a poster-sized canvas) or all virtual (in which case all team members view the canvas on their personal device from their offices). Note that on the right-hand side of Fig. 19.6, there

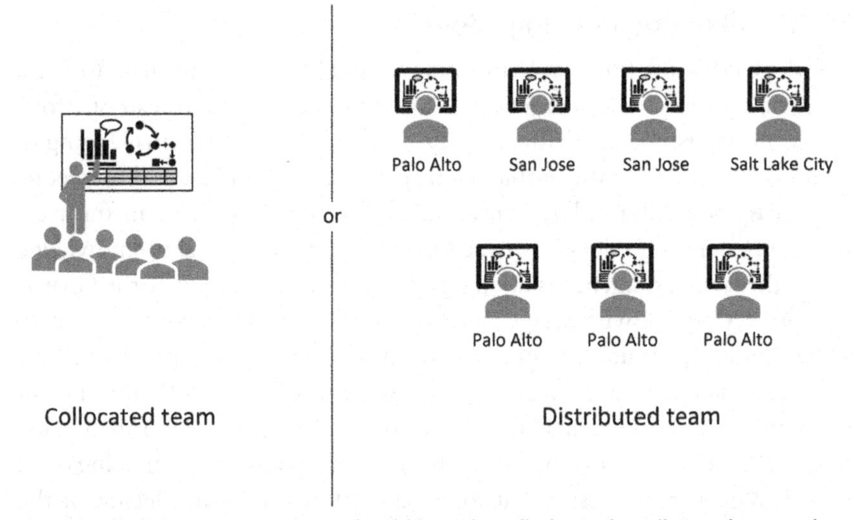

Fig. 19.6 Regular team stand-ups should be either all physical or all virtual, even when a few people are together. Avoid having most in a conference room and some on the phone.

are four people in the Palo Alto office who could have met in a conference room (especially if one of them was the project manager), which implies everyone outside Palo Alto must settle for a speaker phone connection and perhaps a low-resolution video of the room.

Finally, for maximum effectiveness, all participants of virtual meetings are required to be on camera from their personal devices. This asks no more of a person than you would for a face-to-face meeting: be presentable and remain engaged. Use video conferencing software that automatically shows the face of the person talking (for example, Microsoft Teams and Cisco Webex have this feature). It helps quickly identify multitasking where people sign onto audio conference calls and then do email or are otherwise distracted. If some people sense one person can get away with just partial participation, expect others to try the same. Then your meeting will get a fraction done of what it could have done, which isn't fair to the people who engage fully.

19.4 A legend for planning and regular execution

Fig. 19.7 shows a legend for planning and for regular execution. Tasks should be drawn on sticky notes (or their computer application equivalent) with the width equal to the duration of the task; use multiple sticky notes to display long tasks. The critical path task is shown with a red bar (dark gray in the print version), for example, the bars at the top of the sticky notes in

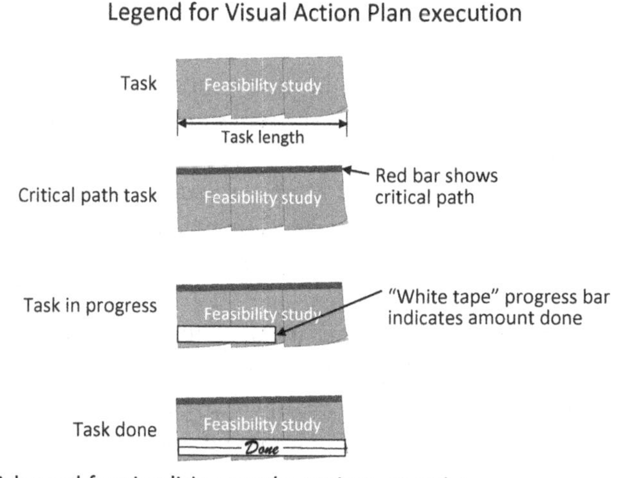

Fig. 19.7 A legend for visualizing regular project execution.

Rows 2, 3, and 4 of Fig. 19.7. Completion is indicated with a progress bar made with, say, a strip of white tape (Row 3); when the task is done, write "Done" on the progress bar (Row 4).

The example in Fig. 19.8 shows a project that is running about 10 days late a month and a half after it started, something you can see because the done flag is a week and a half left of the "Today" arrow. Further, we can see that the feasibility study is the likely cause of the delay; all other tasks that the done flag touches are ahead of the feasibility study. Because the "Done" flag is behind the "Today" arrow, we are late; we have a Stop-Fix alarm, which is the primary topic of the next section.

19.5 Visual Action Plan execution

The regular team stand-up meeting is a series of 2–3-minute conversations that the leader keeps contained in a short meeting—perhaps just 15 minutes. The tone of the meeting is the leader asking closed questions to the team expecting short answers, then building consensus on active and new issues, and then setting up follow-on meetings for topics too deep or narrow for the stand-up. A suggested four-step flow for a schedule-update stand-up meeting is shown in Fig. 19.9:

1. **Update progress bars**
   For each active task in the Visual Action Plan, the owner makes a rough estimate of completion. The project manager draws a line along the bottom of the task to reflect progress (50% done = a line or tape segment from the left to the center of the task block).
2. **Slide "Done" line right**
   Slide the vertical "Done" line to the right as far as the least-progressed task will allow. This sets how much of the project can be counted as "Done."
3. **Measure schedule performance**
   If the "Done" done from Step 2 is greater than or equal to today's date, the project is on time; the regular stand-up meeting is done. If not, go to Step 4. The project in Fig. 19.9 is late and so must proceed to Step 4.
4. **If late, run a play**
   A late project is a Stop-Fix alarm; we need to run a play to get back in the green. The team's ability to hold the schedule depends upon rapidly resolving Stop-Fix alarms. In the simple cases, resolve the issue in the stand-up meeting; if it's too complex to be resolved in a few minutes,

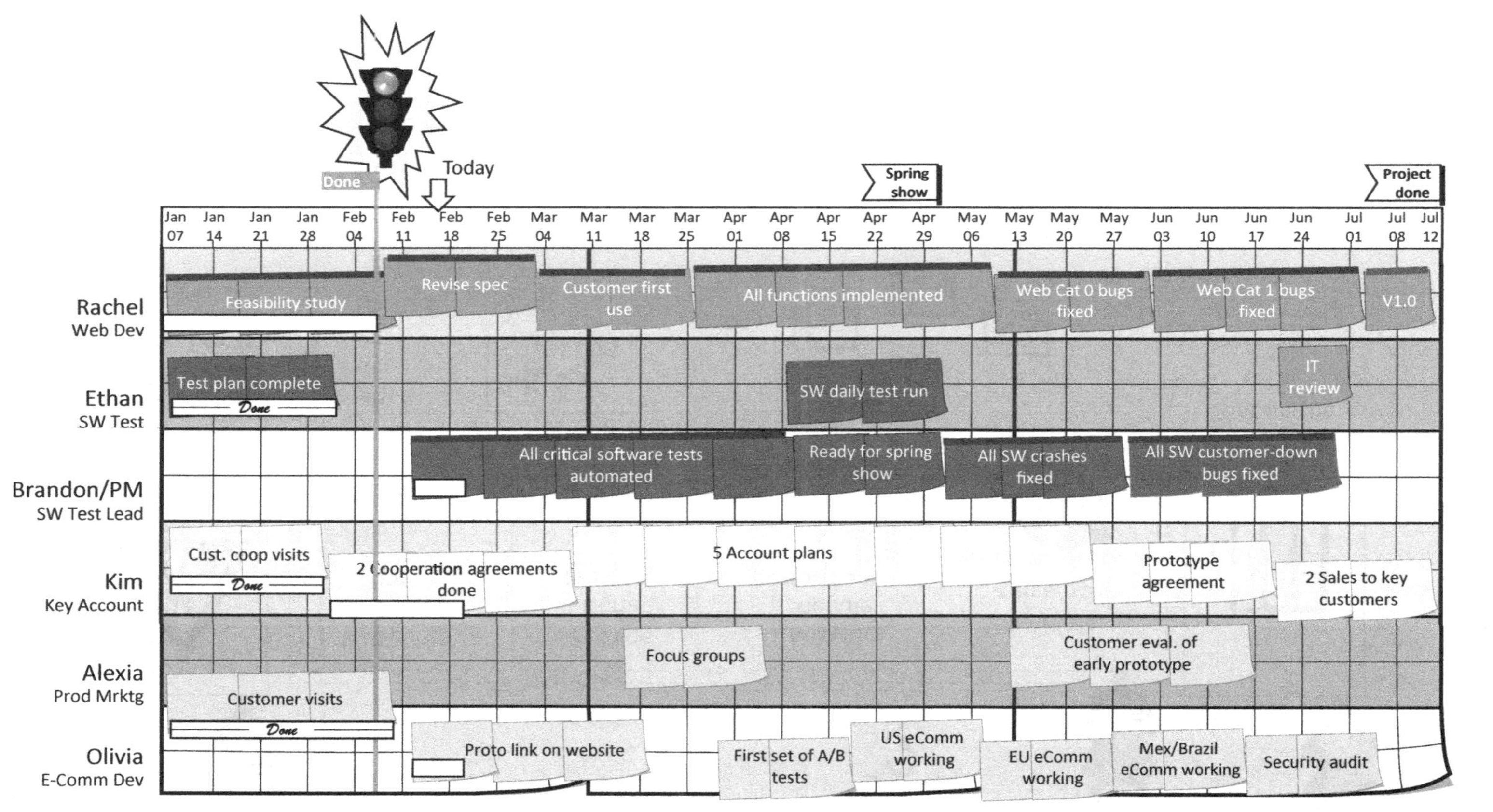

Fig. 19.8 Stop-Fix alarm: Visual Action Plan shows 1.5 weeks of delay between "Done" and "Today."

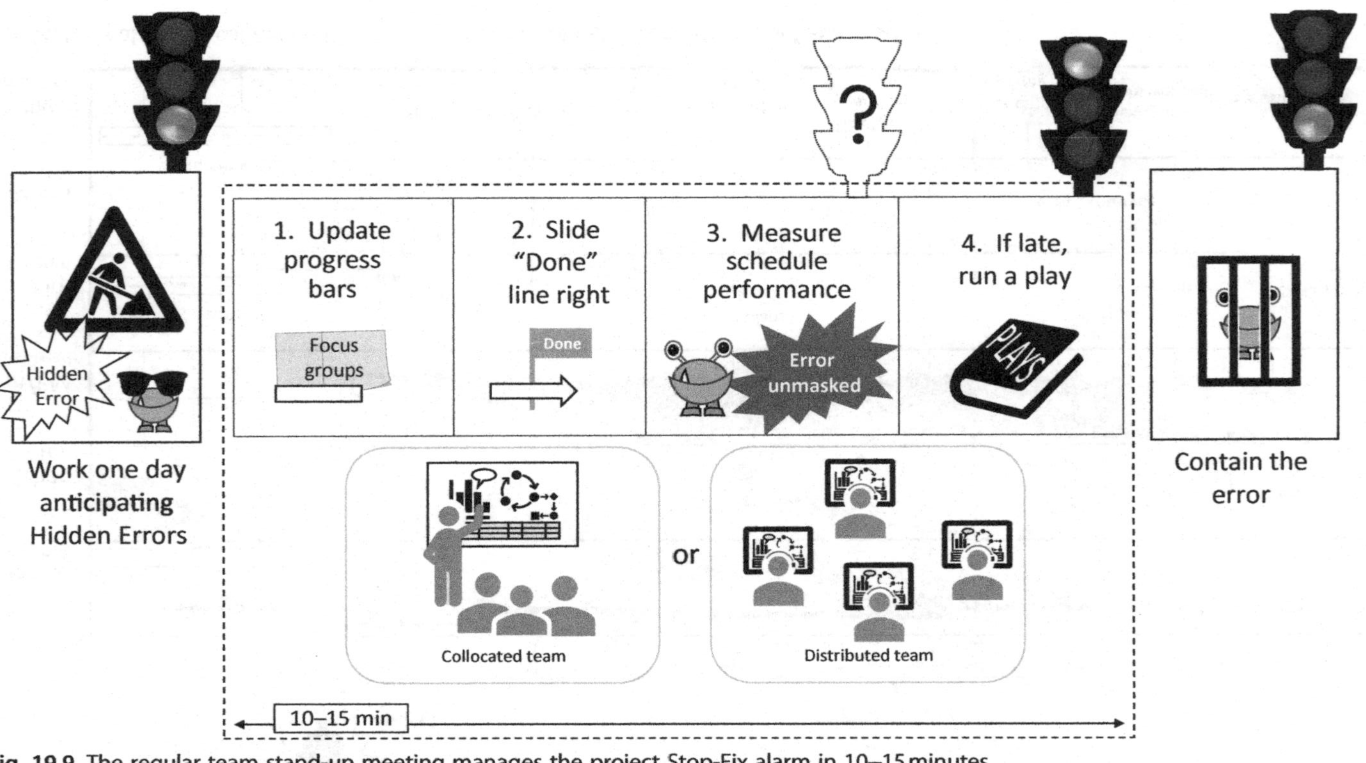

Fig. 19.9 The regular team stand-up meeting manages the project Stop-Fix alarm in 10–15 minutes.

the project manager will likely spend a good part of the day in various meetings to resolve this. Either way, it should be resolved in a day or two.

19.6 Running plays in Visual Action Plans

"Plays" are countermeasures to being late when the regular team stand-up reveals a Stop-Fix alarm. The playbook is short because there just are not that many things we can do to make up schedule. Most plays fall into one of three categories: those that restructure the project to relax constraints, those that increase resources available, and those that reduce the effort required to finish.

19.6.1 Plays that restructure the project

Plays that restructure projects are some of the most innovative steps we can take to make up schedule loss. For example, the team recognizes that a dependency they originally planned on isn't necessary: we thought we couldn't do "Step 2" until "Step 1" was complete. Or we originally planned to send one unit to three customers sequentially; now we realize we can speed up the project by building three units and sending them to three customers in parallel. Or we thought we needed the power supply design to be complete before we could design the product enclosure; now we know a prototype power supply tested to a few critical parameters is sufficient to begin enclosure design. We thought we had to wait for full release of our new Customer Relationship Management (CRM) package, but now we know we can start using the package with an early release in the Americas.

Project restructuring is an "aha" moment that the team experiences when they are diligently managing project performance. These ideas come to mind because of the regular team stand-up meetings, where we have engaged the full creativity of the team. It may take some days for people to digest the problem and discover how to avoid the constraint we originally thought was unavoidable. This path depends on clear requirements and accurate visualization of the current issues. When the whole team understands precisely why a constraint limits team performance, they will naturally shift to finding creative ways to relax those constraints without sacrificing quality or compliance.

19.6.2 Plays that increase resources

Plays that increase resources are those where we can add people to the team, from other parts of the organization or from outside contractors and consultants. They also include where we increase team member focus, for example, releasing key team members from commitments in lower-priority projects. Finally, we can ask for more focus: perhaps a temporary increase in effort or delaying travel. These last examples are the most painful and least sustainable steps we can take. If a temporary burst of energy makes disproportional difference—say, being in time for a show or getting a customer up and running quickly—they may make sense. But you're dipping from a shallow well; most people won't tolerate having their life outside of work disturbed often, nor should they have to. The organization is responsible for adopting workflows that can be executed reliably so that individuals don't regularly have to sacrifice home and family time to make up for lapses.

19.6.3 Plays that reduce effort

Plays that reduce effort include those that reduce the features or performance requirements of the work. They include steps that create multiple releases: a release of a partially done project to meet critical customer needs or start generating revenue sooner. We might launch a bare-bones product and plan to augment it a few months after launch. Or we might remove a requirement altogether. After we have exhausted the other possibilities, we must consider extending the delivery date. Such a step is among the least desirable alternatives, but still superior to the **No-Win Contest** of working to a schedule we know we cannot meet. Once we extend a project, we can no longer claim to be "green"; the "yellow" of working to a renegotiated completion date is the best the team can expect, as discussed in Section 9.5.4 (replanning).

19.6.4 Example: Run a play

Let's continue with the example of Fig. 19.8, where we have a Stop-Fix alarm. We can "see" the problem—the feasibility study was planned for completion ~Feb 7; now it's Feb 16 and there are still a few days of work. Worse, Rachel reveals at the regular team stand-up that she's stuck on a difficult issue, so it could be even longer. In our schedule, we assumed "Revise spec" couldn't start until Rachel finished the "Feasibility study." The rest of the project is on time. If we can join in to help Rachel, we can pull the project back onto schedule.

Notice here that no one is asking Rachel to cut corners. In fact, let's go further and encourage Rachel to finish the work well during the team meeting. Rachel likely is feeling pressure—she doesn't want to let the team down and she wants to do the feasibility study well. She doesn't know how to do both. If the leadership just pounds on, holding the schedule without recognizing the need for quality work, it could send a message that we are not dedicated to outstanding work.

So, let's assume the leader says, "We need to hold the schedule to deliver what our customers expect but we will not compromise our standards... what do we do?" In our hypothetical meeting, there's a minute or two of silence as every team member scans the schedule. There's so much creativity, you can almost feel it.

Ethan breaks the silence: "I can revise the spec. I can push some of my other work out a few days. We don't need the feasibility study to be 100% complete." There's a couple of minutes of healthy conversation as the rest of the team considers Ethan's idea, testing the critical thinking. In this case, the idea looks good, so we'll implement this small restructure to make up the time (Fig. 19.10).

1. Move "Revise spec" from Rachel's swim lane to Ethan's, who will push other work (unrelated to this project) to make time for "Revise spec."
2. Allow "Feasibility study" to slip 1 week to ensure Rachel can finish. Mark the Task-Done to be within 1 week of the end of the slip.
3. Now the "Done" flag is free to slide right all the way to today.
4. Since Done = Today, we can turn off the Stop-Fix alarm.

As we discussed in Section 6.4.3.1, time is the bottleneck in project management. This doesn't mean we only care about releasing on time—that quality, cost, and compliance to regulations don't matter. Of course, they matter very much. But all project issues are reflected in time. If we discover a quality issue that must be addressed, the time to address it comes out of the project team's capacity; that increases delay so that the quality issue can be reflected in units of time. For example, "Our project lost 2 weeks because we discovered we didn't comply with FDA regulations." Similarly, if the cost of providing cloud-based repair tracking is 50% higher than expected because our software supplier just upped their price list, we may need to spend 4 weeks finding a new supplier and adapting our systems to them. So, that cost increase can be reflected as a 4-week delay.

Leaders must emphasize this point: we measure time not because we think all that matters is saying we are finished. Rather, time reflects

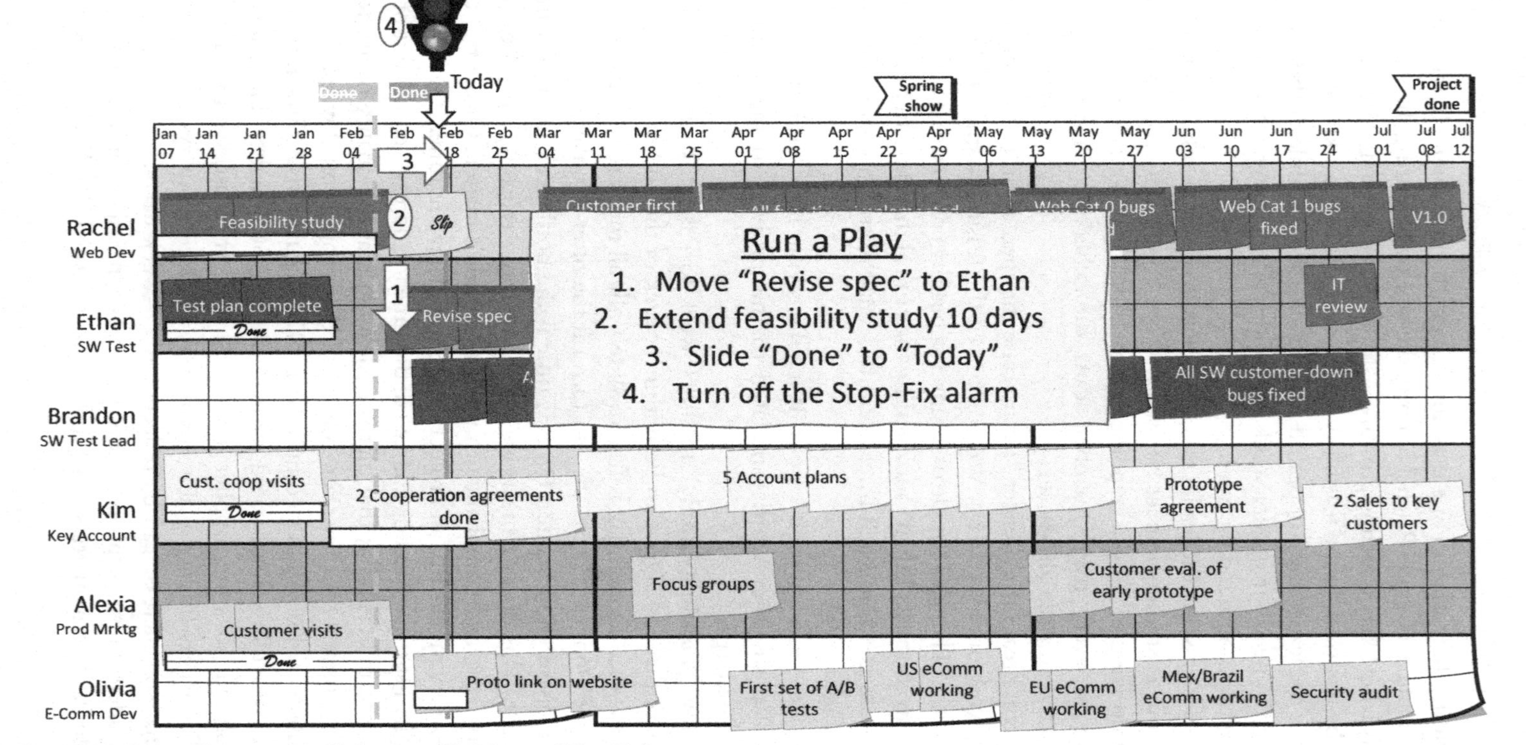

Fig. 19.10 Run a play to turn off the Stop-Fix alarm of Fig. 19.8.

all the things we must do to provide the right product at the right cost and complying with all regulations and organizational requirements. If you fail to communicate this, you risk people feeling pressure to compromise standards, which will create Hidden Errors. It also injures engagement, especially with your "Guardians" who most value protecting the brand.

19.7 Buffer: Smoothing daily management

One of the shortcomings of the method presented here is the requirement to respond so quickly to schedule slips. The method here works well for low-risk, medium-sized projects, the sorts of projects that have few enough unknowns that we can predict the schedule well from the start. When you have projects of more complexity, consider using buffer management to smooth the activity. This will be the topic of the next chapter.

19.8 Knowledge Work Improvement Canvas (KWIC) for visual management

Deploying the Visual Action Plan is hard work. It takes several months to create change and traction, many more months to get all projects compliant, and then years to build the skills in the organization to get the full value. But the initial payoff is fast and the method keeps cutting waste and delivering value for years. Your journey will be unique because each organization has its own needs. So, the details of the visualization, the cadence of meetings, and the playbook will all change. But what won't change is the path to success: gain alignment in the organization about how this will bring value, diligently plan an improvement cycle with the Knowledge Work Improvement Canvas (KWIC) of Fig. 19.11, and execute the plan.

Fig. 19.12 shows how Visual Action Plans and regular team stand-ups **simplify** the workflow of project management. It also **engages**: we can Inspire with the promise of earning trust and providing outstanding performance. We can Connect with cross-functional teams that support each other and Challenge them with a winnable contest: a project they buy into. And we can Protect the brand with standards that we will not violate to make a date. Finally, **experiments** are created through the falsifiable hypotheses to deliver our work reliably on time.

Initiative	Install Visual Schedules	Status	4-Success Map tracking	Last update	12-Dec	Update due	

Guide (Sponsor)

Sponsor	G.M.
Explain need	Our projects and initiatives are systematically late. Our best guess is less than 25% come in on schedule. This causes chaos in the organization. We need to attain 85% or 90% projects on schedule
Need-by date	Need metric in place within 4 weeks, 50% on-schedule projects within 16 weeks, and 90% within 1 year
Owner name	R.E.
Team members	K.N, O.S., R.E, Olivia, Alexia

Focus

Problem or opportunity	Create project management techniques to finish 90% of projects on schedule
Metric	Project completion to date scheduled at project start
Current perf.	About 25%
Targert perf.	90%
Solve method	3-Day Global event with outside expert in project management

Solve

Root causes	Inadequate diligence creating initial schedule Lack of buffer creation at project outset Lack of daily stand-up to rapidly address Stop-Fix alarms
Solution	1) Start projects with full team dedicated to creating a realistic schedule 2) Track progress daily with visual schedules and daily team meetings

Countermeasures		
	1	Create and deploy team event. Define interactions through project, espeically daily stand-up
	2	Define standard way to visualize projects
	3	Define and measure daily stand-up meetings

Action Plan

Deliverable	Next Action	Date	Owner	Plan	Replan	Fcst	Actual
Create training deck for project start event			K.N.	28-Jan	23-Jan		
Define standard project schedule format			K.N.	18-Feb	9-Feb		
Define high-cadence team meet			Pilar	4-Mar	20-Feb		
Buffer				18-Mar	18-Mar		

Test Track

Case	Dates	Project Start Event	Project Date Commit	1 month daily S/U	2 months in Green	Finished on time
Project XYZZY	Plan	1-May	15-May	22-May		
	Replan					
	Forecast					
	Actual					
Proejct Raddleslap	Plan	11-May	25-May	1-Jun		
	Replan					
	Forecast					
	Actual					
Project 1	Plan	19-May	2-Jun	9-Jun		
	Replan					
	Forecast					
	Actual					
Project Hobbgobb	Plan					
	Replan					
	Forecast					
	Actual					

Bowlers

Goal 1	Project on time to schedule											
Date	1-May	1-Jun	1-Jul	1-Aug	1-Sep	1-Oct	1-Nov	1-Dec	1-Jan	1-Feb	1-Mar	1-Apr
Plan	50%	50%	60%	65%	70%	75%	80%	90%	90%	90%	90%	90%
Replan												
Fcst												
Act												

Goal 2	0											
Date	1-May	1-Jun	1-Jul	1-Aug	1-Sep	1-Oct	1-Nov	1-Dec	1-Jan	1-Feb	1-Mar	1-Apr
Plan												
Replan												
Forecast												
Actual												

Fig. 19.11 A Knowledge Work Improvement Canvas (KWIC) to deploy new project management workflow.

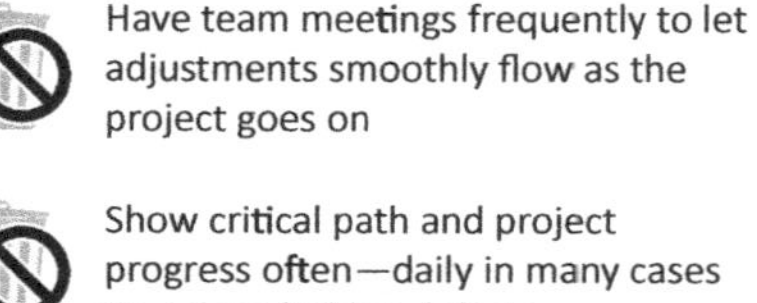
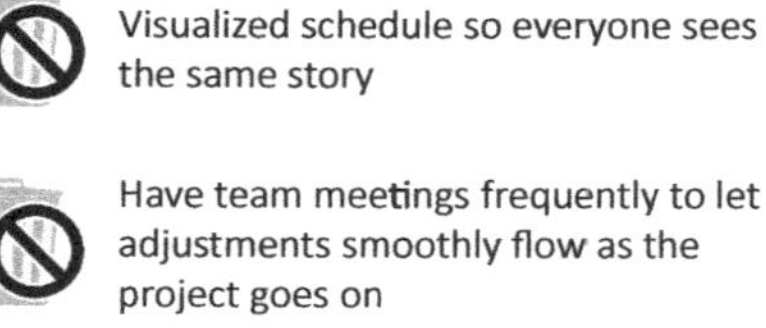

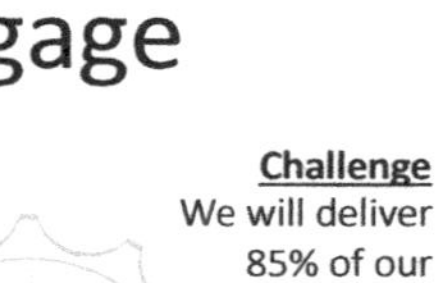
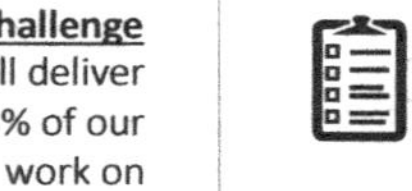
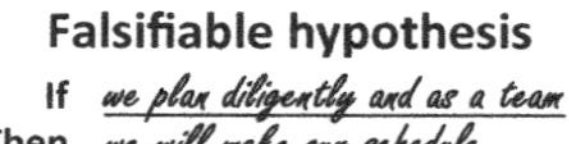

Fig. 19.12 Visual management of projects and initiatives **simplifies**, **engages**, and **experiments**.

References

[1] Quoted by Theodore Kinni, The 50-Year-Old Business Book You Need to Read (or Reread) Right Now, Inc.Com, May 26, 2017. https://www.inc.com/theodore-kinni/the-50-year-old-business-book-you-need-to-read-or-reread-right-now.html.
[2] Huthwaite Innovation Institute, www.barthuthwaite.com (landing page), viewed 2019-Oct-10.
[3] G. Ellis, Project Management for Product Development Leadership Skills and Management Techniques to Deliver Great Products, Butterworth-Heinemann, 2016, p. 107f.
[4] https://www.pmi.org/certifications.
[5] E.M. Goldratt, Critical Chain: A Business Novel (Kindle Location 1920), North River Press, 1997. Kindle Edition.
[6] G. Ellis, Project Management for Product Development Leadership Skills and Management Techniques to Deliver Great Products, Butterworth-Heinemann 108 (2016).
[7] P.R. Williams, Visual Project Management, Think for a Change Publishing®, 2015, p. 23.
[8] What Is a Swimlane Diagram, https://www.lucidchart.com/pages/swimlane-diagram.
[9] Visio, Create a Cross-Functional Flowchart, https://support.office.com/en-us, 2019. and enter "Create a cross-functional flowchart" in the search box.
[10] https://workshopbank.com/project-plan-template.
[11] M. Woeppel, Projects in Less Time. A Synopsis of Critical Chain, Productivity and Quality Publishing Private Limited, 2010, p. 50.

CHAPTER 20

Workflow—Visual management with buffer

The entire bottleneck concept is not geared to decrease operating expense, it's focused on increasing throughput.

Eliyahu Goldratt [1]

20.1 Introduction

This chapter will build upon the last: adding to visual management the concept of project buffer visualized with Eliyahu Goldratt's *fever chart*. We will also build up the Project Schedule Canvas, a single-page view of a project. These techniques are summarized in Fig. 20.1. In addition, this chapter will present Goldratt's portfolio snapshot fever chart, a brilliant visualization technique that allows an organizational leader to take the pulse of dozens of projects in seconds.

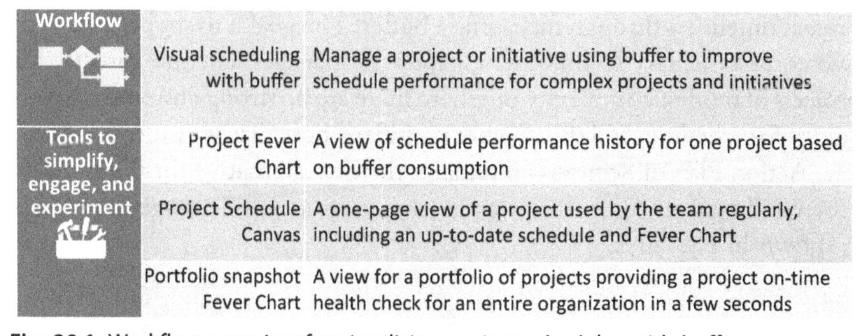

Workflow	Visual scheduling with buffer	Manage a project or initiative using buffer to improve schedule performance for complex projects and initiatives
Tools to simplify, engage, and experiment	Project Fever Chart	A view of schedule performance history for one project based on buffer consumption
	Project Schedule Canvas	A one-page view of a project used by the team regularly, including an up-to-date schedule and Fever Chart
	Portfolio snapshot Fever Chart	A view for a portfolio of projects providing a project on-time health check for an entire organization in a few seconds

Fig. 20.1 Workflow overview for visualizing project schedules with buffer.

Improve
https://doi.org/10.1016/B978-0-12-809519-5.00020-X

Almost all defective projects result from not having the right knowledge in the right place at the right time.

Allen C. Ward and Durward K. Sobek II [2]

Chapter 19 presented techniques to manage a project using time as the bottleneck metric, maintaining on-time delivery through high-cadence stand-up meetings. This method works well for projects of modest complexity or those where schedule commitments are soft. Examples of the latter include when someone uses a Visual Action Plan to manage a goal for personal growth or when a team takes on a challenge to improve their lab or file system. Modest schedule slips are of less concern when there's no customer depending on this work. However, when there is a customer expecting results and the project has significant unknowns, the method of Chapter 19 can demand a great deal of activity to respond to small schedule slips. For example, the team may get hit with multiple issues a week that add a day or two, each generating a Stop-Fix alarm; this can exhaust a team.

In order to smooth the workflow, we can add project buffer at the end of the project. Project buffer is created when the team works to one date, but firmly commits to a slightly later date. The difference between those two is called *buffer*. Buffer then can absorb small changes, say, a day each week, without generating a Stop-Fix alarm. We manage the project by managing buffer, treating it like currency, spending it where necessary to smooth the project but conserving it elsewhere. Time is still the bottleneck metric; buffer presents an alternate way to generate Stop-Fix alarms.

In this chapter, we'll modify the techniques of Chapter 19 to manage project timeliness through managing a buffer. The goal is to supply you with two complementary techniques: Chapter 19 manages schedule directly for projects of modest complexity or where there are no strong customer expectations for schedule, and this chapter considers those that are more complex. The Action Plan of Section 9.3 remains the first alternative for the simplest projects. Together, these three approaches support a wide range of projects as shown in Fig. 20.2.

20.2 The 8 Wastes in unbuffered project schedules

As shown in Fig. 20.3, there are many wastes that come from managing projects with poor techniques. We saw something similar in Chapter 19 where

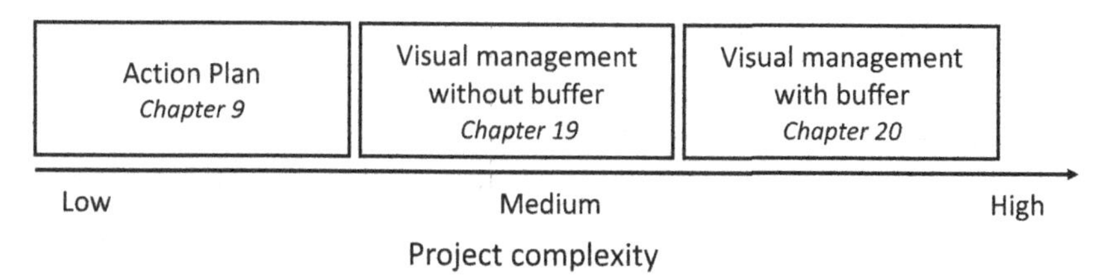

Fig. 20.2 This book presents three project management methods for three levels of project complexity.

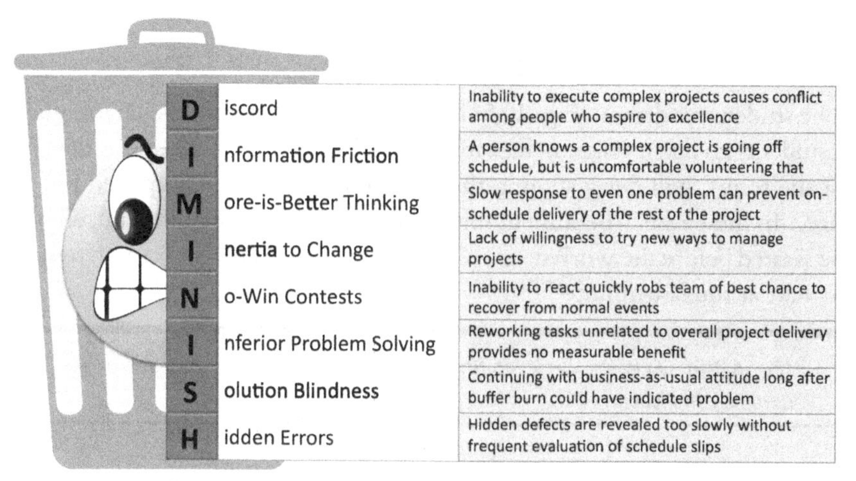

Fig. 20.3 Each of the 8 Wastes affects large-sized project/initiative management.

we compared Visual Action Plans to more pedestrian methods such as long Action Plans and Gantt charts. Here we look at waste reduced by using buffer in projects that need buffer to run efficiently.

Section 20.2 describes visualization techniques that can dramatically reduce this sort of waste, primarily the fever chart. Following that, we'll take up the project Canvas View that will help maintain alignment and flow information daily. We will use buffer consumption from the fever chart to provide an unambiguous indication of success on a weekly or even daily basis, for projects that may be 1 or 2 years long. After that, we'll take up several other methods to simplify tracking projects including the regular team stand-up meeting adapted for buffered schedules and the portfolio snapshot fever chart. Finally, we'll discuss ways to improve standardizing these methods in an organization.

20.2.1 Buffer in project planning

Most people will naturally estimate tasks with a built-in buffer, usually without even realizing it. It's something like how someone who normally has a 10-minute drive to work might leave each day 20 minutes ahead to create "buffer" for unexpected traffic. That works well for one task, but when stringing task after task together as in Fig. 20.4, each with implicit buffer built in, it creates a distributed buffer. Distributing buffer results in loss of schedule performance from two well-known effects: the student syndrome and Parkinson's Law.

20.2.2 The student syndrome

The student syndrome (Fig. 20.5) is named for the common behavior when a student, given an assignment with a great deal of time (e.g., due at the end of the term), will delay starting the work until it can be completed just in time. In these cases, the student initially has a great deal of buffer, but all will be wasted before the work starts. As Goldratt put it, "There is no rush to start so start at the last minute" [3].

The student syndrome: delay starting a task until the time required just allows on-time completion.

20.2.3 Parkinson's Law

In 1958, Cyril Parkinson published "Parkinson's Law" in the Economist [4], a statistical analysis of the British Civil Service from which derives the popular maxim: the time expended for work expands into the time allotted. For example, most people who are allotted 50 minutes for a test will spend all that time. Even if they finish early, they will spend what time remains reviewing

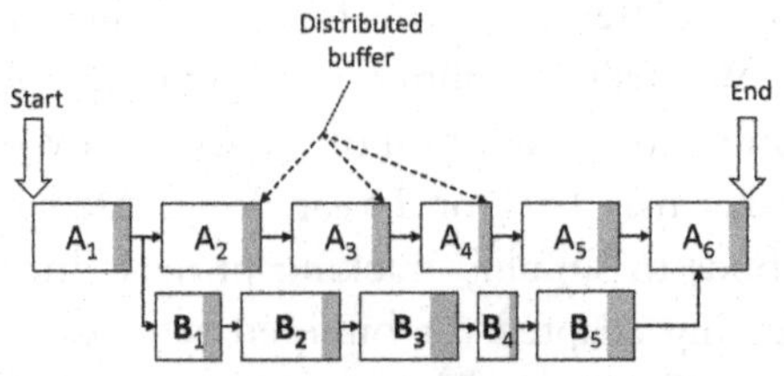

Fig. 20.4 Intuitive task estimations usually distribute buffer among many tasks, a practice that makes staying on schedule more difficult.

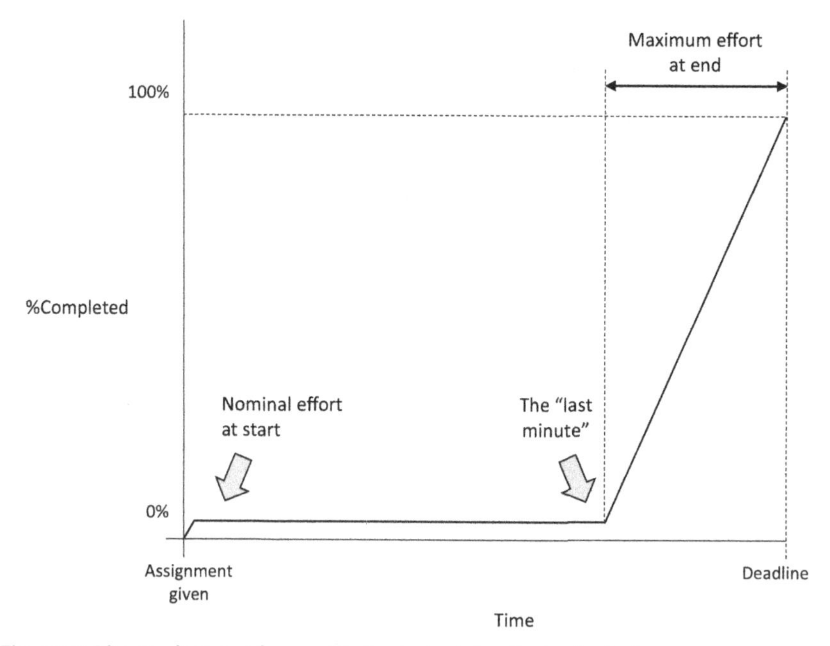

Fig. 20.5 The student syndrome: Start a task at the "last minute."

questions or reworking problems. This is sensible for a 1-hour test, but when people unintentionally apply this thinking to all their daily work, projects slow down even though everyone is busy. If you're allotted 5 days to find a supplier and you find a suitable supplier in 3 days, Parkinson's Law says most people will use the other 2 days to check into a few details about suppliers. It is not necessarily a conscious decision; in fact, most people do this without even thinking about it. Sometimes this is derisively called "gold plating" as if the precious metal was being added in a place a customer didn't value it. As Fig. 20.6 shows, Parkinson's Law can needlessly chew up buffer, burning energy with the work that occurs between "Time required" and "Time allowed."

Buffer collected at the end of a project combats both the student syndrome and Parkinson's Law. As represented in Fig. 20.7, the individual buffer is squeezed out of each task and placed at the end of the project. This fights the student syndrome by minimizing the perceived buffer to start tasks. It fights Parkinson's Law by calling for tasks to be closed as soon as they are ready. Perhaps, most importantly, it provides a concrete leading measure of schedule performance: the consumption of buffer, which will be the foundation of the fever chart; more on that shortly.

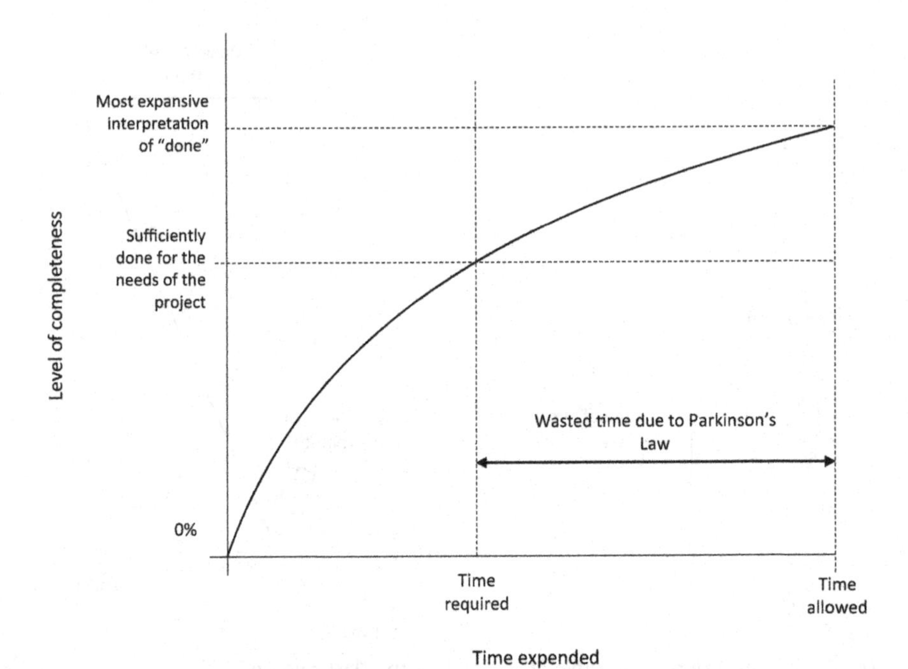

Fig. 20.6 Parkinson's Law: The time expended for work expands into the time allotted.

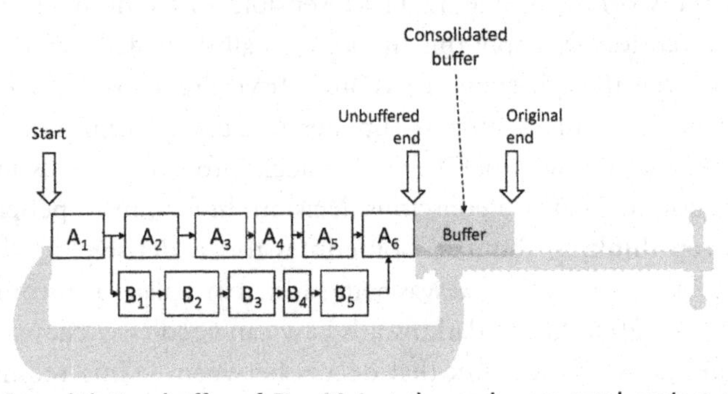

Fig. 20.7 Consolidating buffer of Fig. 20.4 at the end; note total project duration is unchanged.

For our example project of Fig. 19.5, we can squeeze out the buffer from each task so that each sticky note shows, roughly speaking, the task's best-case completion time. This is shown in Fig. 20.8 where 6 weeks of buffer were created in a 28-week project, about 20%. Normally, I used 20%–30% compression for tasks, varying it up or down with increasing or decreasing (respectively) project risk. Notice that buffer is taken from the

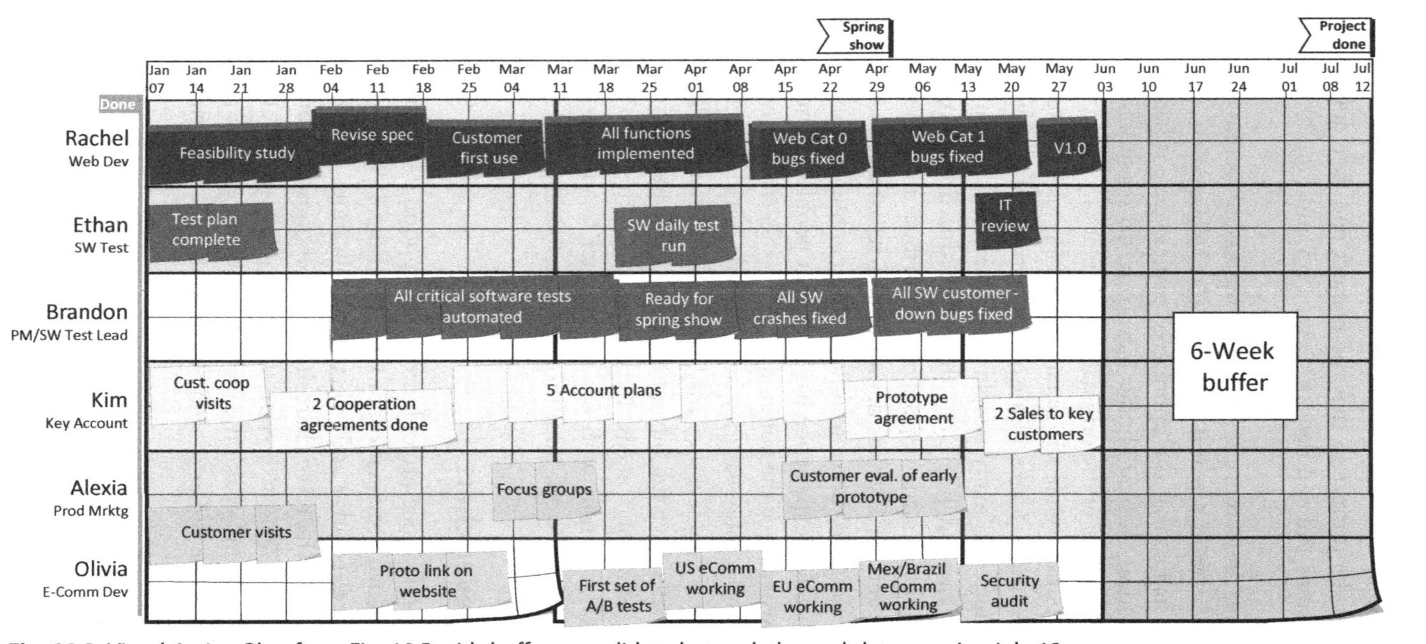

Fig. 20.8 Visual Action Plan from Fig. 19.5 with buffer consolidated at end; the end date remains July 12.

project schedule, not added to it. All of the accumulated buffer is added to the end of the project so that the project end date doesn't change, as can be seen since the end dates for Fig. 19.5 and Fig. 20.8 are the same: July 12.

Goldratt teaches a more aggressive compression technique in his project management method, Critical Chain Project Management (CCPM): cut all task times in half so that the unbuffered project time is half of the original project estimation; then add buffer half again that size [5]. So, the new project time will be 75% of the original estimate: 50% for the compressed tasks and 25% for buffer, while the remaining 25% is taken as acceleration from the original estimate. There is little objective proof of this ratio; moreover, it is counterintuitive to many people and this assumption can be a barrier to adopting the method [6]. Does this mean that 50% compression is too aggressive? Goldratt made unparalleled contributions to the field of project management and CCPM has demonstrated remarkable successes. So, it's certainly a reasonable choice to follow his recommendations. Nevertheless, I've avoided it. My experience has been that people are more accepting of long-term ownership if the buffered schedule reflects their original end date. When things go wrong—as they invariably do—people will act with more determination if they feel they made the commitments willingly. So, if the original estimates have full buy-in from the team and the project estimates meet the organization's needs, I accept them. If they don't, my focus has been to build consensus with the team either to increase resources or to reduce the scope of work until those two factors align.

20.3 The fever chart

The fever chart is a measure of buffer consumption over time. It was part of Goldratt's original presentation of CCPM in the 1990s and remains today one of the most creative and effective visualizations available for schedule performance [7]. The fever chart shows buffer consumption on the vertical axis and project calendar days on the horizontal one.[a] A straight line runs between the bottom left and the top right corners based on a brilliantly simple assumption: the buffer should be consumed no faster than proportional

[a] Goldratt used a different horizontal axis: the amount of critical path "done." For most projects, the graphs are similar. However, there is a significant difference for projects that discover less of the critical path was done that we thought (for example, if a test comes back so bad, we have to restart much of the project). I favor the calendar-based version presented here because it seems more intuitive.

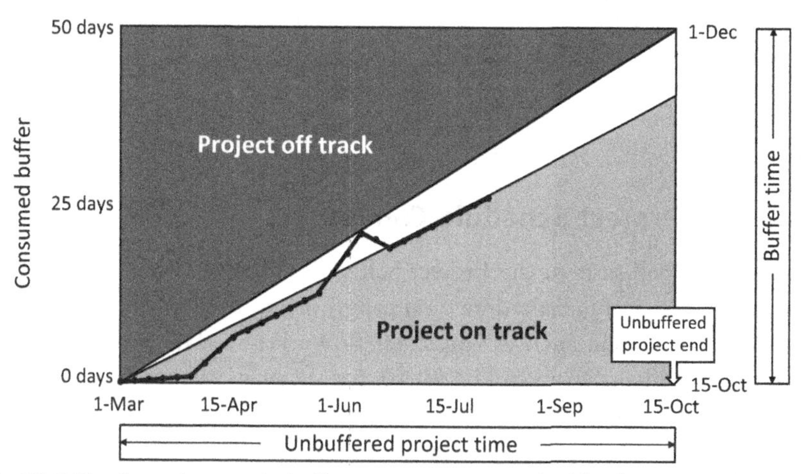

Fig. 20.9 The fever chart tracks buffer consumption over the life of a project. The width of the chart is the project duration excluding buffer. The height is the initial amount of buffer (typically about 25%).

to elapsed time. For example, if we have 50 days of buffer to spend over a 9-month project, no more than half of it (25 days) should be consumed halfway through the project (~July 7); see Fig. 20.9.

The fever chart gives two distinct calls to action: when buffer consumption is in the "yellow" (white in the print version), it directs the team to be prepared to remediate schedule issues. Call that software development contractor who helped you last time to see if they could be available if things worsen in the coming weeks. Push back if someone wants to borrow a resource. Identify potential scope reductions or feature postponements that can help make up time—we may need them next week.

When the fever chart goes into the "red" (dark gray in the print version), it's a Stop-Fix alarm: the team needs to stop and run a "play" from the playbook. Don't just keep pushing the project and hoping for the best. It's time to fix the schedule—reorganize the project to eliminate dependencies, get more resources on the critical path, or find ways to cut scope. The fever chart then gives a leading indicator of schedule performance. Traditional project management is plagued by teams facing up to delays only after due dates have passed; the fever chart along with the discipline of high-cadence regular team stand-up meetings ensures the team will be constantly forecasting delivery dates. This is immensely empowering to the project manager and the team: they have a reliable, relatively precise indicator of urgency. That signal can guide the team to the most important issues as soon as they are identified. We'll see this in action in Section 20.4 where we follow a hypothetical project through its first 4 weeks. But first, let's discuss the project canvas.

20.4 The Project Schedule Canvas

This section will present the Project Schedule Canvas, a single-page view that can be physical (a hand drawn on paper) or captured in almost any presentation application such as Microsoft PowerPoint. The project canvas, shown in schematic form in Fig. 20.10, has six main elements:

1. A title bar at the top
2. A fever chart
3. A static header to the right of the fever chart
4. A date strip that serves as the horizontal axis for both the fever chart and the schedule. It also includes immovable milestones (two, shown as flag icons) above
5. A Visual Action Plan like Fig. 20.8 at the bottom left with a vertical "Done" line; this line is on the far left at the project start.
6. A buffer that is blank at the project start and remains blank unless the project runs over—something we hope will happen only on rare occasions!

Our project from Fig. 20.8 is used to create a Project Schedule Canvas for "Sandy Waters" in Fig. 20.11. We will use this to review how the fever chart works.

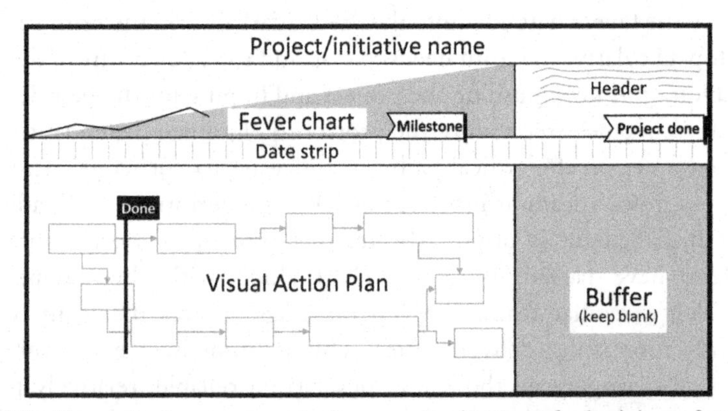

Fig. 20.10 The project canvas is a single-page visualization of schedule performance.

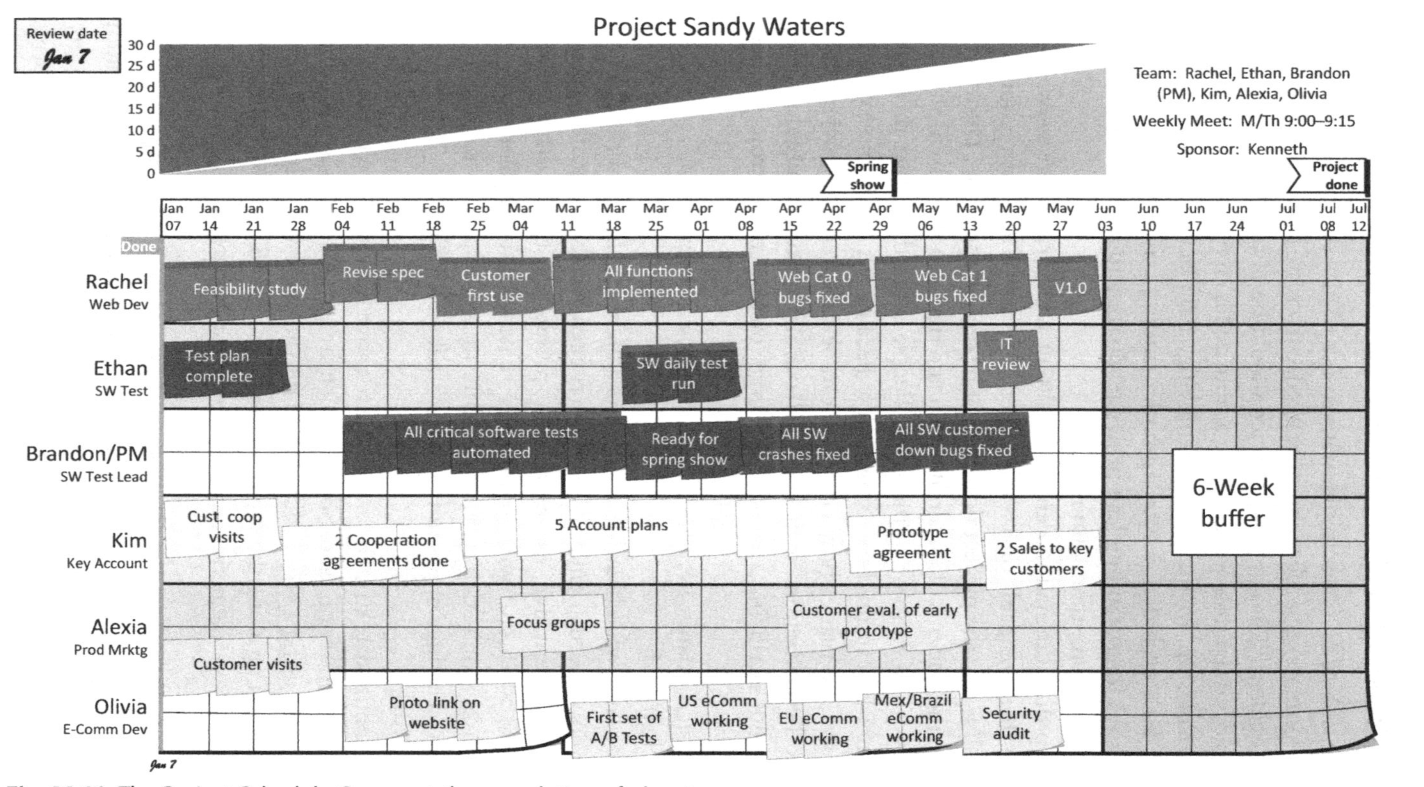

Fig. 20.11 The Project Schedule Canvas at the completion of planning.

20.4.1 Building Project Schedule Canvas boards

For virtual boards, for example, when using display software like Microsoft PowerPoint, most of the techniques needed to create a project canvas should be self-evident. Two pieces of advice should be noted:

- Avoid automation—everything automated is an opportunity for erosion of ownership. Manually sizing, moving, and placing are tedious, but no one feels excluded due to misunderstanding how the software works. This is a long-standing lean principle: do something well manually until you are adept, then automate.
- Keep the information density consistent with a one-page view, using a font that is readable on a single screen. This rule is too often ignored. The result is that the project flow cannot be seen in one view. During meetings, the leader is often scrolling left and right, up and down, and the big story is lost to most of the team. People often think that their project cannot be captured in a single view because it's too complex, but any project, no matter how complex, can be captured in one view; hierarchical visualization (Section 9.3.2.2) can allow more detail at lower levels without sacrificing a view of the larger picture in the Project Schedule Canvas.

For physical boards, here are a few tips that can help:

- Background
  Use a large whiteboard—sticky notes adhere well to that surface. You can put down permanent grid lines with rugged, thin black "gaffer" tape. As a quicker alternative, use six or eight poster-size sticky notes (20″ × 30″, 50 cm × 75 cm) with graph lines; remember the smaller sticky notes for tasks won't stick to the top ~2 in. (5 cm) of poster-size sticky notes due to coatings opposite the poster's adhesive strip.
- Date strip
  The entire date strip should be able to move left relative to the fever chart and Visual Action Plan, both of which are stationary. For a paper display, consider using a set of sticky notes for the dates with a strip of painters' tape running along the top of all the notes to hold them together. Or put the dates on magnetic squares and move all squares left/right as buffer is consumed.
- The fever chart can be cut from construction paper. Also, it can be printed; if you lack a large-format printer, it can be printed on multiple 8.5 × 11 sheets with Microsoft Excel with a red triangle over a green square sized to print on many sheets of paper. Then a pair of scissors and 30 minutes should give a serviceable background for your fever chart for the life of the project.

- Have lots of colors of sticky notes. There never seem to be enough colors!
- Use thin red tape across the top of sticky notes to show the critical path.
- Leave space for the buffer in the canvas, even though it's blank at the start. If the project does run over, you may need the space. More importantly, it communicates to everyone that the team is working to an early date (before buffer), but committed only to a later date (after buffer). Don't break trust with the team by indirectly committing to leadership or customers to meet an unbuffered date.
- The "Done" line (left-hand side of Fig. 20.11) can be a colorful string attached to a sticky note at the top and with a small weight at the bottom to keep the line vertical.

20.4.2 Regular team stand-ups for Visual Action Plans with buffer

As in Chapter 19, the regular team stand-up is a series of conversations that the leader keeps contained to about 15 minutes and then setting up follow-on meetings for topics too deep or narrow for the stand-up. For the Visual Action Plan with buffer, the suggested flow is six steps, as shown in Fig. 20.12 (Steps 1 and 2 are identical to those in Section 19.5):

1. **Update progress bars**

   For each active task in the Visual Action Plan, the owner makes a rough estimate of completion. The project manager draws a line along the bottom to reflect progress (50% done = a line or tape segment from the left to the center of the task block).

2. **Slide "Done" line right**

   Slide the vertical done line as far as the least-progressed task will allow. This sets how much of the project can be counted as done.

3. **Slide date strip left**

   Move the entire strip of dates left until today's date is aligned to the "Done" line.

4. **Update the fever chart**

   Count the number of days of buffer consumed, equal to the days you move the date strip left in Step 3. Update the fever chart with a black dot with the horizontal coordinate set by today's date and the vertical coordinate set by the total number of days consumed in the fever chart. Write today's date at the bottom just beneath the "Done" line to keep a history.

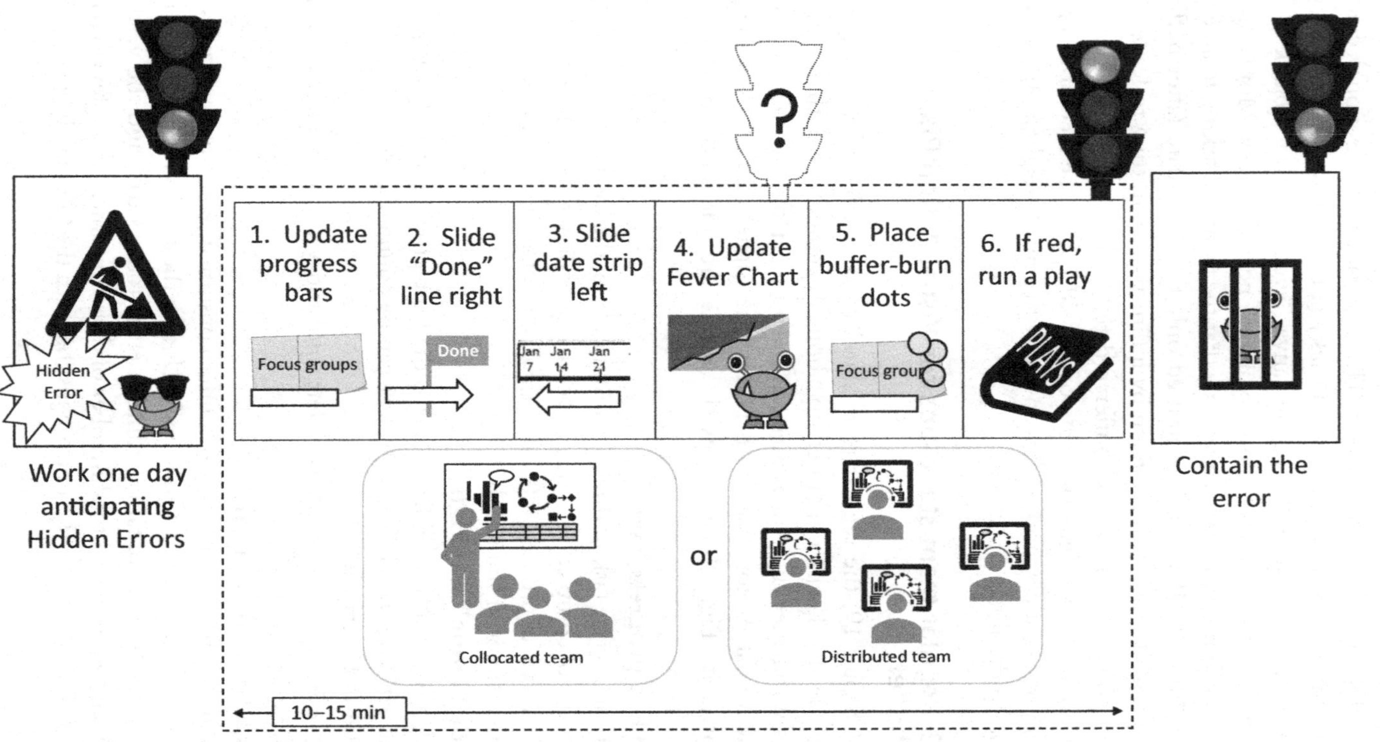

Fig. 20.12 The regular team stand-up is a drill of six steps meant to be completed in 10–15 minutes.

5. **Place buffer-burn dots**
In order to keep a visual history of schedule performance, place buffer-burn dots on the task that is the primary cause for the delay using one dot for each day the task consumed the buffer; horizontally align them next to the current position of the "Done" line. If two tasks combine to consume buffer, split the dots between them as the team agrees. Having a history makes many reoccurring problems obvious and thus contributes to team learning; it also reduces **Information Friction** when explaining the project to others. The total number of dots placed at one stand-up should equal the amount the date strip moved left during that stand-up. It follows then that the total number of dots on the Visual Action Plan will equal the total consumption of buffer as shown in the fever chart.
6. **If the fever chart is "in the red," run playbook**
The chart entering the red is a Stop-Fix alarm. Run a play to get back in the green. The team's ability to hold the schedule depends upon rapidly resolving Stop-Fix alarms. In the simple cases, resolve the issue immediately; if it's too complex to be resolved in a few minutes, the project manager will likely spend a good part of the day in various meetings to resolve this. Either way, it should be resolved in a day or two. We will discuss running plays in Section 20.4.

Kata

Japanese lean thinkers sometimes use the term "Kata" to illuminate the power of short, repeated workflow like the daily or weekly stand-up meeting. It's borrowed from karate exercises, each called a "kata" (recall "wax on, wax off" from the original "Karate Kid" movie). In the regular team stand-up, running the same drill each day for months will eventually allow rapid communication; just a few words will tell a complete story. Let's say that nearly every day at stand-up, Rachel has little emotion; perhaps she has occasional comments like, "Our tasks are on track except 'supplier contracts' is 3 days late." But then 2 months in, she starts with something that communicates high urgency like, "We have a problem." In the context that she rarely brings up problems, she has communicated a large amount by bringing up something at all. Apply the principle of "kata" or "drill" (for musicians, singing or playing scales): do a simple thing again and again the same way, working constantly to get sharper. Over time, the behavior will become so familiar that it becomes part of the organization's culture.

The regular team stand-up demands discipline. First, people must show up and engage in the meeting. That's harder than it sounds. Lack of a quorum will create ineffective meetings and that will likely cause other people—those on the fence about the need for stand-ups—to skip the occasional meeting. And a person need not skip the entire meeting to cause problems. If the stand-up is daily, it may be just 15 minutes long, and being 5 minutes late can prove disruptive. So, the first discipline is that of being present.

The second discipline is staying on the agenda. For a high-cadence meeting to work, it must remain short. There will often appear topics that are outside the scope of a regular team stand-up: meaty topics too deep or too narrow for the full team. To lead a regular team stand-up, the phrase "That's outside the scope of this meeting" must be always at the ready. If the leader allows the meeting to expand well beyond the 15 minutes (and I've seen them go to over 60 minutes), people will stop attending. They will rightly observe that they cannot give 45 or 60 minutes a day to this process.

20.4.3 A legend for visualizing buffer consumption

The legend of Section 19.4 can be augmented to track buffer consumption. This is invaluable to understanding the project history at a glance. Display those tasks that have consumed the buffer through the life of the project, and you'll be better able to predict where new problems are likely to come from, in this project and in future ones. The method here is simple: place one yellow dot for each day of delay from a task (see Fig. 20.13).

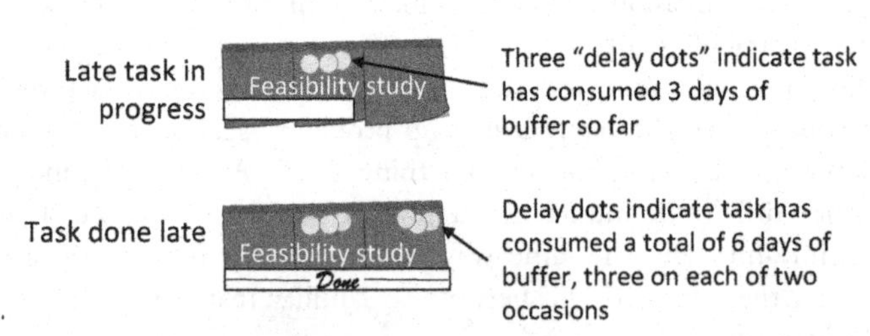

Fig. 20.13 A legend for visualizing buffer consumption.

20.5 Following a project with the Project Schedule Canvas

Let's follow our hypothetical "Sandy Waters" team for the first month of their project. We'll look at the project canvas at a few points along their journey.

20.5.1 Week 2: Everything is on track

Let's say the team met for a day at the end of the previous December to build the plan shown in Fig. 20.11, scheduled to start January 7, which it did. Suppose it's now January 21 as shown in Fig. 20.14 (note date on upper left of canvas) and the four active tasks are running on time. So, the done line is efficiently sliding to the right, aligned to today's date. Thus, no buffer has been consumed. This is normal. In almost every project I've been a part of, the first couple of weeks go according to plan. That may be because the team has especially good insight for planning the early tasks, or because people are more conservative with near-term commitments. In any event, the first 2 or 3 weeks are usually uneventful.

20.5.2 Week 3: An issue comes into focus

On January 28, the team meets and an issue is coming into focus: the feasibility study is lagging (see Fig. 20.15). Rachel has run into several problems and she says 1.5 weeks is the best she can commit to. So, we slide the task progress bar to within 1.5 weeks of the end. And the date done line can move no further to the right than that. The team has a short discussion—often, the task can slide right without affecting the project, especially for tasks that are off the critical path. But in this case, the short discussion confirms what the team thought at planning: the feasibility study is on the critical path and so delays here are likely to have follow-on effects that will delay the project. So, we accept the position of the done line and then slide the date strip until today's date (January 28) is lined up with the done line, which in this case means sliding the date strip 3 days to the left. We now update the fever chart to show 3 days of buffer consumption immediately above January 28; fortunately, the project stays out of the red (dark gray in the print version), if by a hair. But we are in the "yellow (light gray in the print version)," the team is alerted to a looming issue, but decides to continue, ready to react if more issues are discovered at the next stand-up.

20.5.2.1 Visual agreement and canvas automation

Notice from the "3 d" callout box in the upper left corner of Fig. 20.15 that the visual agreement of the delay is represented in three ways: the fever

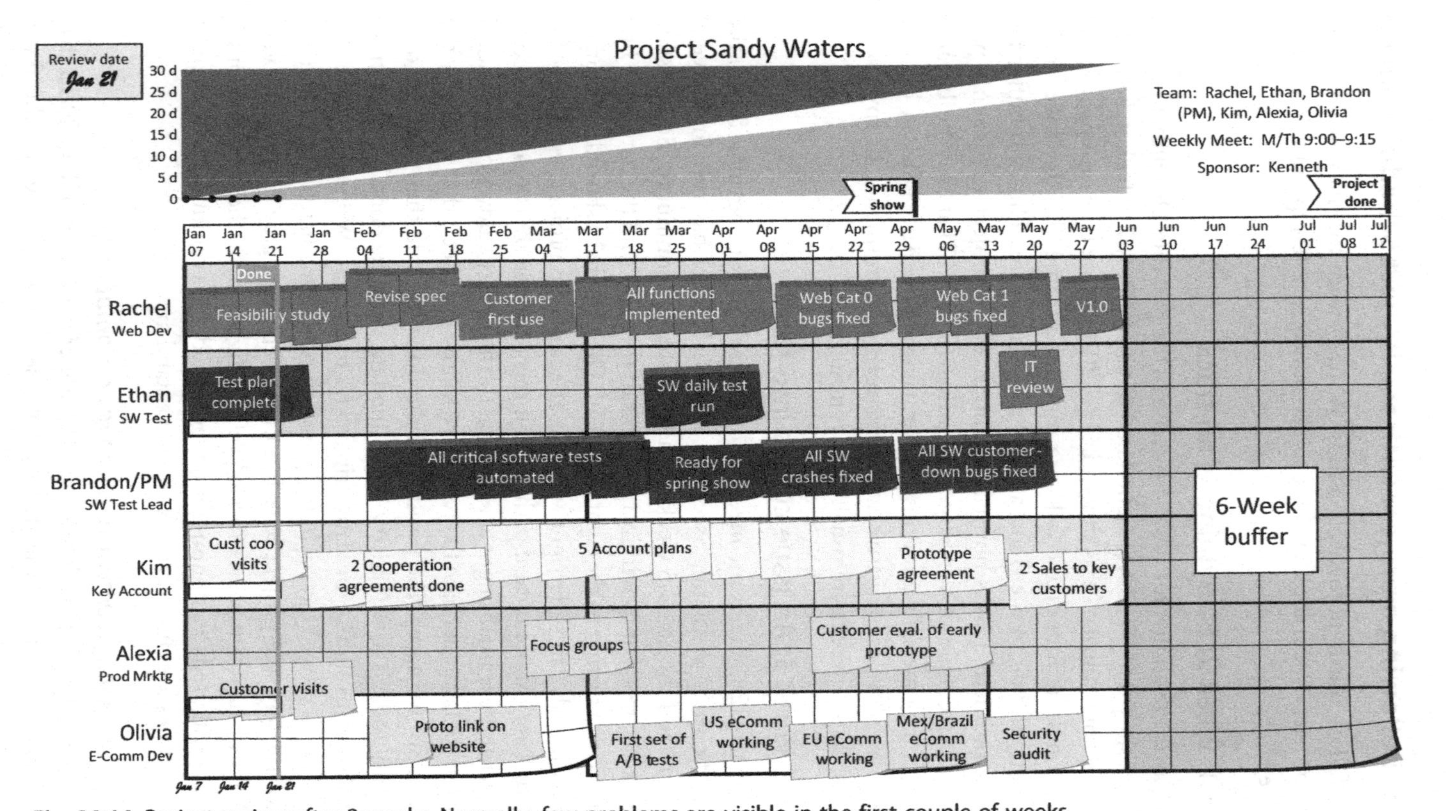

Fig. 20.14 Project review after 2 weeks. Normally, few problems are visible in the first couple of weeks.

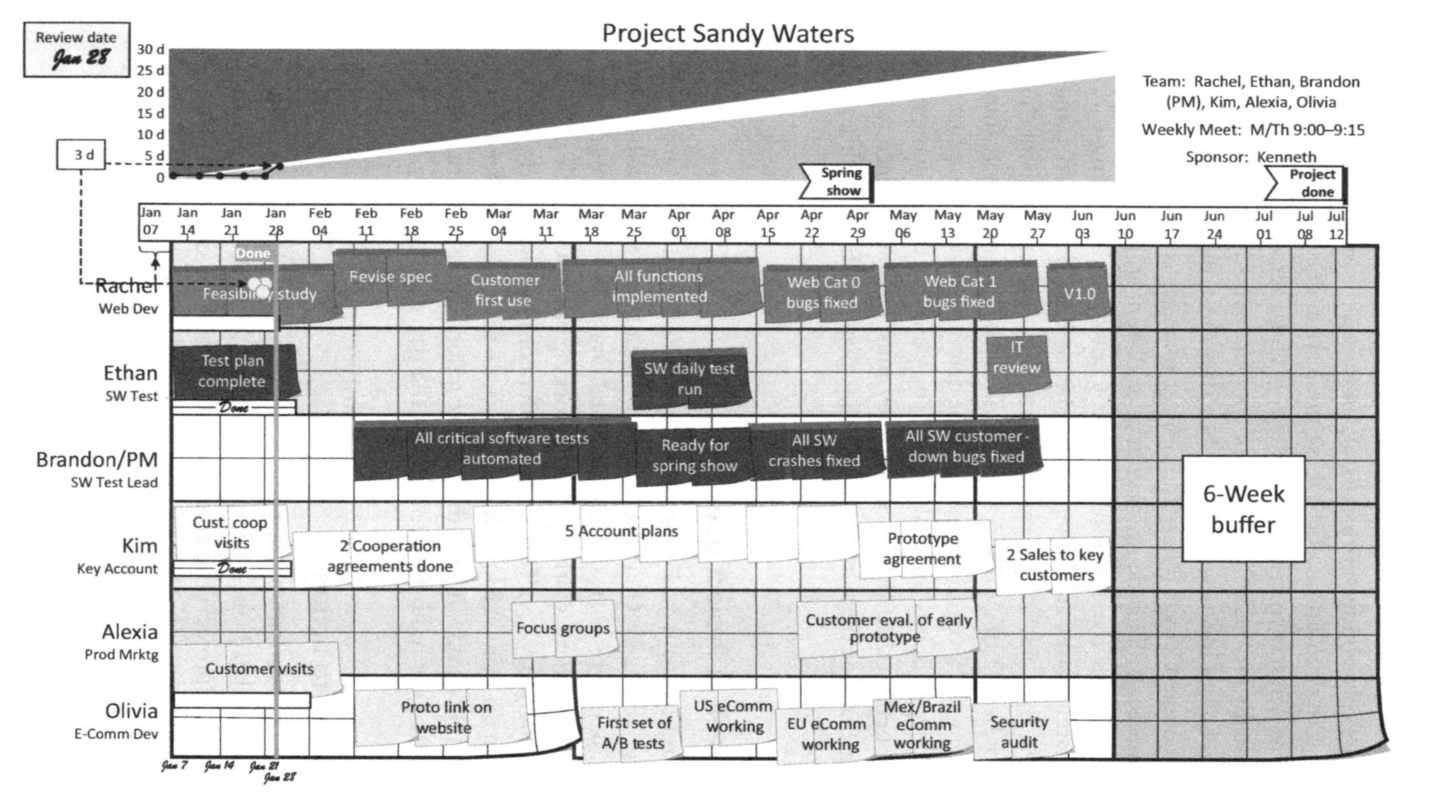

Fig. 20.15 A project review 3 weeks in reveals issues building in the feasibility study. Notice that 3 days of buffer are consumed, shown in three complementary ways: (1) 3 days in the fever chart, (2) 3 days left shift of date bar, and (3) three dots on the feasibility study showing the primary source of the delay.

chart's rightmost "black dot" is set at 3 days, the date strip is shifted left 3 days, and there are three buffer-burn dots, all on the task feasibility study. Visual agreement is the condition where these three are aligned. Well-kept Visual Action Plans will stay in visual agreement. If visual agreement is not maintained, it usually proceeds from a lack of team-meeting discipline, for example, the team is not meeting often or when they meet, they are not using the Project Schedule Canvas (likely the project manager is updating the form after the meeting). This doesn't produce well-run projects; if the project manager maintains the Project Schedule Canvas as a reporting tool but the team rarely looks at it, their ownership will erode. When things go wrong later, that lack of ownership will reduce their engagement. This is the primary benefit of avoiding automation in the canvas such as automatically drawing the fever chart or sliding the date strip: automation that is not understood by all blunts ownership. If at any point the team discovers they are further behind than they thought, they must be able to piece together everything that got them there, something that is difficult with complex automation.

20.5.3 Week 4: Project goes red!

The team meets again on February 4. The feasibility study is continuing to lag (see Fig. 20.16). Rachel knows there is at least a week left, which holds back the task progress bar, the limit for the "Done" line. We must slide the date strip left another 3 working days (6 working days total, a little more than a week) to align to today. That pushes buffer consumption clearly in the red (dark gray in the print version). It also lays down three more buffer-burn dots on the feasibility study. This is the call for action. The project manager sets out to remove the red before the end of the day.

20.5.4 A call to action

This is one of the moments that will define success or failure for this team. Visualization and daily meetings mean nothing if they don't change how the team behaves. A project going into the red is a serious event—all eyes turn to that issue. We get Rachel help, including jumping in to help with the study, taking on other tasks that Rachel is committed to, or working with her supervisor to reduce loading from outside areas temporarily. The fact that three of the four tasks are on time is irrelevant. Only one metric matters for schedule performance: the fever chart is in the red.

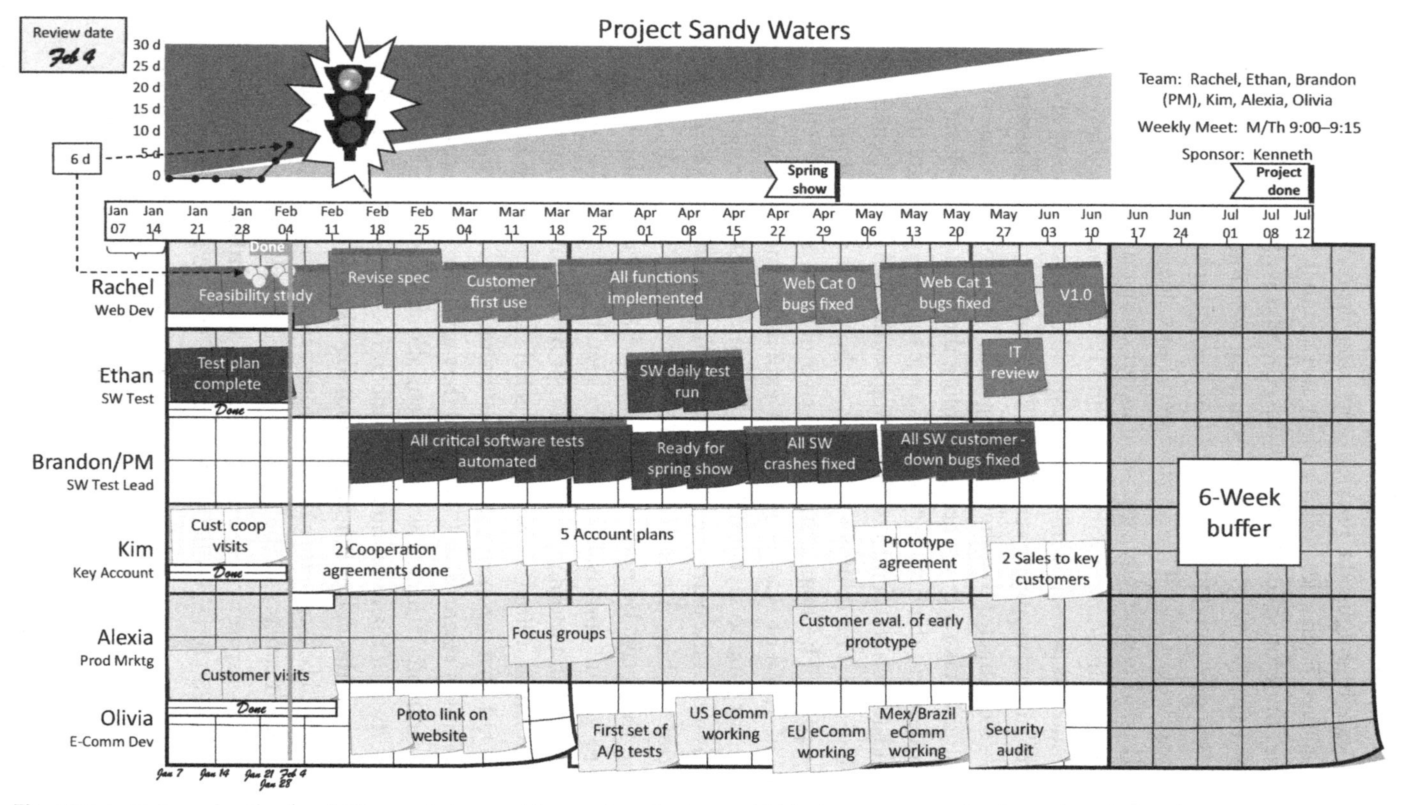

Fig. 20.16 At 4 weeks, the feasibility study has put the project "in the red." It's time to run a play from the playbook.

At the regular team stand-up, the project manager (Brandon, who is also SW test lead) highlights the problem and, after short discussions to see if this can be relieved easily, determines this is an issue that will take several hours to resolve. Brandon schedules follow-up meetings with a few team members in the next few hours and closes the stand-up meeting. Rachel stays back so she and Brandon can go to her supervisor to get Rachel clear of other duties for the next week or so. That's a good step, but it doesn't really help the fever chart—Rachel was already aggressive in her estimate of a week remaining. An hour later, Brandon, Ethan, Olivia, and Rachel meet to see how to relax the schedule. As we see from the critical path, one task follows another so that there's no obvious answer. About 20 minutes in, Ethan suggests a play: he knows a contractor that they can help with "All functions implemented"; that's going to take a few weeks. Doubling the team (Rachel and the contractor) might cut the time by 1 week, so the team plans to hire the contractor in mid-March to spend a week getting set up. Why is the benefit limited to just 1 week? The contractor is new, so he won't work at anything like Rachel's pace. The team assumes the two of them together will work at the rate of 133% of Rachel alone. It's expensive, spending several weeks of contractor time to save just 1 week. But in this case, keeping the project on track is worth the extra spend. The team proposes it that evening to Brandon's boss; the boss is positive and that's enough to adjust the fever chart assuming that play; Brandon and his boss schedule a meeting in a couple of days to get full approval for the extra spend.

Consider for a moment how this might have gone without the Project Schedule Canvas. Often, someone in Rachel's position would keep grinding away at a task that was probably underestimated at the start. She might feel a small amount of resentment ("Why did they give me so little time for something so difficult?") mixed with concern about losing face by being late. These are two of many unintentional incentives for Rachel to avoid transparency about the delay. Another is looking like she's giving up by saying something will be late before the milestone passes. The structure of the Project Schedule Canvas and the regular team stand-ups recognize these disincentives to transparency, seeing them as creating **Information Friction**. Sure, we could chide Rachel to "try harder" or "be more open." But that doesn't work very well. Rather, let's recognize normal human behavior and add mechanisms that account for it. Unfortunately, without such mechanisms, the disincentives to transparency only grow over time—Rachel would be more and more uncomfortable as delay mounts. Eventually, at some dramatic moment in the future, the truth would come out that the project is weeks behind, and at that point, it would be too late to remedy it fully. Rachel would get another check in the lose column; trust from leadership would erode. The project manager would disappoint his boss. Waste would be piled upon waste.

When the team has an effective, rapid means to track schedule performance, they can react before the boss even knows there's an issue. I've watched this work again and again. Teams empowered to take action and trained in methods that reveal schedule issues change their behavior immediately. Perhaps the old behavior was more relaxed, not being overly concerned about a 2- or 3-week delay; now they are more reactive, fighting to recover delay upon seeing just a few days of excessive buffer burn. These methods unlock the ability that was always there. The Project Schedule Canvas works because it enables teams to find issues nearly instantly so that they can apply their creativity and effort to fix the problem at the point most likely to be effective. This is one reason why these methods are so engaging for knowledge staff: they free the staff to apply their skills and experience to win the contest. And winning begets winning. As teams get better at delivering projects on schedule, less time goes to explaining problems and revising schedules. More of their energy goes to the work they want to do most.

20.5.5 Augmenting the Project Schedule Canvas legend

The legend of Fig. 20.13 can be augmented for the cases where the project must be restructured with the techniques shown in Fig. 20.17. Fig. 20.17A is the task as originally planned. If we want to delay the task, we can move the sticky notes to the right and insert a gray sticky note with the word "Push" (Fig. 20.17B); this way, we have a history of tasks being pushed out. If a task slips after it starts, we can add a highlighted sticky (say yellow or neon lime) to the right (Fig. 20.17C) marked "Slip." If we want to accelerate a task by a week, we shorten the task and add a sticky "Accel" where the original task ended (Fig. 20.17D). Fig. 20.17E shows how to push and accelerate to hold an end date, even though the start date is delayed. All of this provides a history as the project proceeds.

20.5.6 Week 4 revisited: Project back in the green

The team meeting the day the project went red proposed to accelerate the "All functions implemented" and thus relax the schedule demand on Rachel's feasibility study. The delivery date is held at a cost of, say, $15k for the consultant, a cost that might seem high to gain just 1 week, but

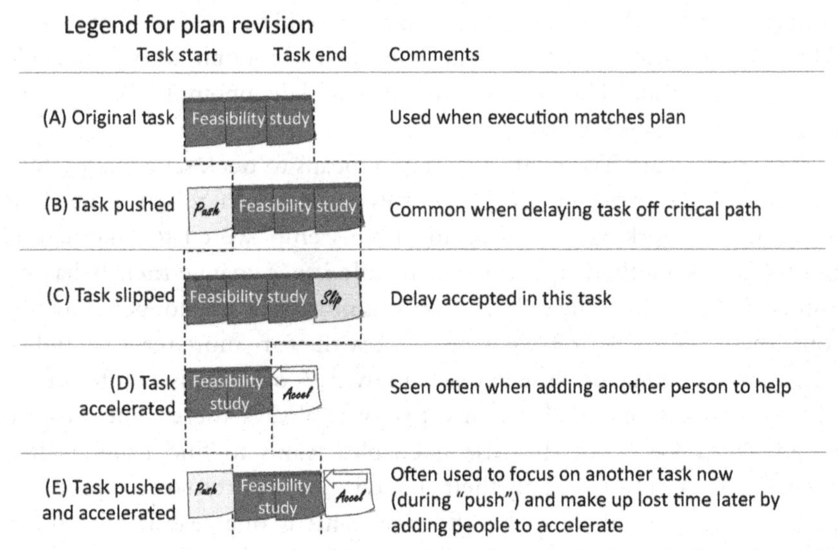

Fig. 20.17 In exceptional cases, tasks must be replanned. This legend allows the agility to replan while maintaining history.

compared to the total project cost it's not significant. Of course, the boss may turn down any play, but at least the team has presented an alternative. My experience is that the boss turns down winning plays, even expensive ones, less often than the team imagines because delivering on time brings so much value to the organization.

The visualization for the restructured plan is shown in Fig. 20.18 in three steps starting from the right and working left:

1. The team ensures the consultant is available and gets an initial go-ahead from the boss. So they push the start of "All functions implemented," but compress it from 4 to 3 weeks and so hold the original completion date.
2. They then push "Customer first use" 1 week; then "Revise spec" is also pushed 1 week.
3. This relaxes schedule demand for the feasibility study, allowing the 1-week slip in the task without affecting completion of the project. As pressure mounted on Rachel, her estimates got aggressive: her estimate of 1 week remaining was questionable. When the team dug in, they saw what she was facing was probably double that. So, Ethan will jump into the feasibility study to help, but the time remaining is still thought to be about 1.5 weeks. Thus, progress is estimated by setting the "Done" flag 1.5 weeks left of the end of "Feasibility study" (including the 1-week slip).

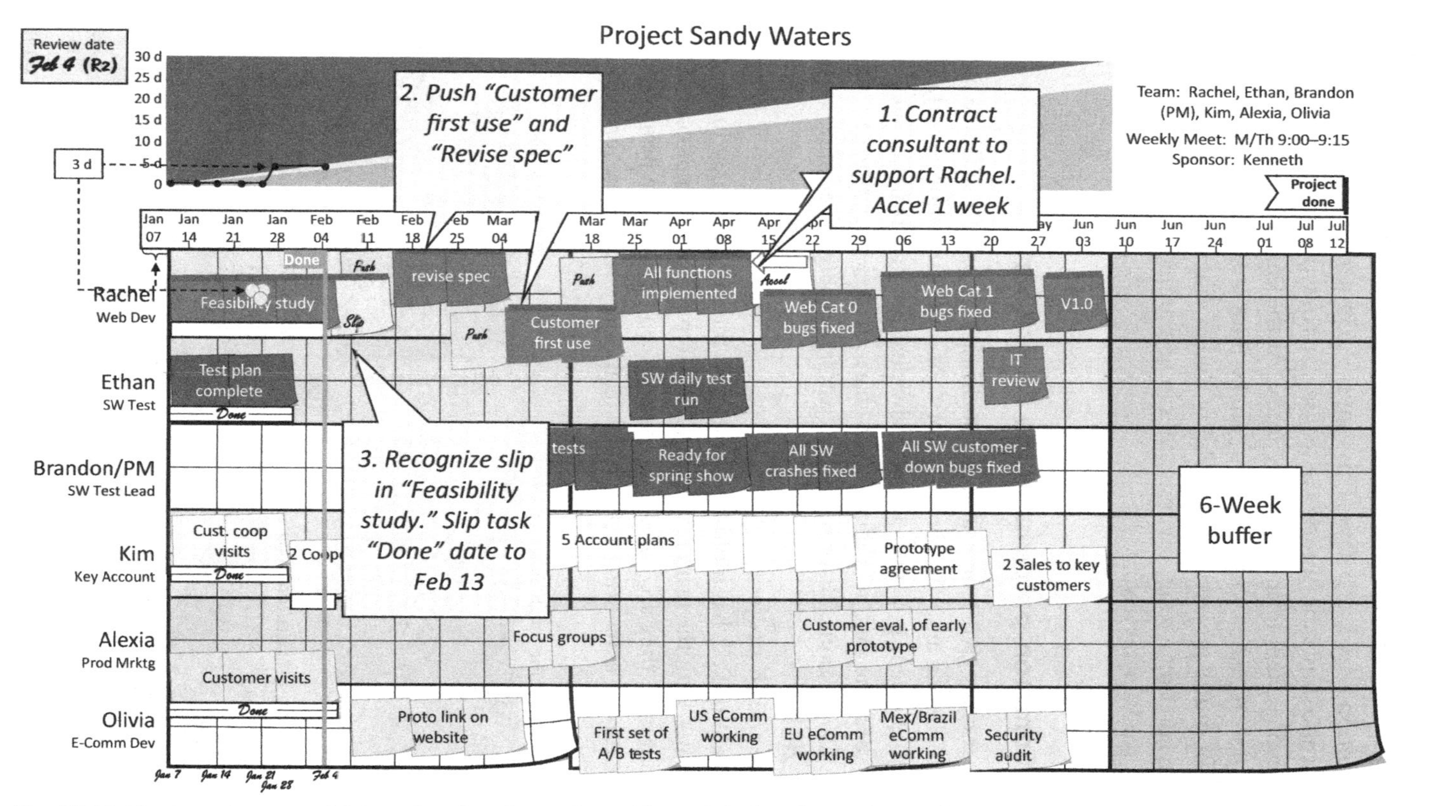

Fig. 20.18 The team is successful in replanning the project without delaying the project end.

Now, we can align Feb 4, today's date, to the done line and we see buffer consumption is back to 3 days. This moves the project on track, back in the green. It also shows the history—in seconds, it's obvious the team is suffering performance issues in the feasibility study. At the end of the project, those yellow dots together with the "Slip" sticky note will help the team learn a lesson for future projects: plan feasibility studies diligently. Nothing creates learning faster than teams owning issues in past projects and capturing them visually for later review.

20.6 Common patterns in the fever chart

The fever chart gives a near-instant visualization of project schedule performance including an overview of history. As an organization uses the fever chart more often, not only do the individual projects benefit, but leaders quickly become deft at reading fever charts to understand project performance. One reason is that a handful of patterns emerge over time, as shown in Fig. 20.19:

A. The tough project

The tough project has multiple swings into the red (dark gray in the print version), with the team reacting quickly to pull it back under control. In the interim, they are just barely in the green (gray in the print version). This can be exhausting because the team must be vigilant for the entire

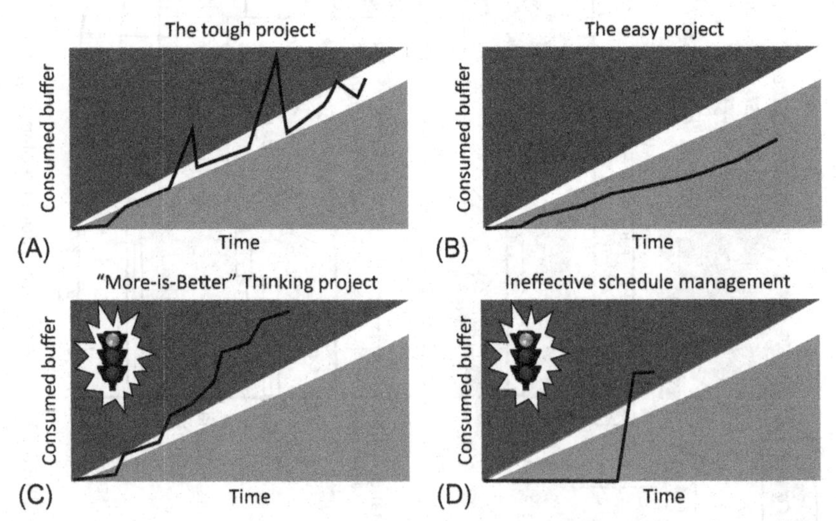

Fig. 20.19 Common patterns in project management provide rapid insight on project history.

term of the project. But it will be a team learning fast. Usually, the tough project comes about, at least in part, because of overly aggressive planning from the team. When that team meets to plan future projects, you can be sure they will work to avoid the same issues.

B. The easy project

The easy project never gets close to the red. Every team deserves a few of these, but if you see this pattern too often, it may be good to ensure the team is setting challenging goals for the benefit of the organization.

C. "More-is-Better" Thinking project

More-is-Better Thinking is one of the 8 Wastes of Knowledge Work. Here the team sees the project is "in the red" and accepts the alarm condition for months. An alarm that never stops ringing is no alarm. There are always reasons: we don't have time to look after this, there's no way to make up time, we need to wait to see what happens, and on and on. But there is always a way to react to the alarm even though sometimes it means extending the project—normally the least satisfying remedy. But if that is true, better to recognize it now and align the organization than to keep working to a date nobody believes. Experience has taught me three things:

a. There is always a way to respond to an alarm quickly.
b. Teams that respond quickly consistently perform well.
c. Allowing a project to live in the red creates poorer performance.

D. Ineffective schedule management

The most concerning issue in Fig. 20.19 is (D) Ineffective schedule management, which often presents with an inactive fever chart that suddenly leaps into the red. The example in Fig. 20.19D says, "Everything was perfect until one day we discovered half the buffer was consumed." In 9 out of 10 cases, the project manager was not managing the project well—perhaps the team wasn't having regular team stand-up meetings or the meetings weren't effective or the team failed to use the Project Schedule Canvas for their routine work. Of course, there are exceptions, for example, if the team one day discovers they were relying on a technique that must suddenly be replaced because it had been patented by a key competitor, that could cause an immediate months' long delay. So there are exceptions, but a pattern like Fig. 20.19D gives cause for a deeper look.

An alarm that never stops ringing is no alarm.

20.7 The portfolio snapshot fever chart

One of the most intriguing variations on the fever chart from Goldratt is the portfolio snapshot fever chart, a helicopter view [8], complementing the ground view of the project fever chart (see Fig. 20.20). Whereas the project fever chart plots a project's history, the portfolio snapshot shows the current state of every project in an organization. Notice that the symbol for the project of Fig. 20.20A (the dark car) appears in precisely the same position in Fig. 20.20B, as it always will. However, the helicopter view of Fig. 20.20B also maps three other projects. The portfolio snapshot allows a department leader to see all projects at a glance. There might be dozens of projects active, but if just one goes to the red in the morning, the department leaders know by lunch and reaction to it can start as soon as that afternoon.

I found the portfolio snapshot invaluable in the years when my team was responsible for 15–20 large projects. Each morning a fresh portfolio snapshot was published. The first months were laborious because there were a lot of projects going "red"; I read the daily snapshot and treated each "red" project as a Stop-Fix alarm. I met with project managers or attended team meetings within a day or two. It took time to bring uniformity on how the many projects in multiple sites reported in, but probably less time than I was spending chasing project data before the snapshot, so I was confident we were on a good path. But things got better—eventually much better. As time went on, people learned the drill; often the first round of questions was resolved between the project managers and the people collecting the daily data before the snapshot was even published. Later still, things improved again. Project teams knew there was an inevitable sequence of events kicked off by going into the red, so they would take the first steps within the teams to resolve many issues among themselves, before their project found its way into the red portion of the published daily snapshot.

Over these months, my hours spent keeping up with projects declined from a couple of hours per day to a couple of hours a week. Better still, over time, the only projects that went "red" were the ones that needed attention; the false positives that were red because of data errors or minor mistakes were more likely to get attention from the project manager before being published. In other words, the correlation between "going into the red" and "a serious project issue" went up to nearly 100%. The

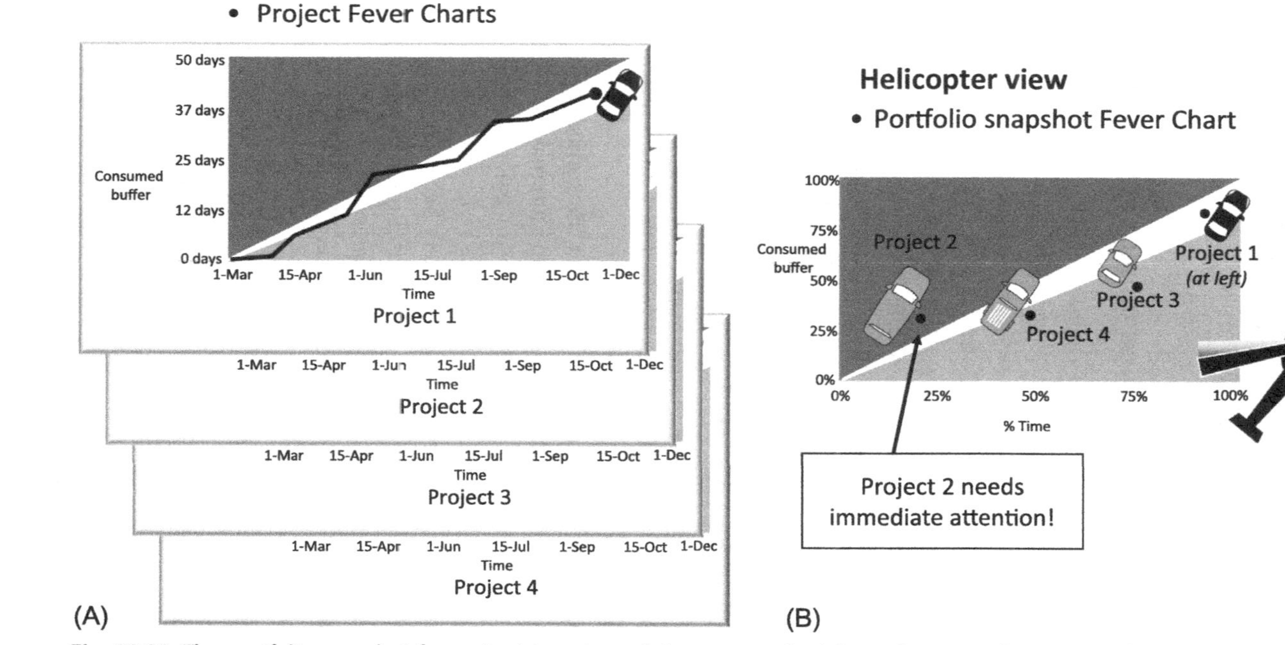

Fig. 20.20 The portfolio snapshot fever chart is a view of short-term schedule performance for every project.

time savings were my personal payoff: it was as if I had an extra day every week to focus on more important and more interesting work. It also benefited the project managers, who spent less time explaining events—not their favorite thing either. But the primary beneficiary was the organization as a whole: our ability to finish projects on time rocketed during that same period. Late projects, commonplace at the start, became the exception in less than a year.

20.8 Creating project management standard work

We can apply the principles of standard work from Chapter 15 to project management as shown in Fig. 20.21. The left-hand side defines how a project (the ground view) is managed and the right-hand side defines how the project portfolio (the helicopter view) is managed. The left-hand side is driven by a number of project managers (perhaps one for each project or two), while the right-hand side is driven by the organization's leadership. The 13 elements of Fig. 20.21 are detailed below.

Ground view (items 1–8)

1 *Process requirements* define tasks as detailed in the organization's internal process requirements, which can define reviews requirements, financial templates, approval processes, and so on.

2 *Tool examples* include the Project Schedule Canvas Template, financial templates, and shared directories.

3 *Training* might be a 1-day class taught at the company for project managers on how to navigate the organization's internal process requirements.

4 *Resourcing* could be the management team assigning five people to the project full-time for 6 months.

5 *Scorecard* is the daily update of the Project Schedule Canvas.

6 *Reporting* could be the Project Schedule Canvas plus a monthly update on finances, customer feedback, and issues/risks.

7 *Review* might be a 30-minute review at a monthly department manager's half-day project review.

8 *Countermeasures* from the review could be captured in a monthly action list.

Helicopter view (items 9–13)

9 *Review workflow* could be set out in a standard quarterly leadership review agenda.

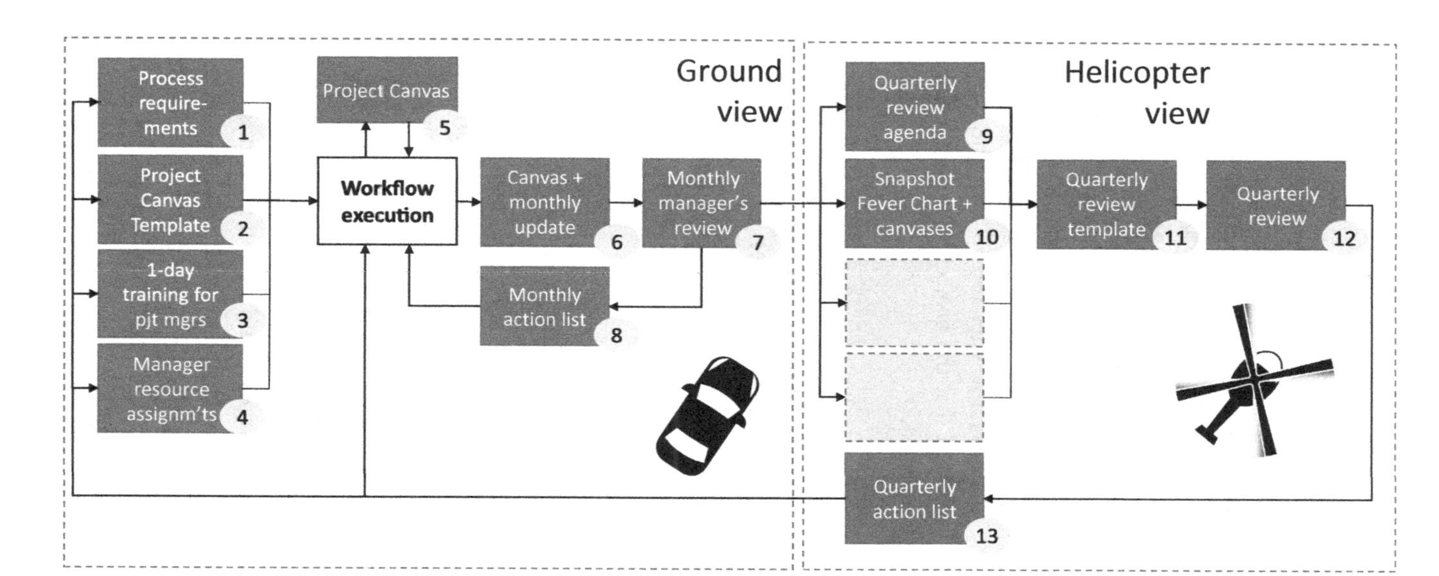

Fig. 20.21 Project management standard work (based on Chapter 15).

10 *Review tools* could include templates for project quarterly reports, portfolio snapshot fever chart, resource allocation template, and financial overview template for the department.

11 *Review information* could be a quarterly project reports, quarterly expense, and resource allocation report.

12 *Review meeting* is the quarterly leadership review for all projects.

13 *Countermeasures* here include three samples. These are the sorts of CMs that would come out of a quarterly review:
- Modify resourcing (add one person to a project team).
- Ask all PMs to sign up for a review on new supplier qualification (perhaps the review detected that several PMs were not sufficiently familiar with the company's process on qualifying new suppliers).
- An instance CM: ask the sales rep to visit the customer (perhaps the review indicated that the project team didn't fully comprehend the customer's needs).

20.9 Knowledge Work Improvement Canvas (KWIC) for visual management with buffer

Please see Section 19.8, the KWIC for visual management (without buffer), which would require a similar improvement cycle.

20.10 Conclusion

Fig. 20.22 shows three of several ways the techniques in this chapter **simplify** the workflow of project management. We can **engage** the team: Inspire with the promise of earning trust and outstanding performance for projects complex enough to demand buffer management. We can Connect with cross-functional teams that support each other in rough times (the way Ethan jumped in to support Rachel in Section 20.5.4) and Challenge them with a winnable game: a project they plan diligently with a realistic buffer and an effective way to manage it. And we can Protect the brand with clear requirements that we will not violate to make a date. Challenge cheers us on to win and Protect reminds us that the win counts only if it's within the rules. Finally, the techniques in this chapter **experiment** by creating falsifiable hypotheses to deliver on time.

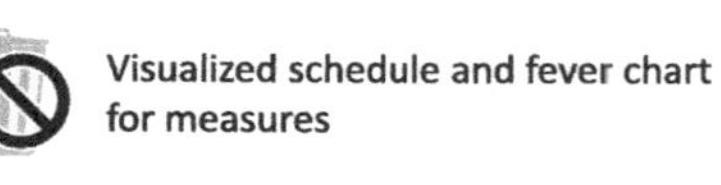
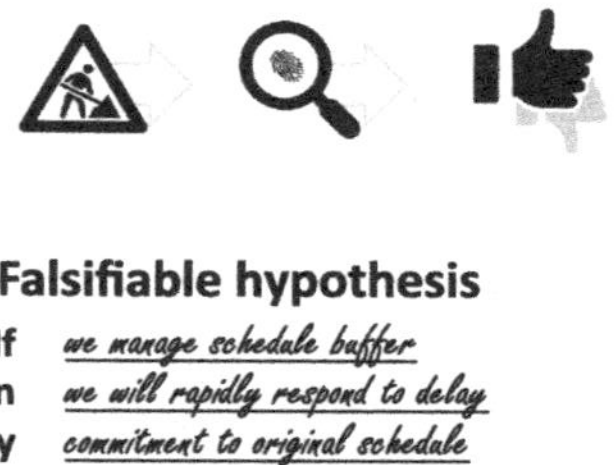

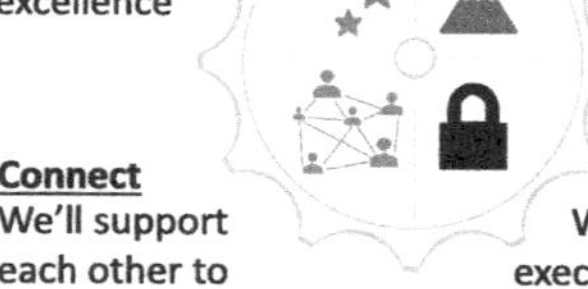

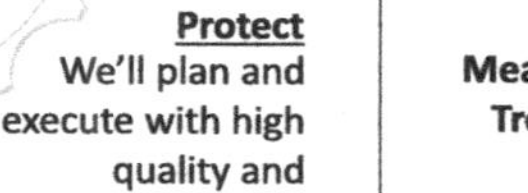
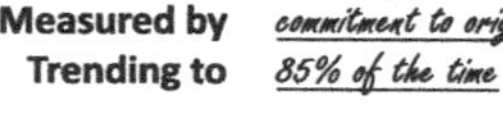

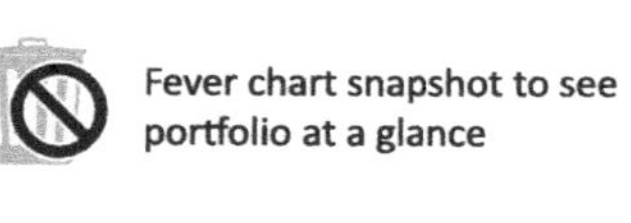

Fig. 20.22 Improving project/initiative delivery simplifies, engages, and encourages experimentation.

References

[1] E. Goldratt, The Goal: A Process of Ongoing Improvement, North River Press, 2004, 298, Kindle Edition.
[2] A.C. Ward, D.K. Sobek II, Lean Product and Process Development, second ed., Lean Enterprise Institute, 2014, 31.
[3] E.M. Goldratt, Critical Chain: A Business Novel, Kindle Edition, North River Press, 1997, Location 1560.
[4] C.N. Parkinson, Parkinson's Law, The Economist, November 1955. Available at: http://www.berglas.org/Articles/parkinsons_law.pdf.
[5] E.M. Goldratt, Critical Chain: A Business Novel, Kindle Edition, North River Press, 1997, Location 1901.
[6] W. Herroelen, R. Leus, On the merits and pitfalls of critical chain scheduling, J. Oper. Manag. 19 (2001) 559–577.
[7] G. Ellis, Project Management for Product Development Leadership Skills and Management Techniques to Deliver Great Products, Butterworth-Heinemann, 2016, p. 128 and 156.
[8] G. Ellis, Project Management for Product Development Leadership Skills and Management Techniques to Deliver Great Products, Butterworth-Heinemann, 2016, pp. 166–167.

CHAPTER 21

Workflow—Kanban and Kamishibai: Just-In-Time Rationalization

A Kaizen [continuous improvement] culture is one in which individuals feel empowered, act without fear, affiliate spontaneously, collaborate, and innovate.

David Anderson [1]

21.1 Introduction

This chapter will address the challenges of managing many smaller tasks. This view divides knowledge work into two categories: projects and initiatives that were the subject of the previous two chapters and the smaller tasks that are too numerous to be managed individually with a project canvas or its equivalent. It's common for people in knowledge work to have one or two main projects or initiatives that occupy a large portion of their time: a salesperson with a key account to win over, a scientist with an invention to validate, or a CFO leading a merger. But for most of us, there are also tens of smaller tasks that must be done: three patents to evaluate for the attorney, reviewing accounts payable for the quarter, and customer sales visits. While any one of these smaller tasks may not be that complicated, the large number of them active at any one time can be overwhelming.

Smaller tasks create complexity in numerous ways beyond just their large number. For one thing, the number of people who can place demands for smaller tasks is often large: HR asking for an evaluation of a new candidate, the boss asking for support for a presentation, Purchasing asking for help with a part that's difficult to procure, Sales asking for a small modification to billing software, and Applications needing technical support for a customer complaint. Another point of complexity is the way scope can change. Operations initially needs a small change in a drawing, but, after the drawing

Improve
https://doi.org/10.1016/B978-0-12-809519-5.00021-1

is changed, it turns out that a certification has to be revalidated and that will take 2 months. Or, you allow a day to scan the competitive landscape and something unexpected turns up; now a product strategy has to be updated.

In many organizations, these smaller tasks are essentially unmanaged: given to knowledge staff to get done somehow. When there's more to be done than there's time for, there's no easy way for staff to know which has priority. Shingo said, "Immature plans and vague instructions are given out and the rest is left to those who implement them. Whenever some blunder or confusion occurs...managers ignore their own shortcomings and bawl out subordinates" [2]. In this case, it's worse because the vague instructions result in the team being assigned tasks that exceed their capacity, what we will call an *unrationalized workload*; there is no path to meet expectations. This is the root of the waste of **No-Win Contests**.

The techniques of this chapter maintain a rationalized workload for a multitude of small tasks as summarized in Fig. 21.1. We will discuss the mindset of "Ruthless Rationalization": never accept work that, when properly done, exceeds the team's capacity. Ruthless Rationalization guides leaders to agree on goals, make plans to meet them, and, when the workload is beyond the capacity of the team, renegotiate either the goals or the resources, or both. Then we will take up two well-known visual techniques for rationalizing smaller tasks: (1) a Kanban board for *irregular* tasks (i.e., tasks that are highly varying in size and timing) and (2) a Kamishibai board for many tasks that repeat at a set cadence like monthly project reviews or quarterly performance reviews; in this chapter, these will be referred to as *regular* tasks.

Workflow	Just-In-Time Rationalization	The process to rationalize workload to available resources in order to minimize multitasking, oversubscription, and ambiguity
Tools to simplify, engage, and experiment	Ruthless Rationalization	The workflow that forces work to fit within the capability of the team by some combination of reducing the goal, increasing the resources, or selecting more valuable projects
	Kanban board	Visual rationalization for **irregular** tasks arrived at by consensus between the knowledge staff and the internal customers
	Kamishibai board	Visual rationalization for **regular** tasks arrived at by consensus between the knowledge staff and the internal customers

Fig. 21.1 The workflow and tools of Just-In-Time Rationalization.

21.2 The 8 Wastes

The Wastes from working on an unrationalized workload are common in knowledge work. Work gets behind and more work gets added. The people doing the work feel constantly overloaded and the people who need the work product feel almost nothing happens unless they "push" with frequent reminders and complaints. Everyone in the organization is losing in a giant traffic jam where tasks on the "highway" are backed up miles and yet new tasks keep being added.

Alexia is a web developer who is enduring a stressful morning because of her unrationalized workload (Fig. 21.2). She comes in focused on Kim's request for website screen updates needed to finalize a print ad that must be placed today. She's not sure how she let it go so long, but she's confident she'll get it done on time—it's only 2 hours of work and she has the whole morning. She sits down at 8:00 ready to get to work. Unfortunately, about 20 minutes later, she gets a call from Kelly, who's upset about a "critical" bug he just found on the website. He says this is top priority. He's convincing and she thinks she can knock that out in an hour so that she puts down Kim's project and works on the bug. Around 90 minutes later, Rich stops by Alexia's desk for an update on the China website. She hasn't started that yet; he's unhappy and talks to her for almost 30 minutes, going over all the times he's asked for it and why she should have started a month ago. Finally, he leaves and, by 10:00, she puts out a quick fix to Kelly's critical bug. Doing a last-minute check, she discovers her fix won't work on the European eComm site. Now she has to start over on Kelly's bug. Right about then, she gets a call that Merix Inc. says the website failed a security compliance check and she needs to go over the details, which takes another 30 minutes. At 12:30, Kim calls for the webpage update, which Alexia completely forgot about. Kim is livid because now the submission deadline has passed; there won't be another chance for at least a month. They spend 10 minutes arguing. When Alexia hangs up, she's so angry she can hardly think about work.

Alexia is a victim of waste from an unrationalized workload. People call and stop by, each jockeying to get their project the highest priority. They push their projects, even when they aren't the most important thing for the business. Alexia will later learn the bug Kelly found could have waited a few days—it was on the website for months without being noticed. Kelly's forceful request won out over Kim's calm nature, but Kim's job was more urgent. But how could Alexia have known that when she decided to drop Kim's job?

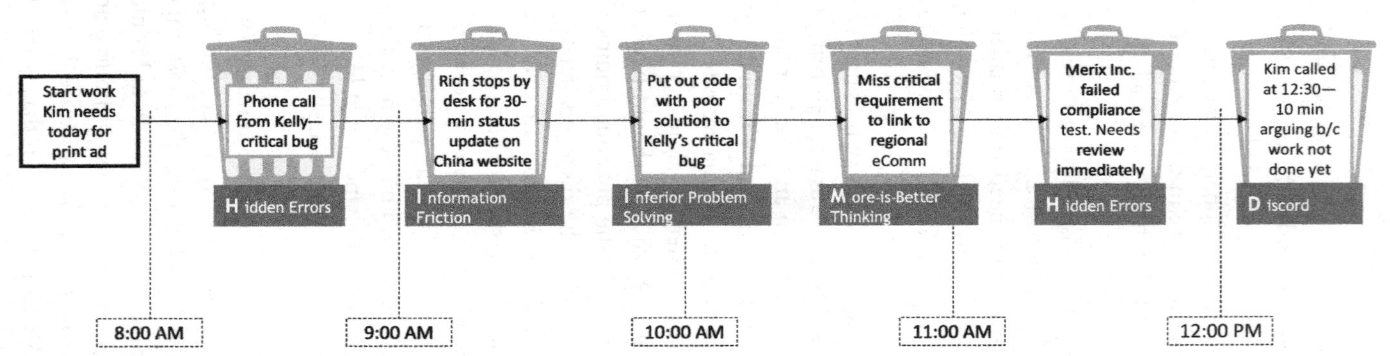

Fig. 21.2 Alexia's morning has plenty of waste from small tasks.

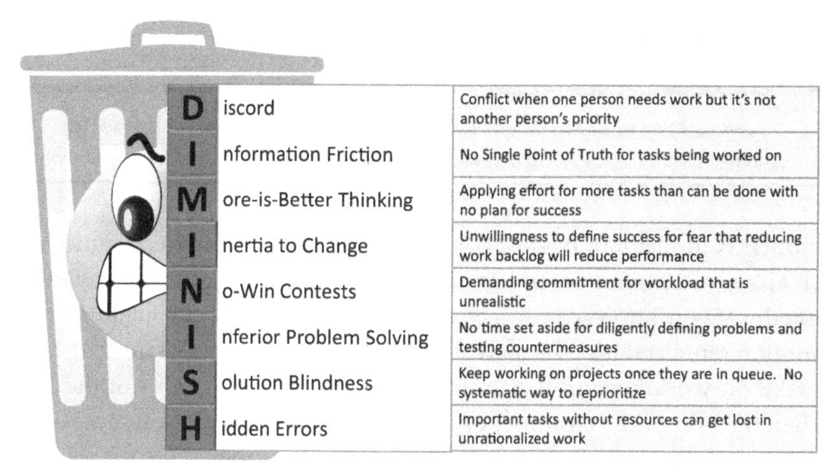

Fig. 21.3 Examples of the 8 Wastes from an unrationalized workload.

Fig. 21.3 shows examples in each of the 8 Wastes. **Discord** in the organization is common because there's not a reliable way to agree on priority of work. Instead, work is doled out faster than the organization can process it, and everyone has to elbow in to get their work done. Tempers can flare when people learn that their work has been sitting idle. **Information Friction** is rampant when the only way to know if the task you need is being worked on is to ask, and you'd better be asking every day to check your task didn't get elbowed out yesterday. **More-is-Better Thinking** takes over because there's no way to meet demand, so let's get work done "as soon as possible." Prioritization often demonstrates **Inertia to Change** because people have learned that the only way to get work done is to push hard; they're likely to be unwilling to take any solution that limits how much they can push. But the pushing is a large part of the waste: the reminder phone calls, the emails, and the explaining, none of which do anything to move work forward. By definition, overloading teams creates **No-Win Contests**. **Inferior Problem Solving** is common in overloaded teams because there isn't enough time to look at issues diligently. An example of **Solution Blindness** is that, once work weaves its way into the prioritized heap of work, there's no easy way to take it out, even if new facts have come to light since the work was accepted. And one of the most damaging **Hidden Errors** is the critical task that got lost in the confusion: the initiative that got dropped that lost us a customer, cost us a compliance fine, or shut down a manufacturing line due to obsolescence. And when there are 20 or 30 tasks stuck in a traffic jam, it's easier to lose track of the critical one.

21.3 Ruthless Rationalization

It is this moment of courage that begins the battle against oversubscription: we will commit to no more than we believe we can do well.

Ruthless Rationalization is just common sense: never agree to do something you know you cannot do. In fact, it's even an outcome of ethics: agreeing to something you know won't happen is a kind of misrepresentation. If it's common sense and it's ethical to avoid, you would think it's rare for people to take on work beyond their capacity. Unfortunately, that's not the case. When the boss or the customer or that always-angry salesperson successfully pressures someone to commit to an unrealistic date, they will probably exit the conversation satisfied that their project is on top. In reality, they are stuck in systemic waste where people are exerting large effort for a momentary win. The organization needs a confident and capable expert who diligently estimates effort and then ruthlessly rationalizes the workload. It is this moment of courage that begins the battle against the waste of oversubscription: "We will commit to no more than we believe we can do well."

Ruthless Rationalization is shown in three steps in Fig. 21.4. It begins by setting a goal: the number of a certain type of task or initiative per week, an amount of revenue secured, or a date by when a service or product will comply with a regulation. Goals are determined in units of value such as those in the value *gem* of Fig. 3.1 such as revenue or on-time completion. Quantify to communicate (Section 6.4.3), understanding the goal will change over time because of the uncertainty in the first iteration. The second step is to estimate the effort to achieve that goal. The third step is to identify the set of work the team will take on. The final check is to ensure the resources are rationalized to the work. If not, we must be "ruthless": either (1) renegotiate the goal, (2) increase the resources, or (3) increase the total value of the workload; the value of the work set is increased by modifying existing tasks and initiatives or finding more valuable alternatives to those we have. Ruthless Rationalization treats an unrationalized workload as a Stop-Fix alarm.

21.4 Just-In-Time Rationalization

The remainder of this chapter will apply Ruthless Rationalization to tasks and small projects. For many teams, everyday work can come from competing customers, both internal and external. This can be harder to

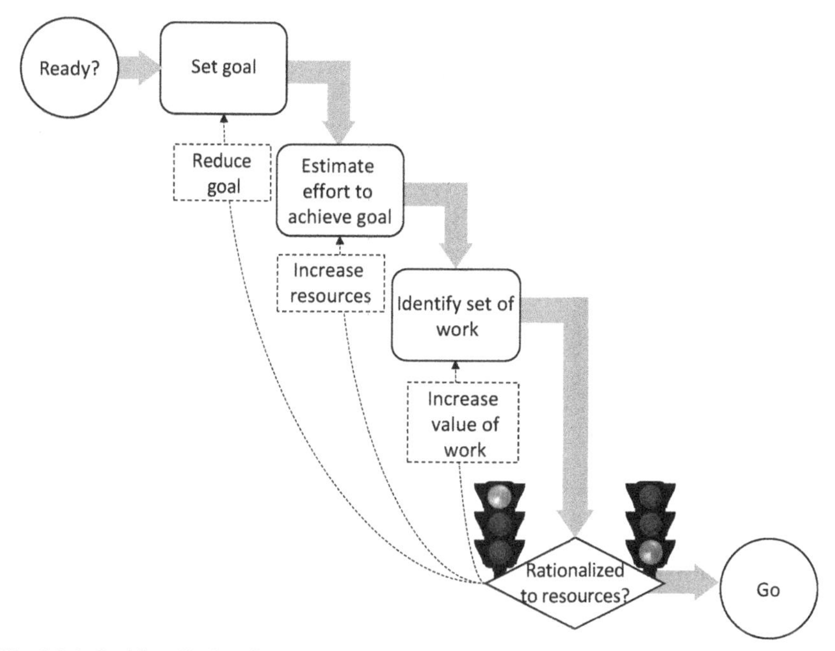

Fig. 21.4 Ruthless Rationalization treats an unrationalized workload as a Stop-Fix alarm.

rationalize than a large project, which is usually a high priority for a dedicated team. Dedicated teams simplify the job of rationalization because there are a smaller number of competing interests, all of which want to see the project successful. In small tasks work, we might take requests from many functions—Product Marketing, Operations, Finance, HR, and General Management—none of which may help weigh their priorities against the others.

21.4.1 Prioritize

The first step in rationalizing is prioritizing, as shown in Fig. 21.5. We start with many tasks, and let's assume it's not possible to do them all (a safe assumption in most knowledge work!). The second step is to prioritize, which begins by filling up the workload with the highest-value projects possible. After getting an initial set of tasks, Step 3 is to rationalize that set against the team's goals. For example, if the primary task of a sales team is to generate in-year revenue, prioritize the tasks by revenue, pick as many as the team can support, and sum the revenue. If it's enough to meet the goals and to satisfy the other requirements of the organization, rationalization is complete. Normally, it's

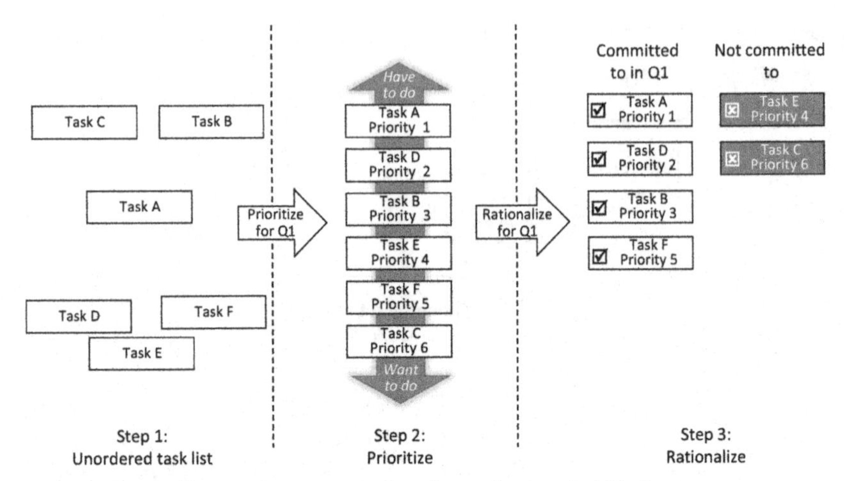

Fig. 21.5 Rationalization is a process that depends on prioritization.

not that simple. Most knowledge work teams support varying needs: taking care of key customers, adding new customers, and meeting regulatory needs are a few examples. So the rationalized set must meet multiple goals.

In Fig. 21.5, the work set was rationalized by selecting Tasks A, B, D, and F; more importantly, the team unambiguously rejected Tasks C and E. There were not enough resources for everything, so the team is transparent, indicating the set they could support. You might have noticed in Fig. 21.5 that Task F (Priority 5) was selected in front of Task E (Priority 4). This is common. It could be that Task B and E rely on one person in the team, so there's not capacity to do both; in such a case, Task F is the best available option. These types of trade-offs after force ranking are normal—force ranking priority must be done over a narrow set of measures, so it almost never tells the full story. Instead, business acumen and technical knowledge must be applied to identify a full work set.

People struggle with this approach at work, thinking it's not possible to rationalize because there are so many differing needs. But we do it every day with our budget at home. Our money satisfies many competing needs, some urgent, some important, some both, and some neither. We all have bills that must be paid, but there are many complex trade-offs: do I save for the college fund or go out to eat? Do I take a vacation or add a bedroom to the house? There is no simple rule, but budget discipline begins by deciding that you will live within a budget, which is an expression of Ruthless Rationalization.

Forecasting workload is complex because there are always unknowns. Work sometimes comes in bunches, something we can call uneven flow; also, domain issues can be unpredictable—a clause in a contract or a detail in an electrical schematic can seem simple at first but turn out to be quite complex. In these cases, experts must willingly shoulder the load to estimate as well as possible, track actual against commitments, and learn to be better the next time. Over months and years, expertise will grow and estimates will steadily improve. Don't be satisfied with stock phrases like "it depends" or "we don't know." Of course it depends. Of course we don't know with certainty. That's why knowledge work relies so heavily on experts.

21.5 Kanban Task Management

Kanban is a bit like the Chinese board game Go—a few moments to learn, a lifetime to master [3].

A Kanban board is a visual way to perform Just-In-Time Rationalization with irregular tasks. It helps teams deal with uneven flow of tasks and creates transparency so internal customers can see what the team is working on. The Kanban board is based on Kanban inventory management, which was derived by Taiichi Ohno and others in Toyota from American supermarkets in the 1950s [4]. Kanban Task Management was applied to knowledge work on a large scale by David Anderson, who helped develop Agile Software Management [5]. The concept can be demonstrated with an ordinary lunch buffet at a pizza restaurant, as shown in Fig. 21.6. When a server (1) sees a

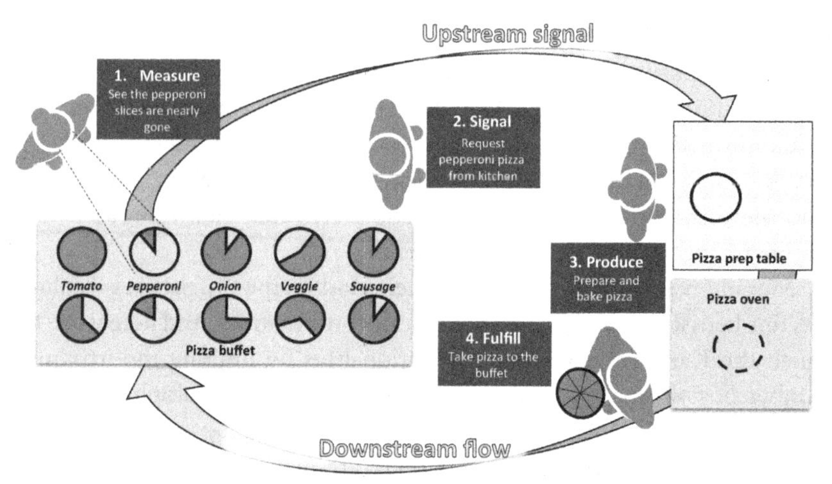

Fig. 21.6 A pizza buffet is a pull system with downstream flow driven by upstream signals.

type of pizza is running low, he or she (2) signals a request to the kitchen for more pepperoni pizzas. The kitchen (3) starts a new pizza and, a few minutes later, (4) delivers it to the buffet table.

The key lean principle in Fig. 21.6 is to see how the upstream signals trigger the downstream flow. These signals help the restaurant respond at just the right rate when, for whatever reason, pepperoni slices are taken at half the rate one day compared to the next. They are the key to the Kanban board [6] of Fig. 21.7. This is a physical board with sticky notes (or it's virtual equivalent) that flow from left to right. Each note represents one task. The backlog of tasks is unlimited; any customer to the process can add work to the backlog, but there's no commitment on when the backlogged work will start—implying, of course, that entering a task in the backlog gives no indication of when, or even if, the task will be worked on.

The unapologetic refusal to commit to tasks in the backlog is probably the most counterintuitive part of Kanban. People seem to expect that how much *effort* they expend to get the project prioritized should be a primary factor in that project's level of priority. In fact, projects ought to be selected by one measure: return, which is how much value we add vs how much effort is required. If return is low, it does not matter how long ago it was requested. Contrast this with the common complaint: "I've been asking for this for months!" It's illogical to think a low-return task requested 6 months ago should supersede a high-return task requested yesterday, but it happens all the time.

21.5.1 How Kanban works

To understand how work flows in Kanban boards, we must start on the right-hand side of the board. The "Done" column is intuitive; when a task is "Confirmed" complete (that is, accepted by the customer), it makes its final move to the right. Perhaps we leave it there a week or two so those who view the board will know it's complete, and then remove it from the board and store it for future reference.

Moving one column to the left, when a task in the "In progress" column gets finished, it can go to the right to await confirmation. And here is the first signal: the Kanban board ruthlessly rationalizes by limiting the maximum number of tasks in progress, here to three. There are multiple techniques to set the queue [7] and the total number is the simplest.

David Anderson points out that there are many ways to visualize tasks that look like Kanban boards. But the test of true Kanban is it must limit

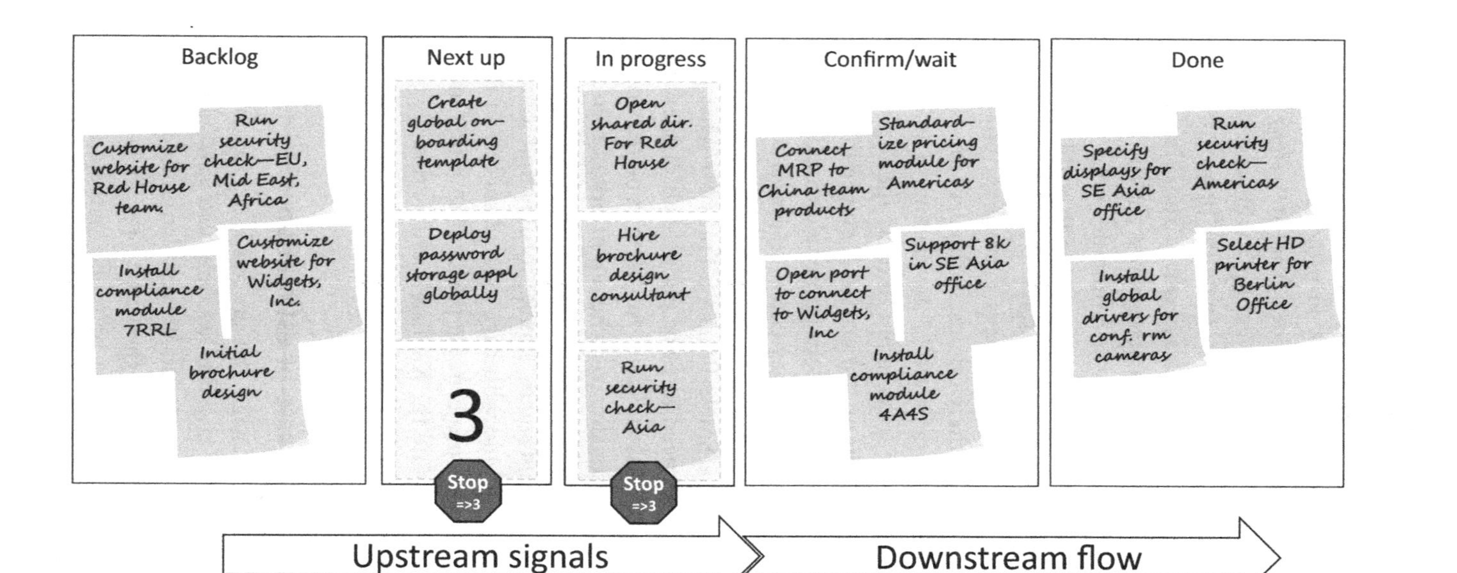

Fig. 21.7 A Kanban board creates a pull system for task management.

"work in progress," shown here as stop signs in "Next up" and "In progress," and the *pull signals* that fill the work to those limits. Anderson says, "...if there is no explicit limit to work-in-progress and no signaling to pull new work through the system, it is not a Kanban system" [8].

The next column to the left is "Next up," which is the next set of tasks that have been prioritized. When a slot opens in "In progress," the team executing the tasks can, at any time, pull a new task from "Next up." "Next up" is filled by customers. A common approach is for the full team including internal customers to meet for 15 or 20 minutes once every week or two to fill "Next up" enough to occupy the team until the next meeting. The Kanban board gives us prioritization and rationalization:

Prioritization: "Next up" ensures customers (usually internal customers) are prioritizing the work just as the decisions are needed. Made too soon, decisions are deprived of the maximum amount of information; made too late, the indecision can delay the team.

Rationalization: Limits on "In progress" ensure the team is constantly rationalized to the workload. Commitments are made on tasks that are at "Next up" or perhaps when they enter "In progress"; no commitments are made for backlog tasks.

The Kanban board is the Single Point of Truth for what tasks are being worked on. There is no need to place phone calls or spew out emails for status; the Kanban board has the status. And there is no reason to use personal networks or force of personality to rachet up priority on your favorite project. Come to the weekly prioritization meeting and make your case with the other internal customers. You'll probably find some people won't accept this, especially people who have created success in their careers by aggressively pushing their projects. Be prepared to intervene with internal customers who work around the Kanban board by directly pressuring team members to work on their projects.

21.5.2 Eliminate the waste of estimation

Kanban "lets the flow manage the process," as Taiichi Ohno is often quoted saying; when one task finishes, the next can start. This cuts the waste of estimation [9], which can be very large indeed. Suppose a task comes into a traditional system—the task customer negotiates a schedule for the task, which if missed must be renegotiated and, if that is missed, renegotiated again. The effort for all estimations except the final one is pure waste.

One practice I've seen work was to do the rough estimates, spending a few minutes on something like T-shirt sizing (small, medium, large) of effort and value of each task. Internal customers then decide what work moved forward based on a quick mental calculation of return based on estimated value and T-shirt size of effort.

The Kanban board works well for independent tasks such as are common in IT, customer service, laboratory testing, and other supporting functions—those lasting a few weeks and coordinating the activities of a few people. It immediately eliminates waste from **No-Win Contests** because "A pull system cannot be overloaded if the capacity…has been set appropriately" [10]. Larger projects need the sort of planning discussed in Chapters 19 and 20. But even for groups that spend most of their time in large projects like R&D, business development, or legal cases, the Kanban board can be used on the slice of resources that goes to small projects and so bring order to that part of the group's responsibilities. For those teams that are almost entirely committed to smaller projects, the whole department may be managed with a Kanban board.

21.5.3 Kanban dashboard

The Kanban board shows flow of tasks but does not visualize performance. For example, measures of value like total revenue generated or %-on-time are not displayed. For that, a complementary visualization that aggregates performance over many tasks is needed such as the dashboard of Fig. 21.8, which has four metrics. (Section 8.5.1 details dashboards.) Anderson suggests a dozen or so metrics that can be used to create continuous improvement [11, 12].

It's important to provide performance measures along with the Kanban board, otherwise internal stakeholders are likely to have poor understanding of Kanban's contribution to the organization over time. The dashboard or other performance measures may not be needed initially, because the Kanban board brings quick, obvious improvement; a lot more gets done and people are happier doing it. But over time, memories fade; the organization habituates to the improvement, forgetting what things were like before the board. And remember from Section 21.5.1 that aggressive internal customer you had to intervene with to stop them from pressuring team members? That same person may complain to the boss that the Kanban system doesn't work. The customer may have a point; for all the waste the Kanban board cuts, it probably is true that the most aggressive people see their influence cut.

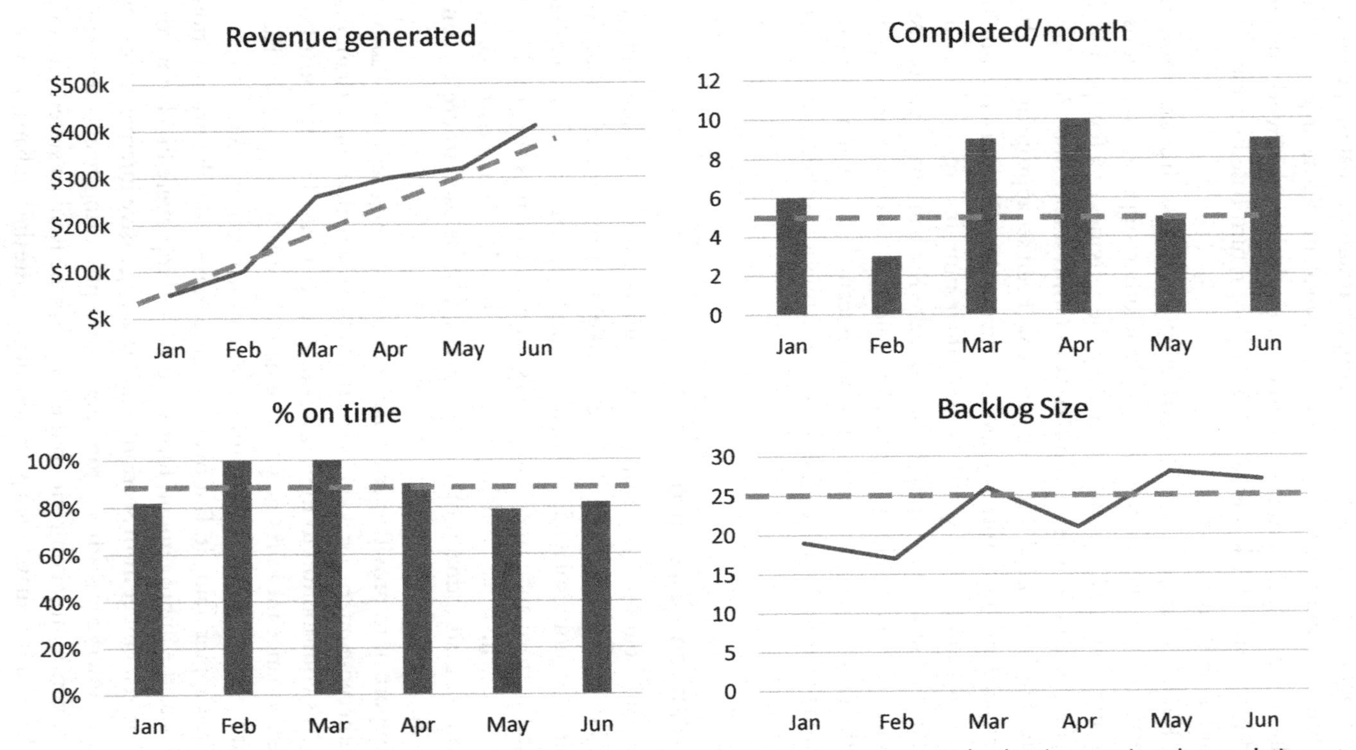

Fig. 21.8 A typical dashboard for a Kanban measure performance from four perspectives: created value (revenue), task completion rate, on-time delivery, and backlog size.

In fact, one of the functions of the Kanban board is to let value drive prioritization in place of listening to the loudest voice. But when that person does complain, you'll want a dashboard at the ready so no one will doubt the increase in productivity that the Kanban board is bringing.

As shown in Fig. 21.9, the Kanban board provides constant rationalization to transact the generation of value. But measuring value over time provides a larger view, always asking if we are doing what the organization most needs, guiding us to transform the Kanban board by fitting and refitting it to current needs.

21.5.4 How to start with Kanban Project Management (KPM)

Perhaps the best advice for those who want to try KPM is to try it first as a board for managing personal daily work [12]. I took that advice and never looked back. I created the Kanban board shown in Fig. 21.10 with four sheets of printer paper and a stack of sticky notes; it was taped up beside my desk so that I could never go more than a few minutes without seeing it. I tried it on a web-based Kanban application, but prefer the accessibility that paper provides. It immediately brought order to my workday. It also helped me cut down on multitasking. When *On deck* or *Do* are overloaded, I return a sticky to *Backlog*. I've made many changes to my Kanban board over the years. For example, I added a Kanban category "Wait for response" that I needed since my Kanban typically flows in a matter of days, but people often take a week or two to respond. Years later, my Kanban board is still constantly changing.

I prefer the Kanban board in physical format when I'm home and convert to digital when I'm traveling by photographing the cards on my physical board and pasting them into my traveling PowerPoint Kanban. Over the last 10 years, I've had many Kanban boards, but I've never gone long without one. After all this time, I'm still surprised how much stress it relieves when I literally see I'm overcommitted and then move a few sticky notes to the left, so that order returns.

21.5.5 Virtual or physical?

One of the first decisions to make is whether the board will be paper or virtual: stored electronically and viewed on a screen. Don't be drawn to the sparkle of a full-feature Kanban software system. Compared to sticky notes and felt-tip markers, they are so polished as to make a physical board seem

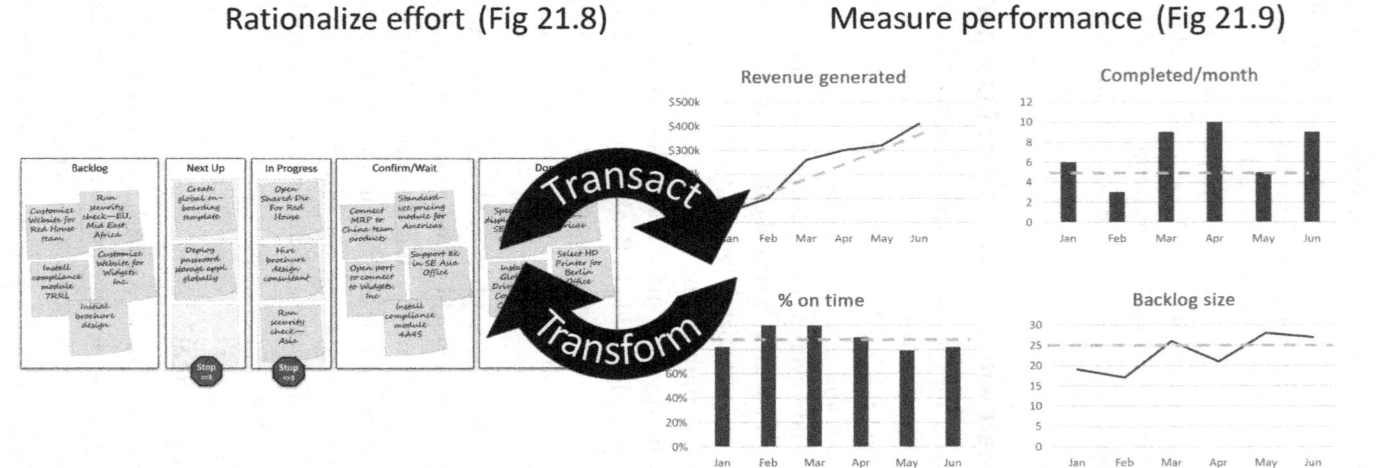

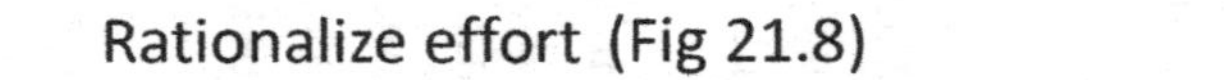

Fig. 21.9 The Kanban board prioritizes tasks to transact our work. The dashboard measures performance to transform how we prioritize.

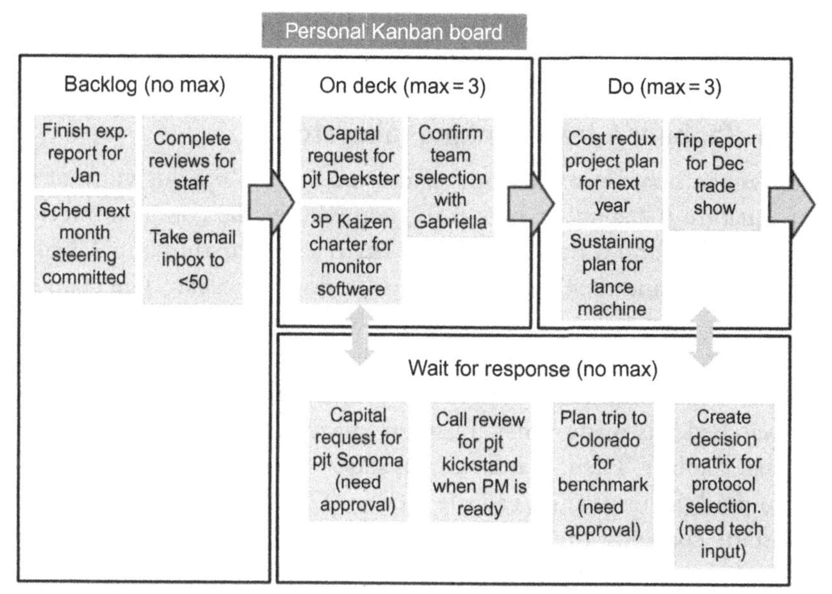

Fig. 21.10 Author's personal Kanban board for daily management.

tired. But the physical board has something almost impossible to replicate with a virtual system: presence. It's always there. Anyone can walk up and touch it. During the week, people start conversations and naturally wander to the physical wall and point at a card… "I think I might know a better way…" Walking in and out of an office area, it's a natural place to stand and contemplate a knotted-up problem. Anderson recounts his early experience: "The physical board had a huge psychological effect compared to anything we got from the electronic tracking tool we used at Microsoft" [13]. He points out how having people standing at the board for a few minutes each day created a dramatic increase in creativity and a high focus on resolving issues.

On the other hand, if your team is highly distributed, you're likely to benefit from a software Kanban board. I've yet to see a video conferencing system that can convey a physical board to remote team members; as a result, those outside the room are always at a disadvantage. Over time, their voices will not be heard as well. I've installed numerous virtual and physical boards, and they've both worked. But the difference between database-optimized, cloud-connected systems and the remarkably low-tech handwritten sticky notes has nothing to do with it. The board should reflect the team: if they are collocated, hang the board near them. If they are distributed, a virtual system will create a flatter playing field.

21.5.6 Kanban: A lean thinking laboratory

The Kanban board provides one of the most unambiguous view of the lean equation (Chapter 4): over time, you will see the complex interplay of value, effort, and waste. It's the ideal lab to learn lean thinking. Changes install quickly, and then produce results (or flop) in weeks. The team can improve as they see gaps. This creates what Anderson calls a "Kaizen culture" by empowering people. "Individuals feel free to take actions, free to do the right thing. They spontaneously swarm on problems..." [14]. I've watched this many times. First, an outside facilitator leads the team to install Kanban. Then, that facilitator stays with the process for a month or two until the team enjoys a little success. Then the team takes over and soon starts changing their board. It never fails—come back a couple of months later and there will be many changes such as modifying the content of the cards, changing the review team, adding metrics, and on and on. Unlike a heavy product development or business acquisition structure where the stakes for any one project are high, Kanban imparts a right and a duty to experiment in the relatively low-risk workflow of small projects.

When implementing Kanban, start by making the fewest possible changes to your existing process. "You must resist the temptation to change workflow, job titles, roles and responsibilities, and specific working practices" [15]. Change nothing that affects people's self-esteem or might injure egos. There's no need. Simply make visible that which is hidden by waste: ambiguous priorities, wasteful multitasking, and persistent barriers.

21.6 The Kamishibai board: Smooth flow for regular work

Where the Kanban board organizes irregular work—tasks and initiatives that vary in size and occur unpredictably—the Kamishibai[a] board organizes regular work that occurs at defined intervals. For example, it can be used to ensure an even flow of project reviews. Let's say we have 12 active projects that should have a 1-hour review at least every

[a] There is an unrelated use of the Kamishibai board for *auditing*, where a random card is picked daily. For example, going to a manufacturing cell each morning and randomly picking 1 card out of a stack of 100 potential topics to audit.

3 months. The Kamishibai board provides visualization to encourage smooth flow of each task so that the reviews are spread out at roughly one per week.

We set up a Kamishibai board for just this purpose and it was invaluable for creating a smooth flow. Before the Kamishibai board, every couple of months we'd realize we had not reviewed any projects, and then we had a raft of reviews to do in a week or so. This created frustration from the stress of having to schedule many reviews quickly; also the reviews during those rushed times seemed to be of lower quality. These are typical wastes of uneven workflow.

Giles Johnston gives a step-by-step guide to building a Kamishibai board, starting with: "The first thing to do is to define the time period that you want to manage" [16]. On the factory floor, the time period is often over a week or a month. In knowledge work, it may be months or even a year. The board will then show all tasks according to whether they are completed so that anyone can grasp the status in a glance.

A Kamishibai board shows a set of cards where each task is written on a two-sided card as shown in Fig. 21.11, an example for project reviews. Note that the text on both sides of the card is identical. All cards are displayed on a single board, with those incomplete (here "unreviewed") showing the light

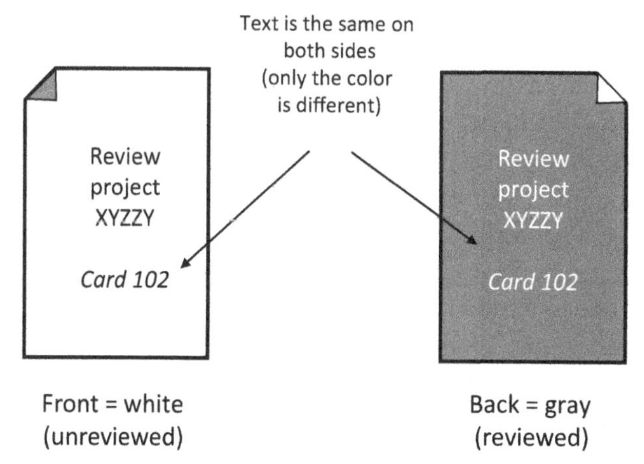

Fig. 21.11 A Kamishibai card has identical text on both sides but with different background colors; the light background shows the task open and the dark background shows it done.

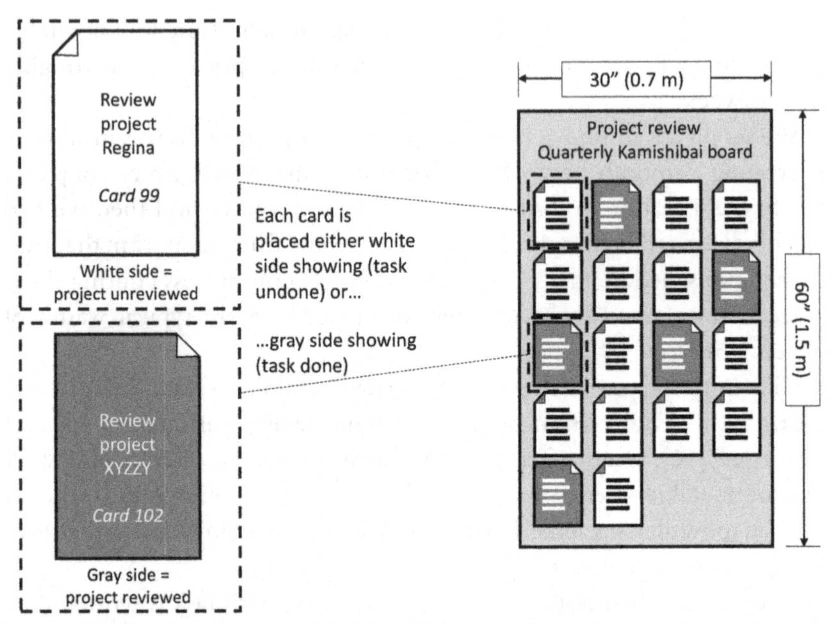

Fig. 21.12 An 18-project Kamishibai board shows each project as either reviewed or unreviewed. Progress over the quarter is apparent in seconds.

background and those complete showing the dark (see Fig. 21.12). Cards can be laminated to be more durable.

Fig. 21.13 shows the flow of the board over the quarterly interval with the left frame showing the board at the start, the center frame showing the board about halfway through the quarter, and the right frame showing the board near the end of the quarter. Progress can be determined in seconds. For example, on June 20 (right side of Fig. 21.13) only 2 projects remain unreviewed of the 18-project portfolio. At least at the first level, this looks about right. Compare this to having to ask every few weeks about each project to get a sense of how evenly the reviews are flowing.

If you'd like to try the Kamishibai board, consider using the sticky-note version shown in Fig. 21.14. The function of the Kamishibai board can be put together in minutes with ordinary office supplies rather than spending weeks to purchase and hang permanent Kamishibai boards and make laminated cards just to try the idea.

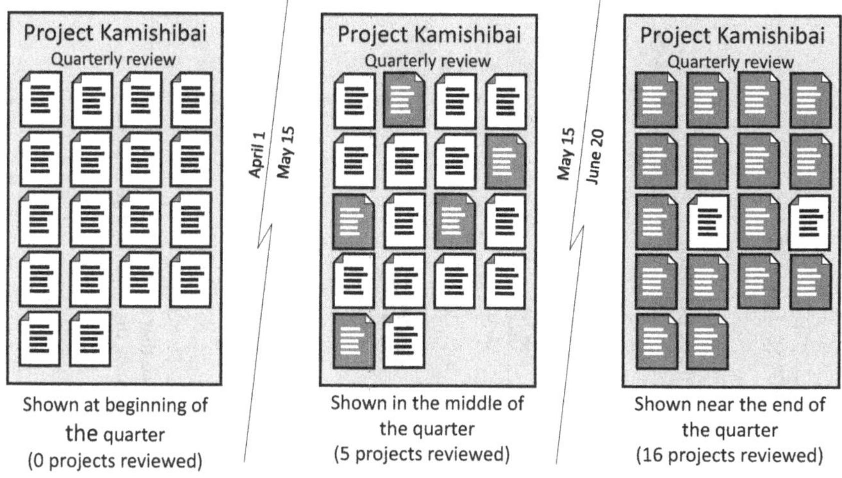

Fig. 21.13 A Kamishibai board shows all cards visualizing progress over the quarter.

21.7 Knowledge Work Improvement Canvas (KWIC) to adopt Just-In-Time Rationalization

Fig. 18.8 shows a Knowledge Work Improvement Canvas (KWIC) to deploy a Kanban Task Management board (Fig. 21.15).

21.8 Conclusion

The Just-In-Time Rationalization **simplifies** the way we prioritize work (see Fig. 21.16). It creates a standard method with SPoT and demands Ruthless Rationalization to eliminate ambiguity. It also **engages** as shown by the Engagement Wheel. Knowing the team can and will delight their internal customers can be inspiring for those who today go home feeling that no matter how hard they work, they are not meeting the organization's needs. It connects the internal customers and the staff who execute their tasks by working together to decide what's most important. It gives a contest that can be won because work is optimally prioritized and constantly rationalized. And it protects the team by ensuring each task is done to completion before starting the next task. The dashboard creates an **experiment**, measuring the value of work and thus creating opportunities to learn and improve.

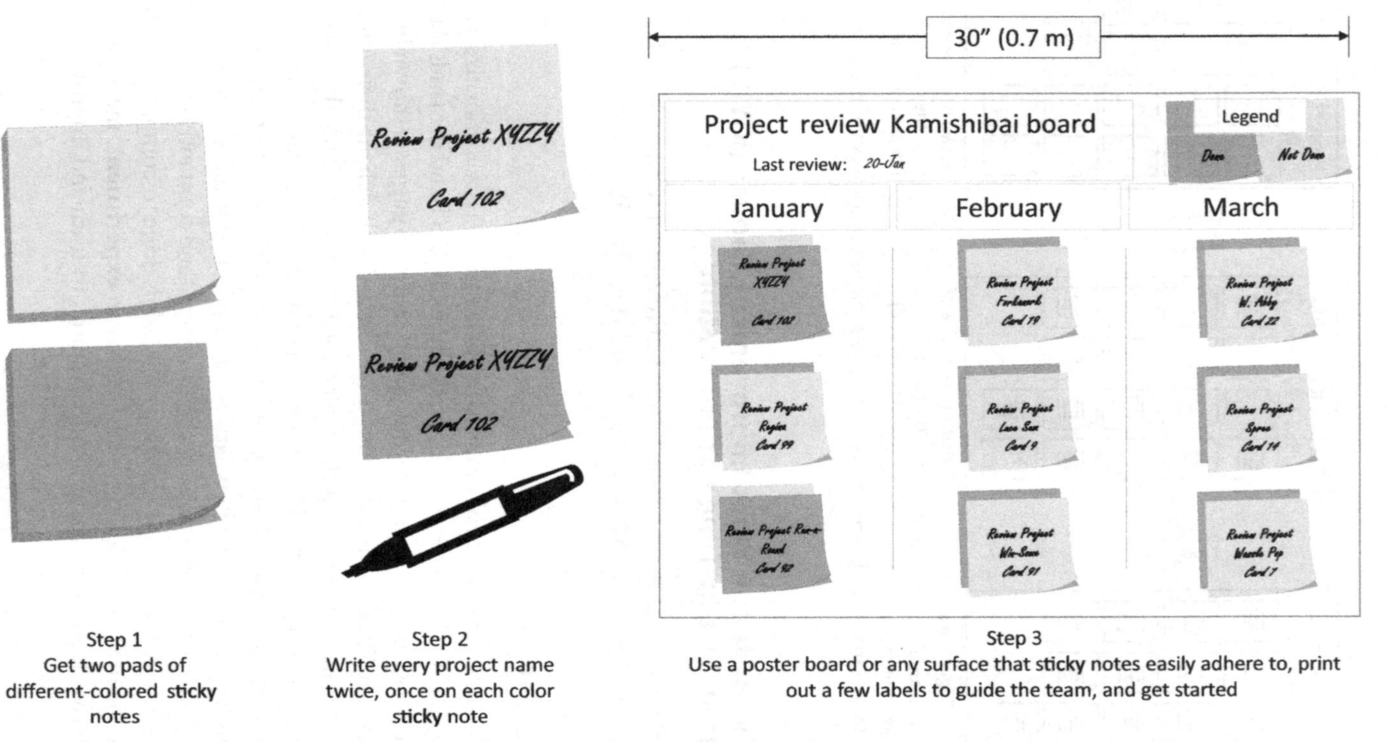

Fig. 21.14 Trystorm a Kamishibai board with sticky notes and a whiteboard.

Initiative	Install Kanban task board	Status	4-Success Map tracking	Last update	12-Dec	Update due	

Guide (Sponsor)

Sponsor	Hannah
Explain need	Our team brings in a large number of small- and medium-sized tasks with no structured means to prioritize. This leads to systematic oversubscription, frustrating the team and disappointing internal customers
Need-by date	2 months fully installed
Owner name	Tucker
Team members	BB Johnson, Urs, Diego, Eyal

Focus

Problem or opportunity	Create global, Single Point of Truth for task acceptance, prioritized by internal customers
Metric	On-schedule delivery, realized value
Current perf.	Unknown
Target perf.	90% on schedule, generating $2M revenue/year
Solve method	Virtual event with full team, 2–3 hours per day, 5–6 meetings over 2 weeks

Solve

Root cause		No standardized means to review and prioritize/rationalize tasks. No standard measures of value
Solution		Install electronic Kanban board for tasks taken by global team; track value delivered annually
Countermeasures	1	Install electronic Kanban
	2	Hold bi-weekly meetings to prioritize new tasks and review old tasks
	3	Report value delivered by Kanban projects

Action Plan

Deliverable	Next Action	Date	Owner	Plan	Replan	Fcst	Actual
Define standard work for Kanban			BB	22-Nov		22-Nov	
Pilot 1			Tucker	29-Nov		29-Nov	
Pilot 2			Tucker	6-Dec		6-Dec	
Train team			BB	13-Dec		13-Dec	
Define standard work for collecting rev by tas			BB	27-Dec		27-Dec	
Buffer				27-Dec		27-Dec	

Test Track

Case	Dates	Meeting held	All projects complete	All projects report rev.	--	--
1st Meeting (12/15)	Plan	1-May	15-May	22-May		
	Replan					
	Forecast					
	Actual					
2nd Meeting (12/18)	Plan	11-May	25-May	1-Jun		
	Replan					
	Forecast					
	Actual					
3rd Meeting (1/2)	Plan	19-May	2-Jun	9-Jun		
	Replan					
	Forecast					
	Actual					
–	Plan					
	Replan					
	Forecast					
	Actual					

Bowlers

Goal 1	Tasks completed to schedule											
Date	1-Jan	1-Feb	1-Mar	1-Apr	1-May	1-Jun	1-Jul	1-Aug	1-Sep	1-Oct	1-Nov	1-Dec
Plan	90%	90%	90%	90%	90%	90%	90%	90%	90%	90%	90%	90%
Replan												
Fcst												
Act												

Goal 2	Revenue earned throughout year											
Date	1-Jan	1-Feb	1-Mar	1-Apr	1-May	1-Jun	1-Jul	1-Aug	1-Sep	1-Oct	1-Nov	1-Dec
Plan	$167k	$333k	$500k	$667k	$833k	$1000k	$1167k	$1333k	$1500k	$1667k	$1833k	$2000k
Replan												
Forecast												
Actual												

Fig. 21.15 A Knowledge Work Improvement Canvas (KWIC) to adopt the Kanban task board.

Simplify

There will be one way to prioritize our small projects

The Kanban/Kamishibai is our SPoT for decisions on small workflows

Ruthless Rationalization removes ambiguity

Engage

Inspire
Our delivery and quality will delight customers

Challenge
We will deliver on time almost every time

Connect
We'll prioritize as a team so we'll stay aligned

Protect
We will take on only what we can do well

Experiment

Falsifiable hypothesis

If *internal customers prioritize*
Then *we'll maximize value*
Measured by *our dashboard*
Trending to *our goals*

Fig. 21.16 Just-In-Time Rationalization simplifies, engages, and encourages experimentation.

References

[1] D.J. Anderson, Kanban: Successful Evolutionary Change for your Technology Business, Blue Hole Press, 2010, p. 60.
[2] S. Shingo, The Sayings of Shigeo Shingo: Key Strategies for Plant Improvement, English translation, Productivity Press, 1987, p. 120.
[3] M. Skarin, Real-World Kanban: Do Less, Accomplish More, Pragmatic Bookshelf, 2015 Forward.
[4] J. Liker, The Toyota Way: 14 Management Principles From the World's Greatest Manufacturer, McGraw-Hill, 2004, 106.
[5] D.J. Anderson, Kanban: Successful Evolutionary Change for Your Technology Business, Blue Hole Press, 2010, p. 60.
[6] P.R. Williams, Visual Project Management, Think for a Change Publishing, 2015, pp. 87–93.
[7] D.J. Anderson, Kanban: Successful Evolutionary Change for Your Technology Business, Blue Hole Press, 2010, pp. 114–116.
[8] D.J. Anderson, Kanban: Successful Evolutionary Change for Your Technology Business, Blue Hole Press, 2010, p. 14.
[9] D.J. Anderson, Kanban: Successful Evolutionary Change for Your Technology Business, Blue Hole Press, 2010, p. 39.
[10] D.J. Anderson, Kanban: Successful Evolutionary Change for Your Technology Business, Blue Hole Press, 2010, p. 13.
[11] D.J. Anderson, Kanban: Successful Evolutionary Change for Your Technology Business, Blue Hole Press, 2010, pp. 139–146.
[12] D. Anderson, Deep Kanban, Worth the Investment? London Lean Kanban Day, https://www.youtube.com/watch?v=JgMOhitbD7M, 2013.
[13] D.J. Anderson, Kanban: Successful Evolutionary Change for Your Technology Business, Blue Hole Press, 2010, p. 55.
[14] D.J. Anderson, Kanban: Successful Evolutionary Change for Your Technology Business, Blue Hole Press, 2010, p. 50.
[15] D.J. Anderson, Kanban: Successful Evolutionary Change for Your Technology Business, Blue Hole Press, 2010, p. 63.
[16] Johnston, Giles. Kamishibai Boards: A Visual Management Tool to Improve 5S and Create Effective Habits (The Business Productivity Series Book 9). Smartspeed Consulting Limited, 2012, Kindle edition.

CHAPTER 22

Workflow—Putting out "fires"

Best efforts are essential. Unfortunately, best efforts, people charging this way and that way without guidance of principles, can do a lot of damage.

W. Edwards Deming [1]

22.1 Introduction

This chapter will discuss how to respond to a "fire," the unfortunate and unexpected, sudden appearance of a negative issue that steals resources from normal work. It takes on many forms. In manufacturing, a design error may cause a quality issue that closes an assembly line. For a legal firm, it may be an unexpected unfavorable ruling that demands an unanticipated filing. For a civil engineering firm, there may be a water runoff issue discovered in the site plan that was already running at the edge of the schedule. For a marketing firm, it may be a customer demand to revise a campaign that was thought to be set.

Whatever your discipline of knowledge work, you're probably sometimes faced with a "fire": an unplanned demand for a large amount of resources to deal with an urgent problem. Handled poorly, these events bring chaos and erode trust in the organization for the team that faces the issue. But a competent and diligent response increases engagement of the team and builds confidence in the organization. Fig. 22.1 summarizes how we will deal with "fires." We will talk about how to plan and execute putting out a knowledge work fire. We will use the Ford Motor Company's 8D (Eight Disciplines) form as a guide to build a Canvas View of a rapidly evolving issue. And we'll introduce the histogram, a powerful tool for knowledge work that is often overlooked.

Improve
https://doi.org/10.1016/B978-0-12-809519-5.00022-3

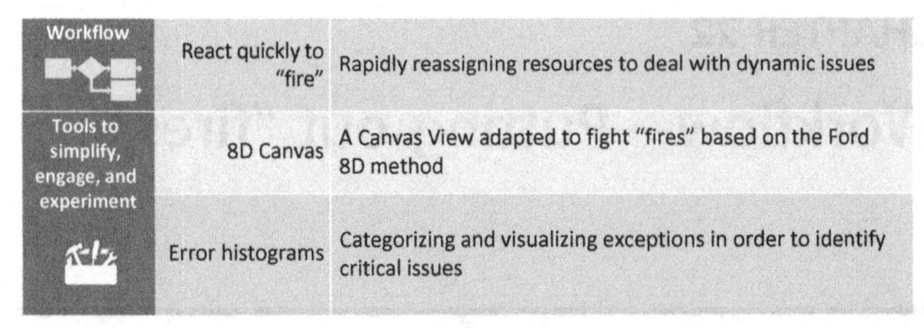

Workflow	React quickly to "fire"	Rapidly reassigning resources to deal with dynamic issues
Tools to simplify, engage, and experiment	8D Canvas	A Canvas View adapted to fight "fires" based on the Ford 8D method
	Error histograms	Categorizing and visualizing exceptions in order to identify critical issues

Fig. 22.1 The workflow and tools of the "firefight," dealing with an urgent, unanticipated problem.

22.2 The 8 Wastes

Fig. 22.2 shows examples in each of the 8 Wastes that occur when responding poorly to a knowledge work "fire." **Discord** can come from unpleasant experiences dealing with customer disappointment; I've lost count of the times a salesperson has complained (legitimately) about how design quality issues erupted, costing many hours when he or she should have been selling. **Information Friction** costs time and frustration when the team responding doesn't deeply understand the issue or when the group that is enduring the consequences doesn't understand the plan to respond: 1000-word emails and multiple calls per day may be necessary just to understand something as simple as what we're doing today. **More-is-Better Thinking** is found in the commonplace response of "doing our best" rather than determining

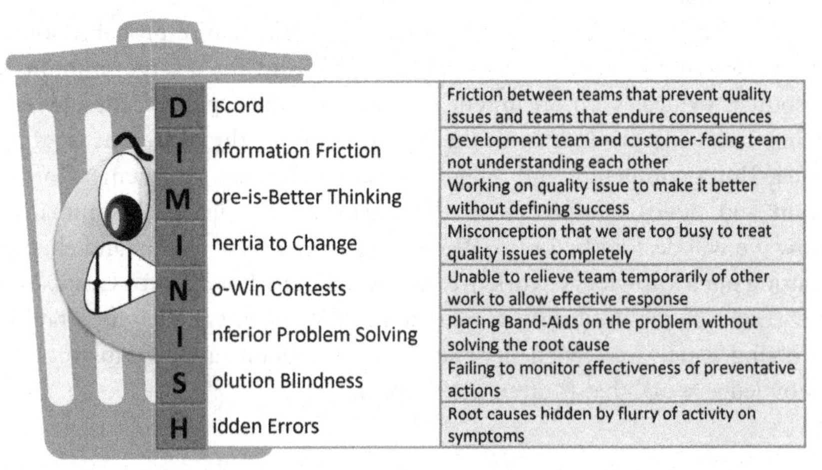

D	iscord	Friction between teams that prevent quality issues and teams that endure consequences
I	nformation Friction	Development team and customer-facing team not understanding each other
M	ore-is-Better Thinking	Working on quality issue to make it better without defining success
I	nertia to Change	Misconception that we are too busy to treat quality issues completely
N	o-Win Contests	Unable to relieve team temporarily of other work to allow effective response
I	nferior Problem Solving	Placing Band-Aids on the problem without solving the root cause
S	olution Blindness	Failing to monitor effectiveness of preventative actions
H	idden Errors	Root causes hidden by flurry of activity on symptoms

Fig. 22.2 Examples of the 8 Wastes experienced when a "fire" must be put out.

what is needed for success and doing it. **Inertia to Change** comes when the team lives continually in the misconception that there's no time to fix the problem well. In fact, treating the issue poorly almost always takes more total time. **No-Win Contests** can be created when there's no way to relax temporarily demands of other work to allow an appropriate response to the fire. **Inferior Problem Solving** occurs when the team places Band-Aids on the problem instead of treating the root cause. **Solution Blindness** is common when the team members assume their solutions are fully effective and don't measure results over time. **Hidden Errors** occur often when putting out fires because the flurry of activity treating the immediate issue can hide root causes.

22.3 A wasteful workflow

An example of a poor experience in a firefight workflow is shown in Fig. 22.3. An unhappy customer calls with an urgent issue that reveals a hidden problem with a product: a broken machine, a connection failure, or an inaccurate blood test. They probably have already spent time trying to figure out the issue, sometimes quite a bit of time. The response to that call is often to start a flurry of activity, but without building the right team. As shown in Fig. 21.4, missing needed team members is a type of **Information Friction**, putting barriers between the problem and the people in the organization who best understand what's causing the problem. Over time, the problem may be fixed; even so, the customers (internal and external) exit the process frustrated. The team is also frustrated, unhappy their product or service is causing dissatisfaction; moreover, they have to juggle their main tasks and the fire with no relief. And, if the root cause is not identified, the problem will occur again, and the wasteful cycle will be back.

22.4 The Ford Eight Disciplines (8D) process

The Ford Global Eight Disciplines [2] or G8D process is a formal problem-solving process customized for quality defects. It separates "containment" responses—the things we do to put the fire out—from preventative actions—the things we do to ensure we don't get a similar fire in future. The 8D is from the factory floor. However, the approach is applicable to knowledge work. As you'll see in the next section, we can apply the canvas approach to this class of problem with just a few modifications. But first, let's review the disciplines required (typically there are nine disciplines with the

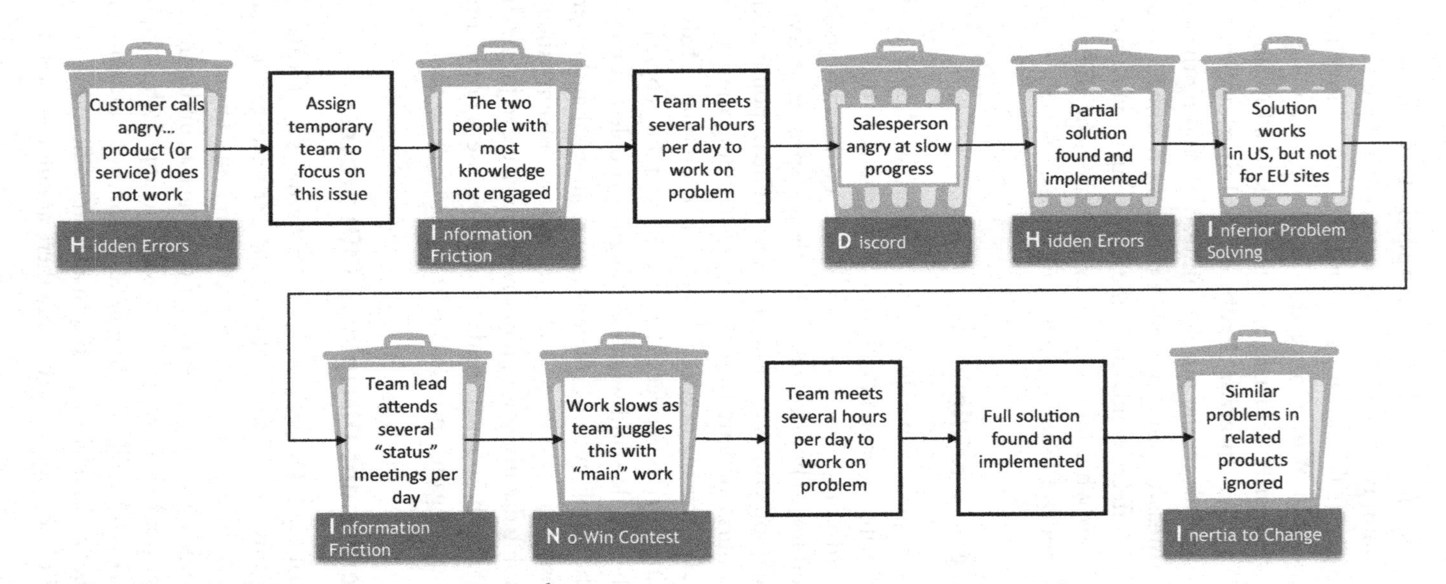

Fig. 22.3 Example of common waste putting out fires.

list going from D0 to D8 [3]). As you'll see, this is an adaptation of the formal problem-solving techniques in Section 12.2.

22.4.1 Plan

Plan an initiative to deal with the problem. Here we will define the sponsor and understand the need and scope.

22.4.2 Use a team

Define a cross-functional team capable of dealing with this problem and a leader.

22.4.3 Define problem

Use the formal problem-solving definition of Section 12.4 to create a problem statement.

22.4.4 Containment

This step holds the primary difference between this canvas and a standard problem solve canvas of Chapter 12: the heightened focus on differentiating containment from preventative measures. As the name implies, firefighting issues can create an unusually high level of damage if not contained rapidly. Accordingly, here there is a great deal of focus on first addressing symptoms.

22.4.5 Root cause

Getting at the root cause is a staple of formal problem solving. Consider a defined method such as the 5-Whys method (Section 12.5.2) or the Ishikawa (Fishbone) method (Section 12.5.3).

22.4.6 Permanent correction

Determine and verify the permanent correction that will address the root causes (similar to Section 12.6).

22.4.7 Corrective action

Take the countermeasures that will implement the permanent corrections.

22.4.8 Verify preventative measures

Verify preventative measures over a period of time long enough to observe that recurrences have been prevented. Here we will use the Success Map as we did in Section 12.8.

22.4.9 Congratulate the team

Recognize the personal effort the team put forth to bring the issue to a successful conclusion.

22.5 The 8D Canvas: A modified KWIC

The 8D Canvas of Fig. 22.4 is based on the Knowledge Work Improvement Canvas (KWIC), modified to separate the containment from the permanent corrective actions. Note that there is a single Success Map for the entire plan, merging together the requirements of D0, D3, and D5. Note also that the Test Track, well-adapted to track individual cases, is used to validate containment (D3), while the Bowler is used to track recurrence (D7).

22.6 The histogram: Systematically tracking recurrence with cause codes

When sustaining improvement over long periods of time, it's difficult to recognize subtle patterns in recurring exceptions. Even during a well-executed response to a "fire," root causes that occur rarely can be obscured by the dominant issue. So, in the moment the most urgent issue is being addressed, it may seem that the issue is fully resolved even though only one of the root causes has received attention. Over the months that follow, it's common for a lower level of issues to occur, often from unidentified lesser root causes. Moreover, different people may discover different issues—say a salesperson talking to a customer, an applications engineer talking to another customer, and an operations person observing an odd response to a production test. The high collaboration in the first weeks of the 8D creation will diminish as the most painful symptoms resolve over the months that follow. We can systematically collect information from a diverse team using *cause codes* to funnel many independent experiences into a few classes of problem and using a *histogram* to visualize the occurrence of these events over time.

22.6.1 Cause codes: Grouping failure

The goal of cause codes is to continuously funnel multiple events into a manageable set of potential root causes. For example, we might have 25 or 30 occurrences of customer complaints that we could drive into 4 or 5 potential root causes. Then, as data accumulates over time, more issues funneling into a given cause code drives more action in that area. When the

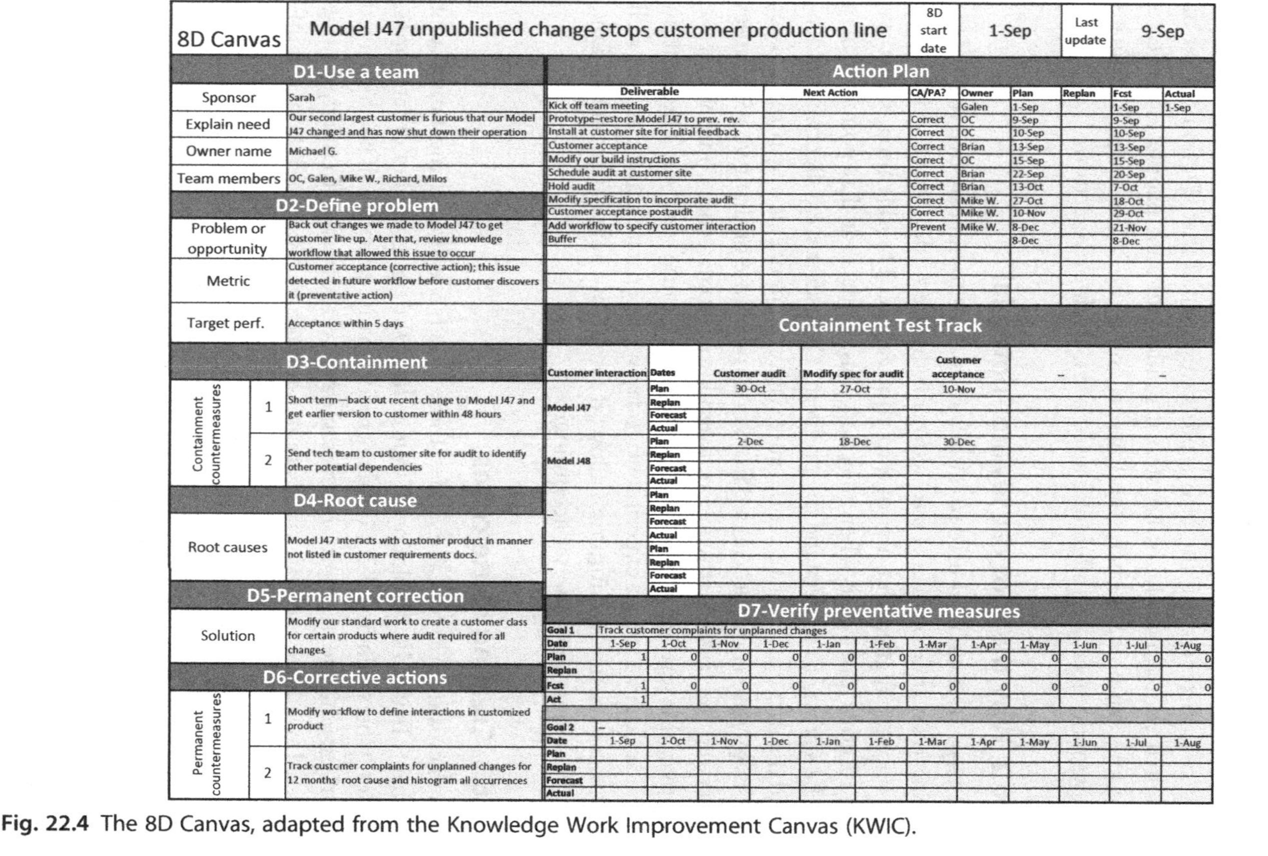

8D Canvas	Model J47 unpublished change stops customer production line	8D start date	1-Sep	Last update	9-Sep

D1-Use a team

Sponsor	Sarah
Explain need	Our second largest customer is furious that our Model J47 changed and has now shut down their operation
Owner name	Michael G.
Team members	OC, Galen, Mike W., Richard, Milos

D2-Define problem

Problem or opportunity	Back out changes we made to Model J47 to get customer line up. Ater that, review knowledge workflow that allowed this issue to occur
Metric	Customer acceptance (corrective action); this issue detected in future workflow before customer discovers it (preventative action)
Target perf.	Acceptance within 5 days

D3-Containment

Containment countermeasures		
	1	Short term—back out recent change to Model J47 and get earlier version to customer within 48 hours
	2	Send tech team to customer site for audit to identify other potential dependencies

D4-Root cause

Root causes	Model J47 interacts with customer product in manner not listed in customer requirements docs.

D5-Permanent correction

Solution	Modify our standard work to create a customer class for certain products where audit required for all changes

D6-Corrective actions

Permanent countermeasures		
	1	Modify workflow to define interactions in customized product
	2	Track customer complaints for unplanned changes for 12 months root cause and histogram all occurrences

Action Plan

Deliverable	Next Action	CA/PA?	Owner	Plan	Replan	Fcst	Actual
Kick off team meeting			Galen	1-Sep		1-Sep	1-Sep
Prototype–restore Model J47 to prev. rev.		Correct	OC	9-Sep		9-Sep	
Install at customer site for initial feedback		Correct	OC	10-Sep		10-Sep	
Customer acceptance		Correct	Brian	13-Sep		13-Sep	
Modify our build instructions		Correct	OC	15-Sep		15-Sep	
Schedule audit at customer site		Correct	Brian	22-Sep		20-Sep	
Hold audit		Correct	Brian	13-Oct		7-Oct	
Modify specification to incorporate audit		Correct	Mike W.	27-Oct		18-Oct	
Customer acceptance postaudit		Correct	Mike W.	10-Nov		29-Oct	
Add workflow to specify customer interaction		Prevent	Mike W.	8-Dec		21-Nov	
Buffer				8-Dec		8-Dec	

Containment Test Track

Customer interaction	Dates	Customer audit	Modify spec for audit	Customer acceptance	–	–
Model J47	Plan	30-Oct	27-Oct	10-Nov		
	Replan					
	Forecast					
	Actual					
Model J48	Plan	2-Dec	18-Dec	30-Dec		
	Replan					
	Forecast					
	Actual					
–	Plan					
	Replan					
	Forecast					
	Actual					
–	Plan					
	Replan					
	Forecast					
	Actual					

D7-Verify preventative measures

Goal 1	Track customer complaints for unplanned changes										
Date	1-Sep	1-Oct	1-Nov	1-Dec	1-Jan	1-Feb	1-Mar	1-Apr	1-May	1-Jun	1-Jul	1-Aug
Plan	1	0	0	0	0	0	0	0	0	0	0	0
Replan												
Fcst	1	0	0	0	0	0	0	0	0	0	0	0
Act	1											

Goal 2	–											
Date	1-Sep	1-Oct	1-Nov	1-Dec	1-Jan	1-Feb	1-Mar	1-Apr	1-May	1-Jun	1-Jul	1-Aug
Plan												
Replan												
Forecast												
Actual												

Fig. 22.4 The 8D Canvas, adapted from the Knowledge Work Improvement Canvas (KWIC).

root cause is left in an open format, people don't reliably drive to actionable causes. Let's say two customers have essentially the same problem recorded differently: one service person records "Customer upset because website sometimes resets" and another records "Occasional data lost on group page." Even though these events might have proceeded from identical causes, a third person reviewing the two reports might miss that. Defining a set of potential root causes helps because it guides different people to drive to the same cause code where the cause is common; it also helps illuminate a newfound need for different cause codes. Both drive action.

22.6.2 Why the root causes are "potential"

The term "potential" is used to modify root cause to indicate that the code is meant to separate problems into actionable buckets. Ideally, we would want the buckets aligned one-for-one to root causes. The dilemma is that we don't know root cause when selecting cause codes—at that point we may just be starting to understand the problem. Only after we collect and understand the data can we state a root cause with confidence. It may turn out later that two cause codes proceeded from one root cause or that two root causes generated the faults that were conflated into a single cause code. When we discover this sort of problem, we can adjust the cause code set. The creation of cause codes starts an experiment: today we'll separate cause codes as well as our experience and intuition allow; if tomorrow new data come to light, we will adjust the code set accordingly.

22.6.3 Creating cause codes

When creating a set of cause codes, there are a few rules, the most important probably being the "Goldilocks" rule to have the right number of categories, typically between 5 and 20. If there are too many, people will be challenged to understand the structure and the same root cause may go to multiple cause codes. If we have too few, there are not enough to distinguish between different root causes; in this case, the buckets we create won't be actionable. Let's return to the issue from the previous sections: we changed the function of a product and unintentionally shut down a production line at the customer "Monmouth Merix." Let's say we ran the 8D of the previous section and found it was our fault: our software was changed without informing the customer. At that point, it's too easy to think that's the cause of all similar complaints. That brings **Solution Blindness**: ignoring data that challenges our beliefs or failing to take reasonable steps to collect such data.

To fight **Solution Blindness**, the 8D form guides us to track recurrence over many months to ensure we solved the full problem. Let's say we put in place our application team for the next 6 months to monitor every complaint from a customer about our changing product function. We'll start with a set of five cause codes, but right now, let's just define the one we know: we made a change without informing customers that depend on our product. So, our initial cause code is pretty simple: the one cause we know and a few "TBDs" to leave space to learn.

Cause codes (initial set)

1. Made change and did not notify customer
2. TBD
3. TBD
4. TBD
5. TBD

Over the next few months, we collect data. The first thing we learn is that the customer is also changing their application that uses our product. In October, the customer calls to complain, but a few days later they realize the problem resulted from their change, not ours. A new customer calls with a similar symptom ("It used to work but doesn't anymore") only to learn our software was installed the second time at that customer site with different parameters. Once the install process is repeated with the original installed parameter set, the problem resolves. And so on. After 4 or 5 months, the cause code set has matured, revealing that there were several causes of this complaint, and just the process workflow change in the original 8D responds to only a portion of incidents.

Cause codes (after 4 or 5 months)

1. Made change and did not notify customer
2. Customer changed their product
3. Install configuration problem
4. End customer user error
5. TBD

22.6.4 Cause code histogram

The histogram in Fig. 22.5 is an ideal way to visualize the occurrence of incidents organized by cause code. Maintaining a ledger format so that the data source is clear, the occurrences are continuously shown on the top half of the screen. The histogram here reveals that months after the 8D form started, the most common problem is configuration errors. This insight can drive action

Customer complaints/uncontrolled product chang	Month	Code	Cause	Countermeasure
Monmouth Merix line shut down from new softwa	Sept	1	Did not notify customer	Run 8D
Monmouth Merix complaint from customer svc	Oct	2	Spec changed	App team reviewed with customer; customer will restore previous firmware
Noonkester Manufacturing complaint	Oct	3	Install config error	Service team to return to customer to reinstall system
Xi Fung customer down	Nov	3	Install config error	Telecon to correct install issue
Monmouth Merix complaint from customer svc	Nov	4	Customer error	Error from customer application team; no response needed
Resilix of Washington State	Dec	4	Customer error	Error from customer application team; no response needed
Xi Fung customer down	Jan	3	Install config error	Telecon to correct install issue
Real Fortuna	Jan	1	Did not notify customer	Using older version (pre 8D). Corrected with 8D

Cumulative occurrences	Code	Cause
2	1	Did not notify customer
1	2	Spec changed
3	3	Install config error
2	4	Customer error
0	5	TBD

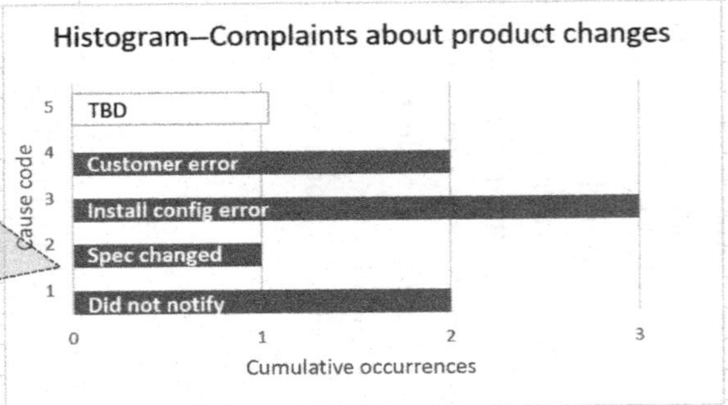

Fig. 22.5 Histogram of customer complaints about uncontrolled product changes.

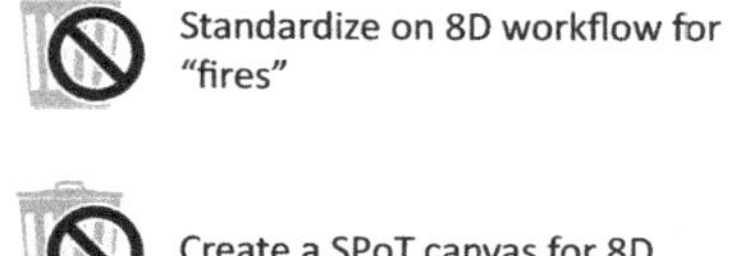

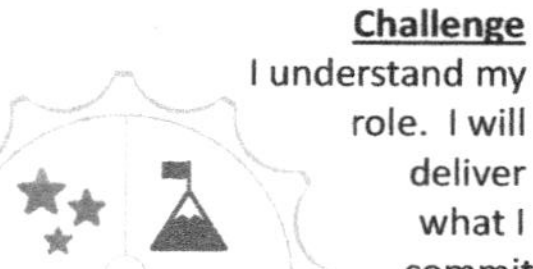
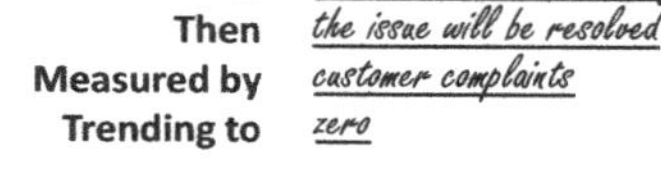
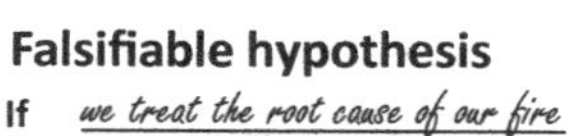

Fig. 22.6 Skillfully putting out "fires" simplifies, engages, and encourages experimentation.

directing the team to dig into the configuration problems. Perhaps the configuration process is too complex or perhaps the instructions are unclear; the root cause is unknown at this stage, but what is clear is that continuing to push on customer notification and audit (from the 8D) alone will be ineffective at eliminating many of the customer complaints.

Histogramming cause codes generates a surprising amount of insight for a modest effort. It challenges our assumptions. And it encourages us to look at nuances in order to increase our understanding of problems over time.

22.7 Conclusion

Effective "firefighting" **simplifies, engages**, and **experiments** as shown in Fig. 22.6. We can simplify the workflow by adapting formal problem solving for the needs of "fires," for example, using the 8D workflow, creating a SPoT canvas for that workflow, and employing histograms to track errors over long periods of time. It increases engagement according to the Engagement Wheel in Fig. 22.6. Knowing the team can reliably extinguish fires is inspiring because they know they have earned the confidence of the organization they work for. And firefighting as a team builds relationships as people experience that they can depend on their teammates. By creating formal problem solving around these urgent situations, the leadership creates contests that each person can win. And the team's "brand" will be protected when the organization commits to fixing things well and preventing similar issues in the future. Finally, effective firefighting is built on experimentation: create a falsifiable hypothesis based on measuring the effects of the errors over time.

References

[1] W.E. Deming, Out of the Crisis, Massachusetts Institute of Technology, Center for Advanced Engineering Study, Cambridge, MA, 1986, p. 19.
[2] Online Course on G8D, Available at: https://www.sae.org/learn/content/pd111860/.
[3] https://asq.org/quality-resources/eight-disciplines-8d.

CHAPTER 23

Workflow—Visualizing revenue gaps

It occurred to me that the same Lean Thinking and methodologies that solved these chronic problems in areas such as manufacturing, product development, customer service, and health care might be used to revolutionize the way sales processes and organizations are designed and managed

Robert J. Pryor, author of Lean Selling [1]

23.1 Introduction

This chapter will focus on two visualizations for revenue gaps that can bring more reliable effort to close those gaps while increasing engagement. The need to close revenue gaps can generate a lot of waste. Instead of the sales team leading the organization to new business, energy can be directed to convincing customers to move up orders already placed and offering discounts to win orders that would have been won eventually. In this chapter, we will talk about closing revenue gaps and two primary visualizations: the single and the double waterfall charts (Fig. 23.1).

Robert J. Pryor worked with CEOs across many industries on the challenges they faced in their sales organizations. He found that in nearly every case, they were disappointed for a few reasons [2]:

- inaccurate revenue forecasts;
- consistently missing revenue commitments (few salespeople met their quota);
- failure to provide the services and experience that users wanted;
- wasted time (expending effort for things the customer didn't value); and
- opaque process (see Fig. 4.2).

Improve
https://doi.org/10.1016/B978-0-12-809519-5.00023-5

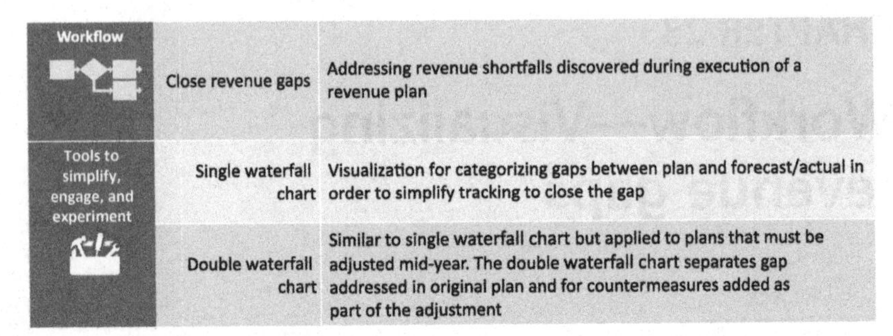

Fig. 23.1 The workflow and tools of closing a revenue gap.

His insightful observation was that these are precisely the reasons leaders are disappointed in other areas of knowledge work.[a] That's no surprise. Selling, especially technical selling, is knowledge work: it requires tacit knowledge and the workflows don't repeat. In this chapter, we're going to take on one workflow from that area: visualizing revenue gaps. That area was selected because it occurs so often in sales organizations and because it's relatively easy to bring improvement to. So, if you're bringing lean thinking to a sales organization, it is a problem area to consider addressing.

23.2 Problem solving for revenue gaps

In this chapter, we will build a problem-solve canvas customized for the issues of closing a revenue gap. The revenue gap is similar to maintaining a schedule on a project: the urgency is around containment. Wherever the issue originated (lost revenue or lost time), we want to recover it from whatever source we can. Lost a major customer? Contain it by finding a new one or increasing orders with an old one. One region down compared to plan? Look to fill the shortfall from another region. The difference between this type of problem and, say, a quality issue is that with a quality issue it's crucial to identify the root cause immediately, because the containment is unlikely to work if we don't understand the root cause. At the other end of the spectrum are problems like revenue gap: the emphasis at the moment the problem is discovered is on finding more revenue. We know more revenue fills the gap, even if we have not fully identified the root cause of the gap.

None of this implies we should not find the root cause. If the organization fails to detect a root cause, the problem will occur again. Long-term success depends on discovering and treating the root cause. But our focus

[a] Pryor focuses on software development, but the logic applies to knowledge work in general.

here will be navigating through the storm. To do that, we'll rely heavily on the problem-solving, visualization, and engagement techniques of earlier chapters. But we'll need to add two visualization tools: the single and the double waterfall charts.

23.3 The 8 Wastes

Fig. 23.2 shows examples in each of the 8 Wastes that can occur when attempting to close a revenue gap. **Discord** is common when revenue gaps are discovered: it's easy for everyone involved to blame someone else. **Information Friction** is common when events or early indicators of unfavorable events are not carried through to the forecast so that unrealistically high numbers remain in the plan. **More-is-Better Thinking** comes when the team loses sight of success and, instead, settles for doing "the best they can." **Inertia to Change** comes in many forms; one common form is when we choose to avoid improving because there's no time for salespeople to do anything but sell. **No-Win Contests** arise when the team is handed down a target that is unrealistic from the start or when events occur that cannot be overcome; either way, there's no identifiable path to success. **Inferior Problem Solving** also comes in many forms, one being the use of unrealistic timing for otherwise effective solutions. For example, if a countermeasure (CM) might generate \$100k/year, it's all too common for a person to forecast credit for the full \$100k the day the CM goes into effect, even though it may take the full year to recognize that amount. **Solution Blindness** occurs when the CMs go

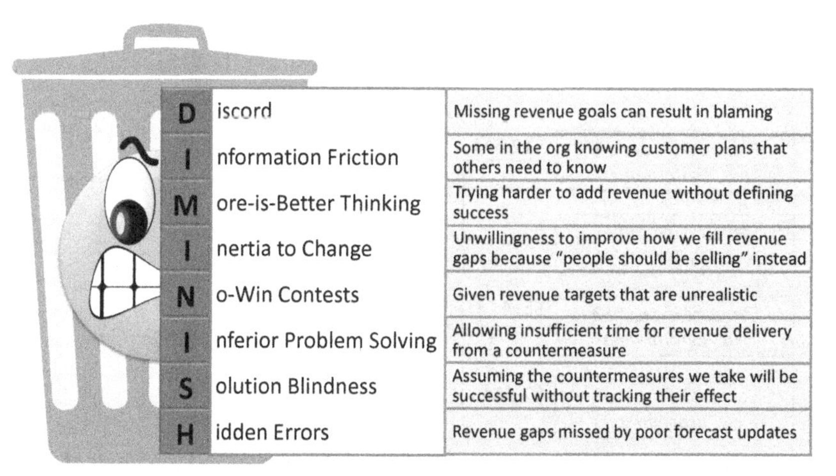

Fig. 23.2 Examples of the 8 Wastes when closing a revenue gap.

unmeasured—essentially, assuming they will work. **Hidden Errors** can happen in many places; one example is when out-of-date forecasts hide an unfavorable revenue trajectory.

23.4 The waterfall chart

The waterfall is "A type of chart used to depict the cumulative effect of sequentially introduced positive or negative values" [3]. It begins with a goal set to meet organization needs; the gap is then the difference between the goal and what we expect to do at the outset (Fig. 23.3). For example, we might use the single waterfall chart to visualize the difference between the plan for the 12 months (as defined at the start of the plan) and the "standard measures" (as we understand them today).

23.4.1 Single waterfall chart

The single waterfall chart visualizes how plan revenue will be built from multiple "standard" measures, which are the things we plan to do from the start of the plan. An example of a waterfall chart is shown in Fig. 23.4. The plan, on the left, is set at $30M in January and does not change throughout the year. The waterfall chart in Fig. 23.4 is shown as it appears in September, 9 months into the plan. Note that the plan is divided into *actual*, the portion that was expected up to September, and *forecast*, what is expected October-December. As can be seen in the chart, the year has not gone as

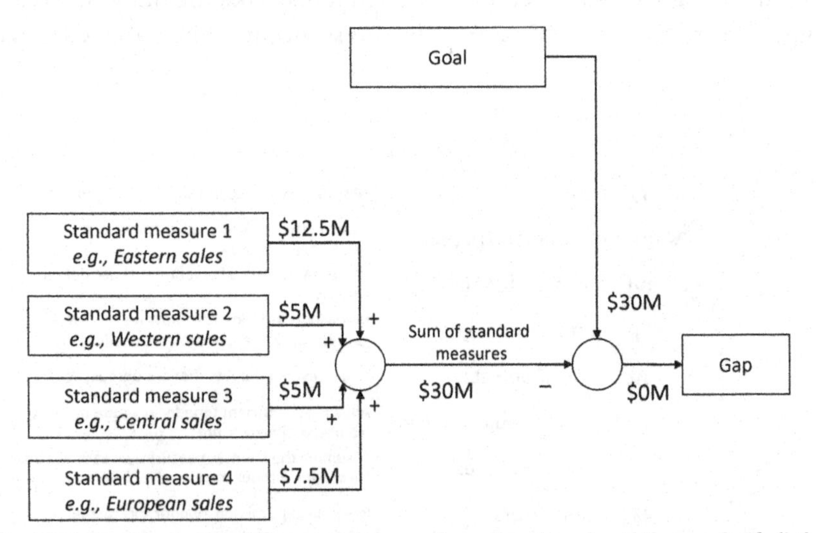

Fig. 23.3 The gap is the difference between the original goal and the result of all the actions we are taking as we understand them today.

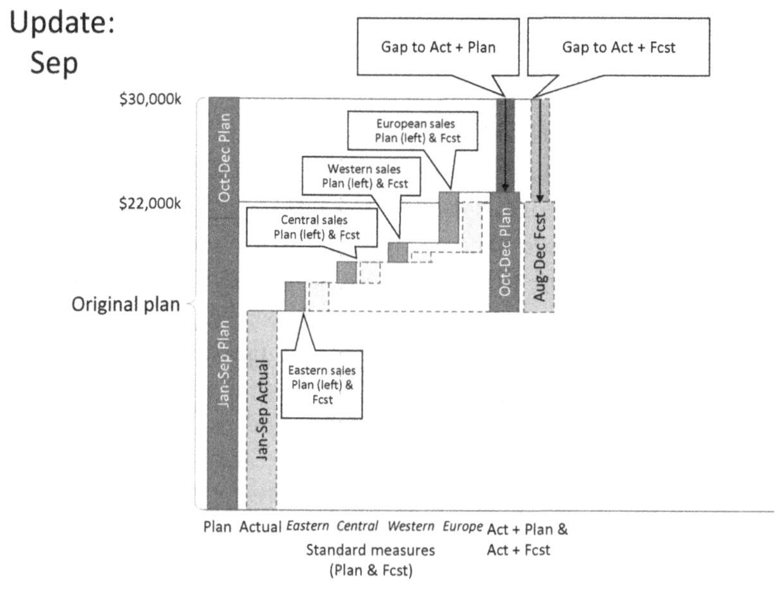

Fig. 23.4 The waterfall chart compares what we know in September (today) with what we thought at the start of the year.

expected, which is immediately visible because the Jan-Sep Actual is well short of the Jan-Sep Plan. There are four sales regions with revenue projections shown in Plan-Fcst pairs to build the Oct-Dec Plan.

Notice that, for the remainder of the year, three of the four regions are on plan, but forecast for the Western Region is now forecast at about half the plan. Perhaps a customer was lost or a key salesperson left; whatever the reason, there is roughly 50% gap in the forecast for the remainder of the year. However, that issue is small compared to the primary problem: the first 9 months of sales were about two-thirds of what was expected. This then creates a gap between the plan (far left) and the best projections available today: "Actual + Plan" and "Actual + Forecast" on the far right. These projections are formed by using the actual revenue for the portion of the year that has passed, and the remainder of the year's plan (and forecast) to project the year's total. Notice how, with just a glance, the waterfall chart compares what we think today to what we thought at the start of the year. The waterfall chart makes it clear that if we don't take strong action, the year will almost certainly finish with a large shortfall. That leads us to take countermeasures, the subject of the next section.

Note that there are many forms of waterfall charts; Fig. 23.4 is just one example. The constant of waterfall charts is the visualization of plan, forecast, and actual against calendar time. The waterfall chart continually asks how what we think today compares to what we thought at the start of the year.

23.4.2 The double waterfall chart

The waterfall chart of Fig. 23.4 visualizes how much we will miss revenue if we don't take action. **More-is-Better Thinking** will have people working hard to sell more, but, looking at Fig. 23.4, there doesn't seem to be an incremental way to close the gap because of the large miss experienced in the months before October. If we want to achieve the year's goal, we need to countermeasure. This requires problem solving to identify meaningful countermeasures, the sort of thing we discussed in Chapter 12. There is no substitute for diligent, expert thinking to find real solutions. But that topic has been covered already in this text, so in this section we'll assume that a competent team applied solid problem-solving techniques and identified the three countermeasures that add a total of $8.6M as shown in Fig. 23.5:

CM1. Discount the product line XXY, a line where sales have suffered since a competitor introduced a lower-cost alternative to generate $2.3M.

CM2. Introduce a new direct marketing campaign in Brazil to generate $4.1M.

CM3. Add a new distributor for Spain/Portugal, which was won over from a key European competitor to generate $2.2M.

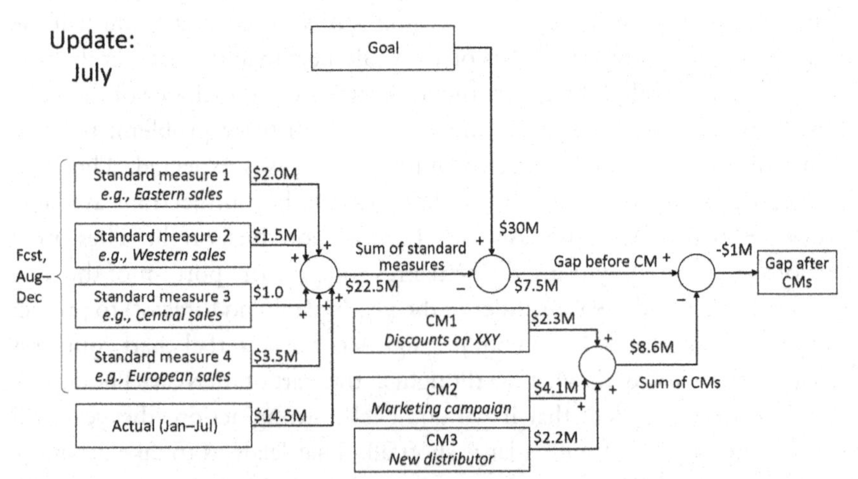

Fig. 23.5 "Gap after CM" adds the countermeasures to close the revenue gap because standard selling is not working.

In preparation for the double waterfall chart, the incremental revenue from these countermeasures is added to the gap to create a metric we will use as the bottleneck: gap after CMs.

The double waterfall chart is needed to separate standard selling from countermeasures, where "standard" are the things we normally do like training, sales calls, incentive programs, and account plans; the "countermeasures" are the steps taken in reaction to this gap because our standard measures are not working. As shown in Fig. 23.6, the double waterfall chart begins with a single waterfall chart on the left. This waterfall chart must be maintained through the period we are measuring revenue. To complete the double waterfall chart, create a set of countermeasures that will make up the gap to the "Actual + Forecast." The implicit assumptions are as follows:

(1) The standard measures alone—just doing what we normally do—will not resolve the gap. We need separate countermeasures.

(2) However, we must maintain the standard measures while implementing the countermeasures.

The standard measures and countermeasures must remain distinct throughout the measurement period. In Fig. 23.6, notice that the countermeasures address the gap between Plan ($30M) and Act+Fcst (the sum of Jan-Sep Actual and Oct-Dec Fcst, $22M). Here, the three countermeasure plans are built to close the $8M gap. There is space to track Plan vs. Fcst/Act for each of the three countermeasures. In this way, the waterfall chart continues to visualize how the original measures are working (left) separately from the countermeasures (right).

23.4.3 Separating standard measures and countermeasures to avoid double counting

For this approach to work, we must avoid double counting, counting the same revenue for a countermeasure and a standard measure. Let's say, for example, that we are having a problem in the Western Region with sales that we will address with XXY discounts (for example, suppose most XXY sales are in the West). Now, when XXY discounts are applied, Western sales will increase. Let's say further that the discount program is global and run by Olivia in Marketing, while the Director of Sales in the Western Region is Lisa. How do we separate the normal heavy lifting of standard selling (which would be credited to Lisa) from the genius of our new XXY discounts (which would go to Olivia)? The answer is, depending on the organization, it could be either, but it needs to be decided unambiguously at the beginning. If it's going to be credited to Sales, the expected

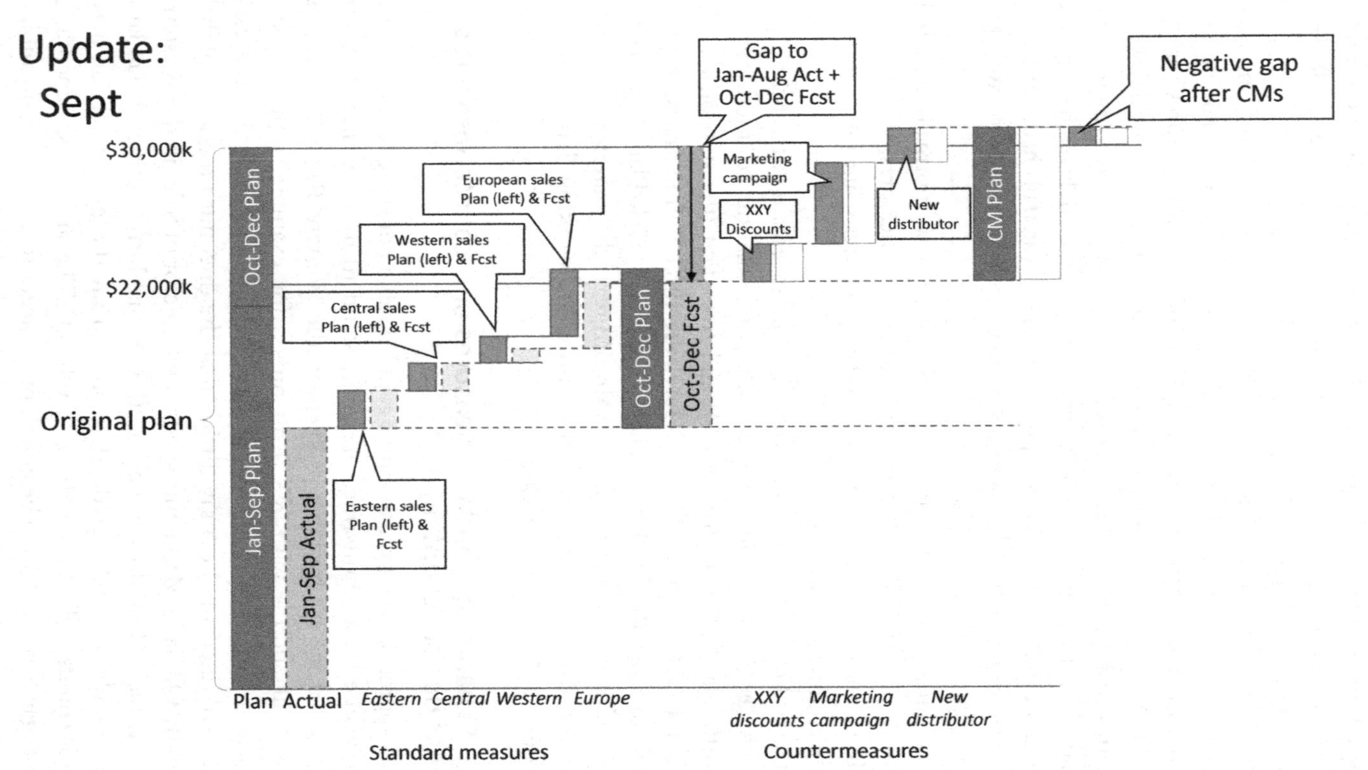

Fig. 23.6 The double waterfall chart starts with the waterfall chart of Fig. 23.4 and adds the effects of the countermeasures on the right.

benefit should be budgeted to the left side of the waterfall chart; if it's going to be managed by Global Marketing, it would be on the right. Implicit in the decision to make this a countermeasure, we must be confident that the source of each affected sale can be separated.

The danger of creating countermeasures outside the region is the question of ownership. As with all task management, unambiguous goals, measures, and ownership are needed for success. With ambiguity comes waste: **Discord** as people argue, blame, and compete for credit, and **No-Win Contests** when there is no path for the team to win.

23.5 Revenue gap Knowledge Work Improvement Canvas

The Knowledge Work Improvement Canvas (KWIC) can be adapted to guide the problem solving of filling a revenue gap. As shown in Fig. 23.7, the Test Track is replaced with the double waterfall chart. So, referring to Fig. 9.21, *change* is still managed with the Action Plan and *value* with the Bowler; *traction* is measured with the double waterfall chart. The left side is the same as the standard KWIC used throughout this text.

23.6 Conclusion

As shown in Fig. 23.8, the skills to close revenue gaps **simplify**, **engage**, and **experiment**. We simplify first by applying the formal problem solving of Chapter 12 to revenue gaps. We simplify further by using the single or double waterfall chart as part of a canvas to solve gap closure measures and sustain them over time. The Engagement Wheel shows that, through effective management of initiatives to fill revenue gaps, engagement can be increased. Knowledge staff almost always understand the inherent need to manage revenue to create an environment where we can reach our goals. Using cross-functional teams to address revenue gaps creates connection in the organization. Having confidence that revenue gaps will be addressed with clear and achievable goals creates a contest the team can win. Finally, knowing we can respond in the storm creates a sense of protection. We create experimentation by measuring gaps and tracking them over time; if the gaps don't close, we know we must learn to create new and more effective countermeasures.

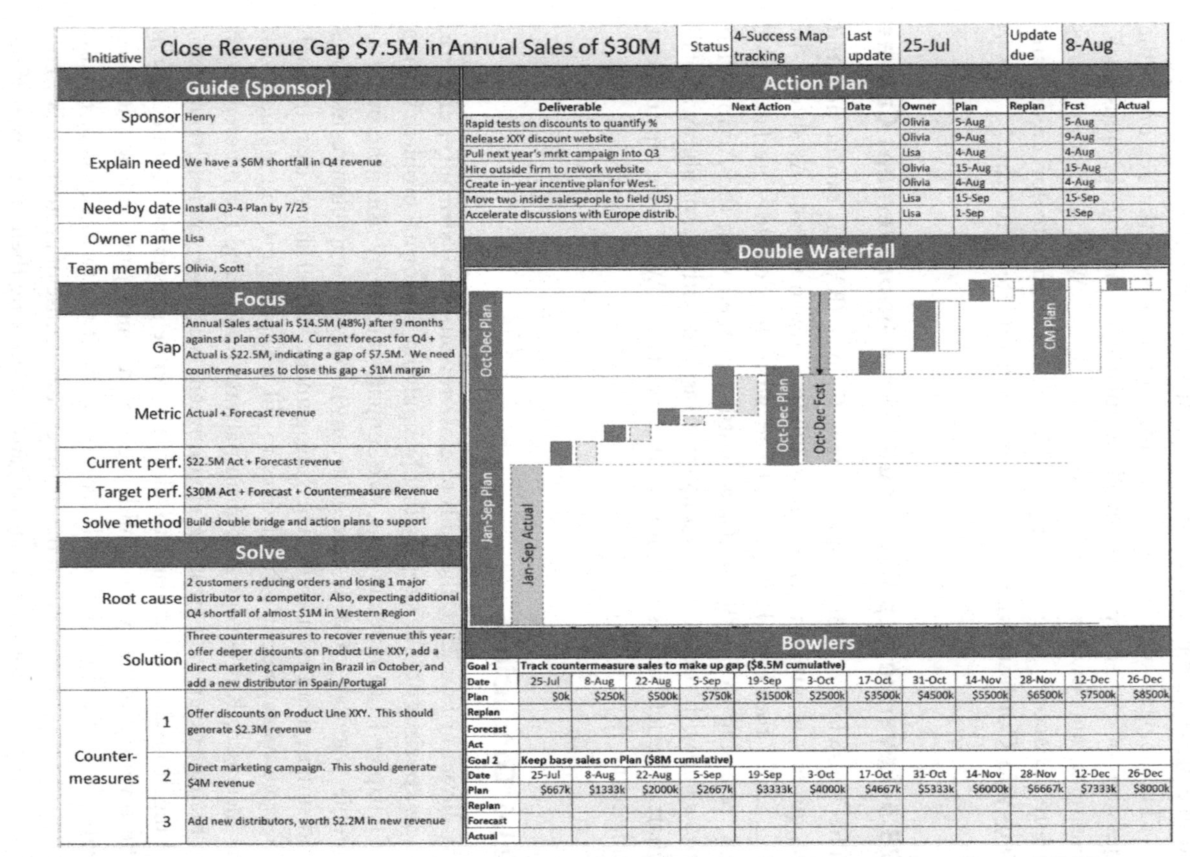

Initiative	Close Revenue Gap $7.5M in Annual Sales of $30M	Status	4-Success Map tracking	Last update	25-Jul	Update due	8-Aug

Guide (Sponsor)

Sponsor	Henry
Explain need	We have a $6M shortfall in Q4 revenue
Need-by date	Install Q3-4 Plan by 7/25
Owner name	Lisa
Team members	Olivia, Scott

Focus

Gap	Annual Sales actual is $14.5M (48%) after 9 months against a plan of $30M. Current forecast for Q4 + Actual is $22.5M, indicating a gap of $7.5M. We need countermeasures to close this gap + $1M margin
Metric	Actual + Forecast revenue
Current perf.	$22.5M Act + Forecast revenue
Target perf.	$30M Act + Forecast + Countermeasure Revenue
Solve method	Build double bridge and action plans to support

Solve

Root cause		2 customers reducing orders and losing 1 major distributor to a competitor. Also, expecting additional Q4 shortfall of almost $1M in Western Region
Solution		Three countermeasures to recover revenue this year: offer deeper discounts on Product Line XXY, add a direct marketing campaign in Brazil in October, and add a new distributor in Spain/Portugal
Counter-measures	1	Offer discounts on Product Line XXY. This should generate $2.3M revenue
	2	Direct marketing campaign. This should generate $4M revenue
	3	Add new distributors, worth $2.2M in new revenue

Action Plan

Deliverable	Next Action	Date	Owner	Plan	Replan	Fcst	Actual
Rapid tests on discounts to quantify %			Olivia	5-Aug		5-Aug	
Release XXY discount website			Olivia	9-Aug		9-Aug	
Pull next year's mrkt campaign into Q3			Lisa	4-Aug		4-Aug	
Hire outside firm to rework website			Olivia	15-Aug		15-Aug	
Create in-year incentive plan for West.			Olivia	4-Aug		4-Aug	
Move two inside salespeople to field (US)			Lisa	15-Sep		15-Sep	
Accelerate discussions with Europe distrib.			Lisa	1-Sep		1-Sep	

Double Waterfall

Bowlers

Goal 1	Track countermeasure sales to make up gap ($8.5M cumulative)											
Date	25-Jul	8-Aug	22-Aug	5-Sep	19-Sep	3-Oct	17-Oct	31-Oct	14-Nov	28-Nov	12-Dec	26-Dec
Plan	$0k	$250k	$500k	$750k	$1500k	$2500k	$3500k	$4500k	$5500k	$6500k	$7500k	$8500k
Replan												
Forecast												
Act												

Goal 2	Keep base sales on Plan ($8M cumulative)											
Date	25-Jul	8-Aug	22-Aug	5-Sep	19-Sep	3-Oct	17-Oct	31-Oct	14-Nov	28-Nov	12-Dec	26-Dec
Plan	$667k	$1333k	$2000k	$2667k	$3333k	$4000k	$4667k	$5333k	$6000k	$6667k	$7333k	$8000k
Replan												
Forecast												
Actual												

Fig. 23.7 An initiative to close a revenue gap can be managed with a Knowledge Work Improvement Canvas (KWIC).

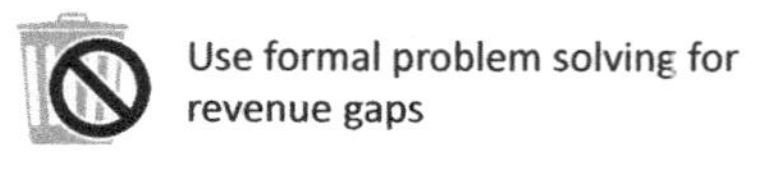
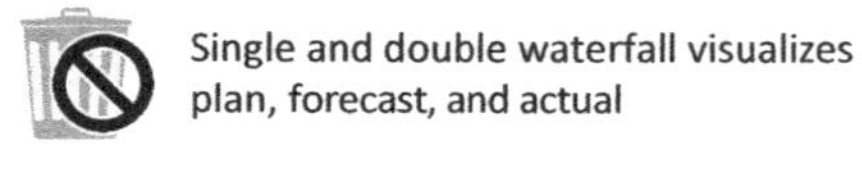
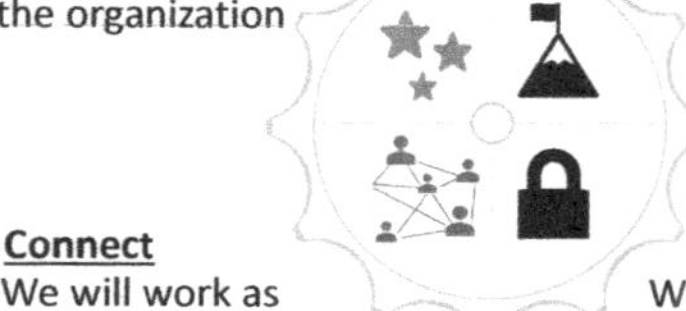

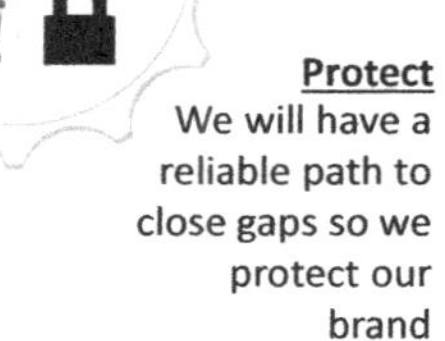
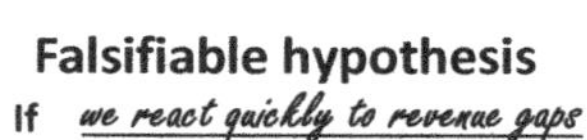

Fig. 23.8 Ability to close revenue gaps simplifies, engages, and encourages experimentation.

References

[1] R.J. Pryor, Lean Selling: Slash Your Sales Cycle and Drive Profitable, Predictable Revenue Growth by Giving Buyers What They Really Want, AuthorHouse, 2014, Kindle Edition, 14%.
[2] R.J. Pryor, Lean Selling: Slash Your Sales Cycle and Drive Profitable, Predictable Revenue Growth by Giving Buyers What They Really Want, AuthorHouse, 2014, Kindle Edition, 12%.
[3] P.R. Williams, Visual Project Management, Think for a Change Publishing, 2015, p. 41.

CHAPTER 24

Workflow—Leadership review of knowledge work

There are three kinds of leaders. Those that tell you what to do. Those that allow you to do what you want. And lean leaders that come down to the work and help you figure it out.

John Shook [1]

24.1 Introduction

In this chapter, we will discuss a few techniques that can improve leadership review of knowledge work. This review is typically carried out by a team of leaders who together bring a great deal of experience and acumen, though they normally have less domain expertise than the people carrying out the work they are reviewing. For example, a product development team might have their work occasionally reviewed by a management team composed of functional leaders from Operations, Finance, Customer Service, Applications Engineering, Sales, and General Management. Those people cannot normally direct the expert in their area of expertise, but they can test the critical thinking by comparing work to their experiences: has the customer been involved enough? Is the value of the work clear? Are we doing enough to ensure compliance with regulatory bodies? So, the review is a partnership between knowledge staff, who bring expertise and transparency, and the leaders, who collectively bring decades of experience to the table.

While a sense of accountability should pervade virtually every aspect of organizational life at a great company, the place where it must be demonstrated and addressed most clearly is meetings [2].

Improve
https://doi.org/10.1016/B978-0-12-809519-5.00024-7

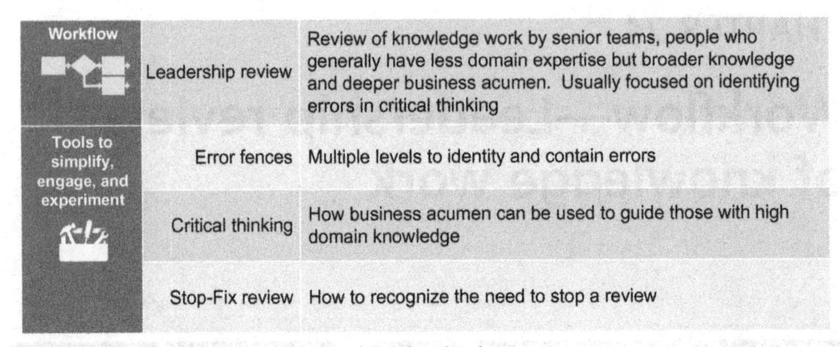

Fig. 24.1 The workflow and tools of the leadership review.

Done well, leadership review identifies gaps that might be very expensive to find later; this increases the odds of success for the entire organization, and it does so by growing and building up people. It also increases accountability by reviewing what was done against organizational standards and individual commitments. Done poorly, reviews can create a great deal of unnecessary work, provide unintentional disincentives for transparency, and leave knowledge staff feeling attacked or ridiculed [3].

This chapter introduces a few techniques for leadership reviews (Fig. 24.1). First, we'll look at the concept of error fences: how a well-designed leadership review connects into a system of countermeasures an organization can take to reduce the likelihood of errors escaping to the customer. This mindset helps focus the review time on the areas of greatest value and helps those having their work reviewed understand the value of the review. We'll talk then about critical thinking and how to pressure test knowledge work. Finally, we'll apply the principles of Stop-Fix to leadership reviews.

24.2 The 8 Wastes

Fig. 24.2 shows examples of each of the 8 Wastes that are common in leadership reviews of knowledge work. **Discord** arises when different leaders evaluate work to conflicting goals. **Information Friction** can come from unnecessarily long narratives from knowledge staff during reviews: rambling discussion or thick slide decks, which can unintentionally obscure what's critical. **More-is-Better Thinking** comes about when reviews turn into progress reviews vs evaluations of success and critical decision points. **Inertia to Change** can come when experts resist being transparent, becoming

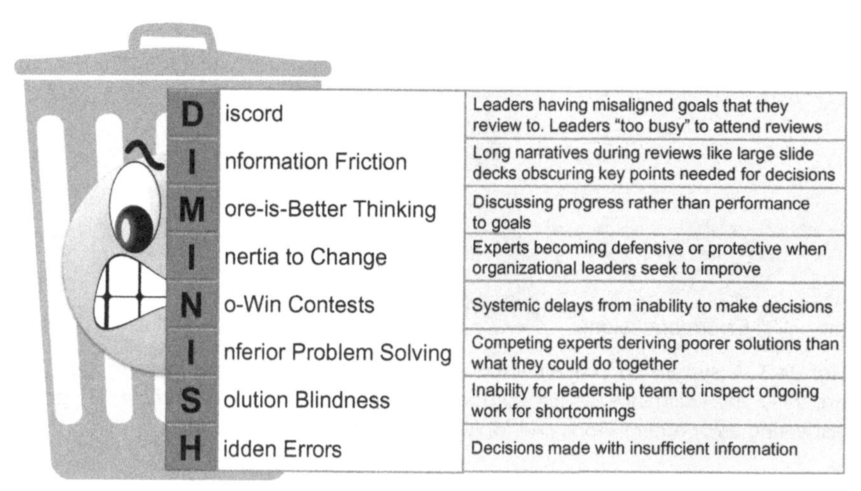

Fig. 24.2 Examples of the 8 Wastes in leadership reviews.

dismissive or defensive when their work is evaluated for potential improvement. **No-Win Contests** result when leaders are unable to make decisions, delaying work that would otherwise be successful. **Inferior Problem Solving** commonly comes when experts compete with each other so that they have a poorer understanding of the problem than they would if they worked together. **Solution Blindness** comes about when the leadership team is unable to review ongoing work in light of new information, for example, because work has gone too far to stop. **Hidden Errors** occur when decisions are made with insufficient information because critical data is obscured.

24.3 A system of barriers to defects

We are doomed to failure without a daily destruction of our various preconceptions.

Taiichi Ohno [4]

Leadership review done well brings large value at a reasonable cost. Much of the cost is concrete: people-time spent in meetings, time spent preparing, and time spent managing the outcome (e.g., administrative time to manage meeting minutes and actions). But there are abstract costs as well—for example, employees who feel ignored or attacked may have lower engagement.

The value of leadership review is more challenging to measure. The primary value is identifying errors and omissions in work that must be remedied, such as increasing customer feedback points, deepening operations investigation,

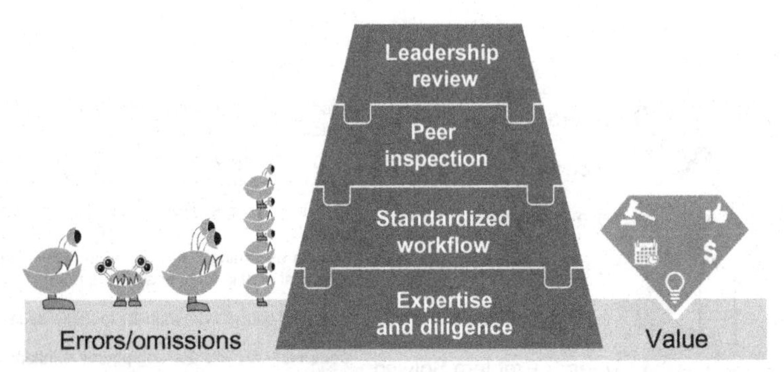

Fig. 24.3 A system to protect knowledge work from inevitable defects: (1) high expertise and diligence in knowledge staff, (2) standardized workflow, (3) peers inspecting each other's work, and (4) leadership review.

applying for more intellectual property protection, expanding the customer service plan, and reevaluating financial what-if scenarios. These might ultimately create a great deal of value, but there is rarely a practical way to compare what ultimately happened to what might have happened without the review.

Leadership review is part of a system to prevent defects from escaping into the flow of value (see Fig. 24.3). To start, recognize that knowledge work is complex and some level of errors and omissions are unavoidable. We cannot prevent their creation, but we can stop most of them from escaping into the value streaming to our customers. This view helps prevent blaming. It communicates to the knowledge staff: we know **Hidden Errors** are unavoidable, so let's work together to find them.

Leadership time is limited, so work product needs to arrive at the review ready. Referring to Fig. 24.3, this includes: (1) having been completed with high diligence and expertise, (2) having followed the standards erected to prevent common errors from the past, and (3) having been reviewed with peers and teammates, for example, through regular team stand-ups. These steps should detect most errors. Only when these requirements have been met is work ready for leadership review.

24.3.1 Avoiding asking leadership teams to inspect standard work

Leadership review detects errors and omissions that other techniques miss. Expertise and diligence work well inside the domain; they work also in the uncountable connections with other functions in an organization, but not quite as reliably. Standardized workflow can be outstanding at catching errors that have occurred in the past, but it doesn't do well with issues the organization has never seen or has seen so rarely that they are not captured in the standards. Team inspection helps, but a typical knowledge team has nothing close to the breadth of experience and diversity of viewpoints of a strong leadership team. Leadership review is far too expensive to find all errors, but it works well as part of a system.

As a practical example, consider standard work. I've witnessed many times people wanting leadership teams to do detailed reviews as to whether work meets a standard like a technical review; that's unrealistically time-consuming for most leadership teams. Instead, consider creating simple Stop-Fix alarms in the standard to reflect whether someone has met the standard vs expecting a senior leader to inspect for that level of detail. A simple example is a requirement for releasing a new product or service: the head of Operations could sign off on a checklist regarding production readiness ahead of the leadership review. Then treat this as a Stop-Fix alarm: if the checklist isn't signed off, the review is postponed. This is something like when a police officer checks your right to drive a vehicle on a roadside stop. If you don't have a license, you are simply unqualified to drive; the police don't give you an impromptu driving test. Similarly, standards like operations plans and financial requirements can provide unambiguous indications if basic requirements have been met so that leaders can focus on those things that only they can review well. We can apply thinking from Deming to knowledge work: let the workflow build quality into the work in order to lessen dependence on inspection and review.

Cease dependence on inspection to achieve quality. Eliminate the need for inspection...by building quality into the product in the first place.

W. Edwards Deming [5]

As knowledge staff come to understand the lesson of Fig. 24.3, they are more likely to understand the value that leadership review brings. As leaders focus on those areas where they bring the most value, they will reinforce those lessons. They will also recognize where the standards lack unambiguous indications and, accordingly, direct effort to correct that. Most knowledge staff will welcome effort that results in building up their work. That will naturally create more transparency and more diligent preparation for review.

24.4 Critical thinking

The key ingredient is to have the courage to face the inconsistencies between what we see and deduce the way things are done. The challenging of basic assumptions is essential to breakthroughs.

Eliyahu Goldratt [6]

Critical thinking is a mysterious skill. I remember first witnessing it on a regular basis almost 20 years ago. Our General Manager at Kollmorgen, Kevin Layne, held monthly all-day meetings to review the most complex projects in the organization. He and a few others had an almost-magical ability to ask questions that drove quickly to the trouble areas. It might be an unpredictable customer asking for us to perform a complex technical analysis. Or it might be an operations issue that the presenter had put to rest too quickly. It might be a technical area or intellectual property issue or a flawed contract clause. I made mental notes where he predicted a customer would back out or a manufacturing line would stall. I was (and still am) impressed with his near prescience.

I didn't know it at the time, but that was my first regular exposure to critical thinking and, two decades later, it would be the most difficult part of my job. I've gotten better through practice. Some people seem to have an innate skill; by contrast, I learn by the mistakes I made. Whether you come by it naturally or through the school of hard knocks, you can recognize the pattern: a person leading the questions marries curiosity and skepticism with a firm understanding of how something will affect the organization over many facets. They are patiently building a mental model of how this work product will create value, finding some areas to tap the brakes and others to press the accelerator, while constantly asking if this vehicle needs to be parked. This is what organizational leaders can do in a way no one else can, and it is the review process that connects knowledge work directly into that skill.

Critical thinking is a broad topic. Let's start with a definition from the Foundation of Critical Thinking:

> *Critical thinking is the intellectually disciplined process of actively and skillfully conceptualizing, applying, analyzing, synthesizing, and/or evaluating information gathered from, or generated by, observation, experience, reflection, reasoning, or communication, as a guide to belief and action [7].*

Ken Petress, a communication scholar and journalist, divides critical thinking into seven categories of evaluation [8]:

- Sufficiency—is there enough evidence to support the claims?
- Relevance—is the evidence pertinent to the issues?
- Reliability—do the sources have a good track record? Did they have sufficient expertise to draw the conclusions they did?
- Consistency—are the elements of evidence consistent with each other and with other experiences and sources?
- Timely—is the evidence recent enough to be applicable?
- Accessible—are supporting materials available for verification?
- Objectivity—are supporting materials fair and sufficiently free from bias?

One of the most accessible places to observe critical thinking in business is ABC's television show Shark Tank [9], where entrepreneurs pitch for about 10 minutes to five "sharks," venture capitalists from whom they seek an investment in their company, generally between $50k and $5M [10]. The entrepreneurs present the value proposition of their product or service along with information on varying topics like market acceptance, competitive response, operational performance, intellectual property, and profitability. The show is fascinating. Over the years I've made a mental model of the critical thinking on the show and unscientifically split the stated reasons the sharks reject a business as follows (see Fig. 24.4): leadership, market access, performance to date, value prop, competition, and operational issues. I've noticed they focus a lot on leadership, often investing in ideas they are unsure about because they like (that's the word often used) the entrepreneur, and just as often losing interest because they don't. And if they sense anything shady, most back out fast. I often recommend the show to aspiring leaders to increase their exposure to critical thinking.

Helen Lee Bouygues advises that those who want to improve their critical thinking need to start by practicing three habits [11, 12]:

1. Question assumptions

   Chapter 1 of this text has a provocative title: "30% of what you think is wrong," meaning that a large portion of the time, ideas, insights, and solutions that seem amazing at first will eventually be proven wrong—not dead wrong, but wrong enough to cause disappointment. Bouygues' advice here is the logical conclusion of this mindset: we constantly need to be challenging our assumptions because a large portion of them are wrong.

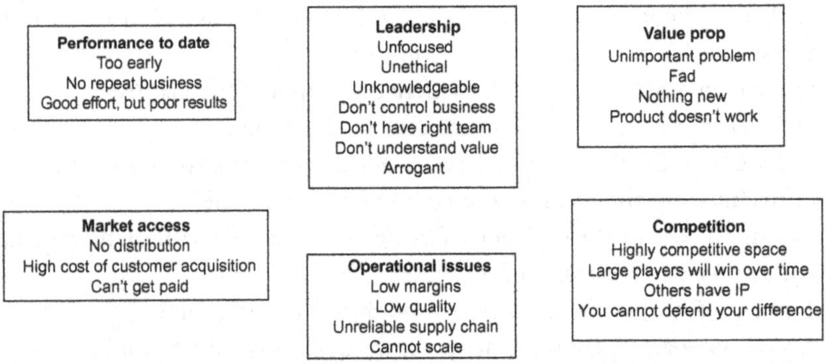

Performance to date
Too early
No repeat business
Good effort, but poor results

Leadership
Unfocused
Unethical
Unknowledgeable
Don't control business
Don't have right team
Don't understand value
Arrogant

Value prop
Unimportant problem
Fad
Nothing new
Product doesn't work

Market access
No distribution
High cost of customer acquisition
Can't get paid

Operational issues
Low margins
Low quality
Unreliable supply chain
Cannot scale

Competition
Highly competitive space
Large players will win over time
Others have IP
You cannot defend your difference

Fig. 24.4 Shark Tank television show demonstrates how to find gaps using critical thinking.

2. Reason through logic

   "Critical thinking starts with logic. Logic is the unnatural act of knowing which facts you're putting together to reach your conclusions, and how" [13]. Where "question assumptions" has you constantly challenging the evidence you are relying on, "reason through logic" asks if you are linking evidence to build reasonable conclusions.
3. Seek out diversity of thought and collaboration

   Each of us has bias and gaps that we can reduce, but never eliminate. However, a diverse review team helps balance out these biases and gaps. In fact, research shows that when people listen only to people who agree with them, they become more polarized [14]. On the other hand, "opposite styles can balance each other out" [15]. This is why the most effective management reviews are cross-functional: Operations, R&D, Sales, and Finance bring valid insights, but incomplete views; collaboration among them joins those insights to bring comprehensive oversight.

24.5 Critical thinking: Finding the pillars that support value

> *...one of the most important things that we do is establish this notion that nothing is unchallengeable. You don't take anything for granted and you test and pressure test everything.*
>
> ***George Yancopoulos, co-Founder and Len Schleifer, CEO of Regeneron [16]***

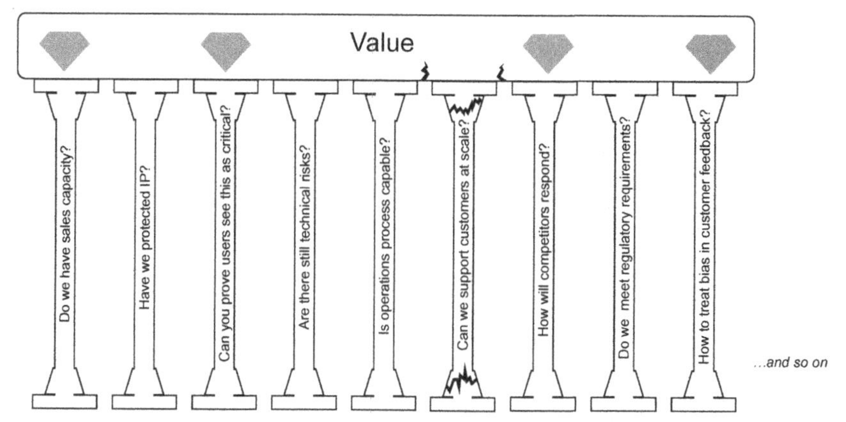

Fig. 24.5 A model of critical thinking: Failure in just one support can be ruinous.

A visual model of the value of critical thinking is shown in Fig. 24.5. The ability to support a flow of value to a market is almost always a heavy burden that must be supported in many ways. Hidden flaws in just a single support can be ruinous for the delivery of value. The example in Fig. 24.5 highlights customer support among eight other supports; if all elements are in place except customer support, value will not be supported. This is what makes leadership review so valuable; where domain experts generally focus on one area of their expertise, leaders are usually broader, considering the areas knowledge staff may have little or no exposure to.

Tom Gilb, an author known for the development of software metrics, carries with him "12 Tough Questions" [17]. These questions are a sort of general-purpose list to aid critical thinking. Gilb put them on the back of his business card to invite people to use the questions on him. I put them on my computer's wallpaper to help commit them to memory (Fig. 24.6).

24.6 Bias and critical thinking

Each of us has our own unique biases. You may be biased for action—ready to leap on a problem quickly—or biased for analysis so that you are not satisfied with shallow thinking. Those biases are probably what got you to your position, be it a general manager, a development engineer, or a doctor. But those same biases can injure your ability to think critically. There are hundreds of cognitive biases, but a few of the most commonly discussed biases are [18]:

- belief bias—logic appears stronger when the conclusion is more believable;
- clustering illusion—seeing patterns where none exists;

1. How is the value quantified?
2. How is the risk quantified?
3. Are you certain about this?
4. Where did you get this? Do I have access?
5. What goals does this affect?
6. What are the critical elements?
7. Where have you seen this work?
8. Does this meet the full requirements and expectations?
9. Are we doing the most important things first? How do you know?
10. Who owns success?
11. What experiments are we running to know this works?
12. What are the costs of failure?

Fig. 24.6 Critical-thinking questions based on 12 Tough Questions by Tom Gilb.

- confirmation bias—interpreting information to confirm preconceptions;
- loss aversion—the disutility associated with losing an object is seen as greater than the utility associated with acquiring it;
- recency bias—the tendency to overvalue recent events;
- sunk-cost fallacy—honoring costs that are lost and cannot be recovered[a];
- availability heuristic—overvaluing those things most available in memory, which elevates the most vivid examples; and
- overconfidence bias—the tendency to overvalue one's judgment.

Most of us are largely unaware of our biases. However, having diverse groups evaluating complex work usually cancels biases held by a minority of the reviewers. The leadership review works to minimize bias by relying on a team of experienced reviewers who firstly are presumably less biased than less experienced people, and secondly meet as a group to help cancel the biases they do have.

Simply meeting as a team doesn't ensure the elimination of bias. In fact, one bias that is amplified by team review is the "bandwagon effect" or group think. The Innovator's DNA LLC lists five dangers of groupthink [16]:

1. Rationalizing

   Justifying behavior and results with plausible reasoning, even if untrue.

2. "Satisficing"

   Accepting an inferior solution because it's good enough.

3. Pressure to conform

   The innate push to join onto the consensus of the group.

[a] This reference does not contain *sunk-cost fallacy*.

4. Default to the status quo
   Leaving things as they are is the easy decision.
5. Minimizing risk rather than seeking excellence
   Overemphasis on protecting what we have.

24.7 Unambiguous requirements

In this section, we will discuss the need for unambiguous requirements for reviews. These requirements are meant to set a floor that all review work must meet before it earns a turn at leadership review. This type of requirement will be largely based on Fig. 24.3: that before a meaningful leadership review, the work must have been executed by a diligent expert, it must meet standardized work requirements, and it must have been reviewed by an appropriate group of colleagues. The reason to protect the review is twofold: first, leadership review is an expensive way to root out errors and so should be reserved for that work that has benefited from the other (presumably less expensive) parts of the framework. Second, the freer work is of more obvious defects, more likely the review will uncover subtler ones. Put another way, the less time leaders must spend finding errors that could be found other ways, the more time they can invest in finding those errors they are best positioned to spot.

Creating unambiguous requirements takes a modicum of effort. Let's consider a simple example. Suppose we have an initiative to open a manufacturing site in another region of the world, and further, let's suppose the organization has a standard process to follow when opening new sites. We might state, "You must meet process X." But this is ambiguous. It's highly unlikely that process X is so clear that any person who might ever use it can evaluate whether their work fully "meets" Process X requirements. There will be statements like "Ensure we meet local regulations" and "Obtain competitive shipping quotes," but how does a person determine when they have done enough to meet the spirit of those directions? You intend for leaders to focus on strategic issues around the new site, but can wind up spending most of the meeting evaluating tedious details that could have been settled better elsewhere.

There are two steps that remove ambiguity with modest effort. The first is to provide templates with clear sign-off and specify expert review on tactical issues ahead of leadership review. In other words, create an unambiguous sign-off. For our new manufacturing site, the requirement could be considerably less ambiguous with a small modification: "You must obtain sign off from the regional Operations leader that you have met process X at least 3 days before the review."

Fig. 24.7 Stopping reviews when the team is unprepared can be the best way to prevent defects from being created.

Another place to improve leadership review is requiring the A3 format for all problems that don't fit a standard issue. In other words, presentations are limited to standard templates and A3 canvases (Section 12.9). The improvement in this simple requirement can be dramatic. Better and faster decisions are made, which helps organizational performance and engages knowledge staff.

Changes of this sort bring value over time. But some benefits will come quickly to those who "(1) identify the errors we cannot tolerate and then (2) don't tolerate them" (Section 14.3.1). The first puts more burden on leadership and the second on those presenting. But leaders should assume that whatever limits they create will be tested. And this is where the Stop-Fix approach works to reinforce the requirements: "not tolerating the errors" usually means stopping the review. If there is a requirement for sign-off that is not met or a specified format is ignored, stopping the review is often the best reaction (see Fig. 24.7). Actually, it's the only logical reaction if we believe the requirement is necessary for high-quality review and the organization is committed to high-quality reviews. Grinding through slipshod work will pass errors on. It also tells the team that published requirements are not required in practice.

24.8 Project review Knowledge Work Improvement Canvas

The Knowledge Work Improvement Canvas (KWIC) can be adapted to guide the installation of a project review meeting. Referring to Fig. 24.8, the Test Track was adapted for the monthly meeting by entering a series of Yes/No questions to score process integrity. Normally, we use the Test Track to monitor a few early cases. Here there is no need to monitor a subset of cases carefully; there are only 12 meetings per year, so we can look at all of them. Here, we will evaluate process integrity as our indication of traction, filled out as a team in the last few minutes of each meeting. The other elements of this KWIC are similar to those used in earlier chapters.

Initiative	Install Monthly Project Review	Status	4-Success Map tracking	Last update	25-Oct	Update due	8-Dec

Guide (Sponsor)

Sponsor	Regina (CEO)
Explain need	Our project reviews are failing to deliver the value we need. Reviews are constantly running over so decisions are rushed if they are made at all
Need-by date	1-Jan
Owner name	Jiri
Team members	Olivia, Greg, Alexia, Christian, Gabriella

Focus

Gap	We have insufficient estimated revenue in our project portfolio
Metric	Primary metrics: estimated revenue of 1) projects started this year and 2) projects completed this year
Current perf.	Unknown (we don't measure this)
Target perf.	$6M / year for both
Solve method	1-day offsite for leadership team

Solve

Root cause		Out review process is chaotic. We do not reliably review projects for critical success factors
Solution		Create a standard process project review that ensures timely review with the right people for every project start and completion
Counter-measures	1	Standardize approval process for project start and completion
	2	Track estimated revenue from approvals
	3	Track KPIs for monthly meeting effectiveness

Action Plan

Deliverable	Date	Owner	Plan	Replan	Fcst	Actual
Modify standard work for reviews		RME	15-Nov		15-Nov	
Create standard templates for finance		RME	15-Nov		15-Nov	
Require canvases for all review issues		RME	15-Nov		15-Nov	
Define required attendees		RME	15-Nov		15-Nov	
Track projected revenue from approvals		JJ	15-Dec		15-Dec	
Use agenda buffer to improve schedule		AA	1-Jan		1-Jan	
Set up Process Integrity Scorecard for monthly meeting		Olivia	15-Dec		15-Dec	

Process Integrity Scorecard

Month	Meeting held	All required people present	All projects conformed to standard templates	All project starts reviewed; clear approval/rejection	All completions reviewed; clear approval/rejection
January review					
February review					
March review					
April review					
May review					
June review					
July review					
August review					
September review					
October review					
November review					
December review					

Bowlers

Goal 1	Planned revenue for projects approved to start this year											
Date	1-Jan	1-Feb	1-Mar	1-Apr	1-May	1-Jun	1-Jul	1-Aug	1-Sep	1-Oct	1-Nov	1-Dec
Plan	$500k	$1000k	$1500k	$2000k	$2500k	$3000k	$3500k	$4000k	$4500k	$5000k	$5500k	$6000k
Replan												
Fcst												
Act												

Fig. 24.8 An Initiative to install a new leadership review—a Knowledge Work Improvement Canvas (KWIC).

Fig. 24.9 Leadership reviews can simplify, engage, and encourage experimentation.

24.9 Conclusion

This chapter has presented several techniques to serve those providing leadership review of knowledge work. Leadership teams bring experience and acumen, and though they may have less domain expertise than knowledge staff, they can nevertheless bring substantial increased likelihood of success by identifying **Hidden Errors**. Doing these reviews well not only can improve the performance of the organization by stripping out expensive errors and omissions, but also can increase engagement with knowledge workers by teaming up with them to improve their work. As knowledge staff engage in the process, transparency and diligence will increase, improving the effectiveness of the review over time.

Part of the benefits of good leadership review is to simplify preparation with clear requirements and standard format (see Fig. 24.9). The Engagement Wheel for leadership review shows that strong leadership review inspires a team who knows their work will be honed by senior people finding errors before those errors escape. They can feel connected to leaders who spend their time and energy to improve the work of their organization. Tough reviews will create a meaningful challenge—for example, to pass the first time—so long as reviews are fair. They will protect the organization by preventing errors from escaping that could damage profitability, quality, and customer relationships. Experimentation with leadership review can be done by measuring the value of projects approved, for example, quantified by estimated future revenue of the work.

References

[1] Chris Anstey, Embracing Lean-finding a new approach for food factories". A Kimberly-Clark Professional Whitepaper. Available at: https://www.kcprofessional.co.uk/media/58701737/Embracing_Lean_Whitepaper_from_Kimberly-Clark_Professional.pdf.
[2] P. Lencioni, Overcoming the Five Dysfunctions of a Team, A Field Guide, Jossey-Bass, 2005, p. 67.
[3] P.R. Williams, Visual Project Management, Think for a Change Publishing, 2015, p. xiv.
[4] T. Ohno, Taiichi Ohnos Workplace Management, McGraw-Hill Education, 1988, p. 164. Kindle Edition.
[5] W. Edwards Deming, https://deming.org/explore/fourteen-points.
[6] E.M. Goldratt, The Goal: A Process of Ongoing Improvement, North River Press, 2004. Kindle edition, Introduction to first edition.
[7] A statement by Michael Scriven & Richard Paul, Presented at the 8th Annual International Conference on Critical Thinking and Education Reform, 1987. http://www.criticalthinking.org/pages/critical-thinking-where-to-begin/796.
[8] ATEEC, KEN PETRESS Professor Emeritus University of Maine at Presque Isle Copyright Project Innovation, ProQuest Information and Learning Company,

Spring 2004, p. 4. https://docplayer.net/11412657-Critical-thinking-an-extended-definition-petress-ken-http-www-findarticles-com-p-articles-mi_qa3673-is_200404-ai_n9345203.html. All rights reserved.
[9] Shark Tank, https://abc.go.com/shows/shark-tank.
[10] N. Nichols, Inside the Shark Tank with Mark Cuban, D Magazine (March 2014). https://www.dmagazine.com/publications/d-magazine/2014/march/inside-shark-tank-with-mark-cuban/.
[11] H.L. Bouygues, 3 Simple Habits to Improve Your Critical Thinking, Harvard Business Review, May 6, 2019.
[12] H.L. Bouygues, Improve Your Critical Thinking at Work, HBR Ideacast, July 23, 2019. https://hbr.org/ideacast/2019/07/improve-your-critical-thinking-at-work.
[13] S. Robbins, The Path to Critical Thinking, Harvard Business School Working Knowledge, May 30, 2005. https://hbswk.hbs.edu/archive/the-path-to-critical-thinking.
[14] C. Pazzanese, Danger in the Internet Echo Chamber. To Combat Endless Feeds of One-Sided Data, Sunstein Suggests an 'architecture of serendipity', Harvard Law Today, March 24, 2017.
[15] S.M. Johnson Vickberg, K. Christfort, Pioneers, Drivers, Integrators, and Guardians, Harvard Business Review, March–April 2017.
[16] C. Lefrandt, Building an Innovative Culture & Environment, Innovator's DNA LLC, 2019. Keynote Presentation.
[17] T. Gilb, Twelve Tough Questions. VERSION 0.23, July 18, 2002. http://concepts.gilb.com/dl24.
[18] M. Battersby, S. Bailin, Critical Thinking and Cognitive Bias, OSSA, May 10, 2013. https://scholar.uwindsor.ca/ossaarchive/OSSA10/papersandcommentaries/16/.

Index

Note: Page numbers followed by *f* indicate figures, *t* indicate tables, and *b* indicate boxes.

M

T

U

V

Printed in the United States
By Bookmasters